The C# Player's Guide

Second Edition

RB Whitaker

RB Whitaker
rbwhitaker@outlook.com

ISBN-10: 0985580127
ISBN-13: 978-0-9855801-2-4

Contents at a Glance

Table of Contents

Part 2: The Basics

Part 3: Object-Oriented Programming

Part 4: Advanced Topics

Part 5: Mastering the Tools

Part 6: Wrapping Up

Acknowledgements

The task of writing a book is like writing software. When you start, you're sure it's only going to take a few *weeks*. *It'll be easy*, you think. But as you start working, you start seeing that you're going to need to make changes, and lots of them. You need to rearrange entire chapters, add topics you hadn't even thought about, and you discover that there's not even going to be a place in your book for that chapter called *Muppets of the Eastern Seaboard*.

I couldn't have ever finished this book without help. I'll start by thanking Jack Wall, Sam Hulick, Clint Mansell, and the others who wrote the music to the Mass Effect trilogy. (You think I'm joking, don't you?) I listened to their music nearly incessantly as I wrote this book. Because of them, every moment of the creation of this book felt absolutely epic.

I need to also thank the many visitors to my XNA tutorials site, who provided feedback on the early versions of this work. In particular, I want to thank Jonathan Loh, Thomas Allen, Daniel Bickler, and Mete ÇOM, who went way above and beyond, spending hours of their own personal time, reading through this book and provided detailed critique and corrections. With their help, this book is far more useful and valuable.

I also need to thank my mom and dad. Their confidence in me and their encouragement to always do the best I can has caused me to do things I never could have done without them.

Most of all, I want to thank my beautiful wife, who was there to lift my spirits when the weight of writing a book became unbearable, who read through my book and gave honest, thoughtful, and creative feedback and guidance, and who lovingly pressed me to keep going on this book, day after day. Without her, this book would still be a random scattering of Word documents, buried in some obscure folder on my computer, collecting green silicon-based mold.

To all of you, I owe you my sincerest gratitude.

-RB Whitaker

Introduction

In a Nutshell

- Describes the goals of this book, which is to function like a player's guide, not a comprehensive cover-everything-that-ever-existed book.
- Breaks down how the book is organized from a high-level perspective, as well as pointing out some of the extra "features" of the book.
- Provides some ideas on how to get the most out of this book for programmers, beginners, and anyone who is short on time.

The Player's Guide

This book is not about playing video games. (Though programming is as fun as playing video games for many people.) Nor is it about *making* video games, specifically. (Though you definitely can make video games with C#.)

Instead, think of this book like a player's guide, but for a programming language. A player's guide is a popular kind of book that is written to help game players:

- learn the basics of the game,
- prevent them from getting stuck,
- understand how the world they're playing in works,
- learn how to overcome the obstacles and enemies they face,
- point out common pitfalls they may face and locate useful items,
- and master the tools they're given.

This book accomplishes those same goals for the C# programming language. I'll walk you through the language from the ground up, point out places where people get stuck, provide you with hands-on examples to explore, give you quizzes to ensure you're on the right track, and describe how to use the tools that you'll need to create programs. I'll show you the ins and outs of the many features of C#, describing *why* things work the way they do, rather than just simple mechanics and syntax.

My goal is to provide you with the "dungeon map" to direct you as you begin delving into C#, while still allowing you to mostly explore whatever you want, whenever you want.

I want to point out that this book is intentionally *not* called *Everything you Need to Know about C#,* or *The Comprehensive Guide to C#*. (Note that if books with those titles actually exist, I'm not referring to them specifically, but rather, to just the general idea of an all-encompassing book.) I'm here to tell you, when you're done with this book, you'll still have lots to learn about C#.

But guess what? That's going to happen with *any* book you use, including those all-encompassing books. Programming languages are complex creations, and there are enough dark corners and strange combinations that nobody can learn everything there is to know about them. In fact, I've even seen the people who designed the C# language say they just learned something new about it! For as long as you use C#, you'll constantly be learning new things about it, and that's actually one of the things that makes programming interesting.

I've tried to cover a lot of ground in this book, and with roughly 350 pages, anyone would expect that to be quite a bit. And it is. But there are plenty of other books out there that are 800 or even 1200 pages long. A book so heavy, you'll need a packing mule to carry it anywhere. That, or permanently place it on the central dais of an ancient library, with a single beam of dusty light shining in on it through a hole in the marble ceiling. Instead of all that, the goal of this book is effectiveness and clarity, not comprehensiveness. Something that will fit both on your shelf and in your brain.

It is important to point out that this book is focused on the C# programming language, rather than libraries for building certain specific application types. So while you can build desktop applications, web pages, and computer games with C#, we won't be discussing WPF, ASP.NET, DirectX, or any other platform- or framework-specific code. Instead, we'll focus on core C# code, without bogging you down with those additional libraries at first. Once you've got the hang of C#, heading into one of those areas will be much easier.

How This Book is Organized

This book is divided into six parts. Part 1 describes what you need to get going. You'll learn how to get set up with the free software that you need to write code and make your first C# program.

Part 2 describes the basics of procedural programming—how to tell the computer, step-by-step, what to do to accomplish tasks. It covers things like how information is stored (in variables), how to make decisions, loop over things repeatedly, and put blocks of code that accomplish specific tasks into a reusable chunk called a method. It also introduces the type system of the C# language, which is one of the key pieces of C# programming.

Part 3 goes into object-oriented programming, introducing it from the ground up, but also getting into a lot of the details that make it so powerful. Chapter 18, in my opinion, is the critical point of the book. It is where we get into the details of making your own classes, which is the most powerful way C# provides for building your own data types. Everything before this point is giving us the building blocks that we need to understand and make classes. Everything that we do after is simply providing us with more ways to use these custom-made types or showing how to use other classes that have been made by others.

Part 4 covers some common programming tasks, as well as covering some of the more advanced features of C#. For the most part, these topics are independent of each other, and once you've made it past that critical point in Chapter 18, you should be able to do these at any time you want.

Part 5 changes gears, and covers more details about Visual Studio, which you use to create programs in C#, additional information about the .NET Framework, and some tools, tricks, and information you can use as you program.

Finally, Part 6 wraps up the book with some larger scale programs for you to try making, a chapter on where to go next as you continue to learn C#, and a glossary of words that are defined throughout the book, which you can use as a reference when you run across a word or phrase that you are unfamiliar with or have forgotten about.

Try It Out!

Scattered throughout the book are a variety of sections labeled *Try It Out!* These sections give you simple challenge problems and quizzes that give you a chance to play around with the new concepts in the chapter and test your understanding. If this were a class, these would be the homework.

The purpose of these *Try It Out!* sections is to help you get some real world practice with the new information. You can't learn to drive a car by reading the owner's manual, and you can't learn to program without writing any code.

I strongly encourage you to spend at least few minutes doing each of these challenges to help you understand what you're reading and to help you be sure that you've learned what you needed to.

If you have something else you want to explore with the new concepts instead of the challenges I've provided, all the better. The only thing better than playing around with this stuff is doing something with it that you have a personal interest in. If you want to explore a different direction, go for it!

At the end of the book, in Chapter 46, I have an entire chapter full of larger, tougher challenge problems for you to try out. These problems involve combining concepts from many chapters together into one program. Going through some or all of these as you're finishing up will be a great way to make sure you've learned the most important things you needed to.

The most important thing to remember about these *Try It Out!* sections is that the answers are all online. If you get stuck, or just want to compare your solution to someone else's, you can see mine at **starboundsoftware.com/books/c-sharp/try-it-out/**. I should point out that just because your solution is different from mine (or anyone else's) doesn't necessarily mean it is wrong. That's one of the best parts about programming—there's always more than one way to do something.

In a Nutshell

At the beginning of each chapter, I summarize what it contains. These sections are designed to do the following:

- Summarize the chapter to come.
- Show enough of the chapter so that an experienced programmer can know if they already know enough to skip the chapter or if they need to study it in depth.
- Review the chapter enough to ensure that you got what you needed to from the chapter. For instance, imagine you're about to take a test on C#. You can jump from chapter to chapter, reading the *In a Nutshell* sections, and anything it describes that you didn't already know, you can then go into the chapter and review it.

In Depth

On occasion, there are a few topics that are not critical to your understand of C#, but they are an interesting topic that is related to the things you're learning. You'll find this information pulled out into *In Depth* sections. These are never required reading, so if you're busy, skip ahead. If you're not too busy, I think you'll find this additional information interesting, and worth taking the time to read.

Glossary

As you go through this book, you're going to learn a ton of new words and phrases. Especially if you're completely new to programming in general. At the back of this book is a glossary that contains the definitions for these words. You can use this as a reference in case you forget what a word means, or as you run into new concepts as you learn C#.

Getting the Most from This Book

For Programmers

If you are a programmer, particularly one who already knows a programming language that is related to C# (C, C++, Java, Visual Basic .NET, etc.) learning C# is going to be relatively easy for you.

C# has a lot in common with all of these languages. In fact, it's fair to say that all programming languages affect and are inspired by other languages, because they evolve over time. C# looks and feels like a combination of Java and C++, both of which have roots that go back to the C programming language. Visual Basic .NET (VB.NET) on the other hand, looks and feels quite different from C# (it's based on Visual Basic, and Basic before that) but because both C# and VB.NET are designed and built for the .NET Framework, they have many of the same features, and there's almost a one-to-one correspondence between features and keywords.

Because C# is so closely tied to these other languages, and knowing that many people may already know something about these other languages, you'll see me point out how C# compares to these other languages from time to time.

If you already know a lot about programming, you're going to be able to move quickly through this book, especially the beginning, where you may find very few differences from languages you already know. To speed the process along, read the *In a Nutshell* section at the start of the chapter. If you feel like you already know everything it describes, it's probably safe to skip to the next chapter.

I want to mention a couple of chapters that might be a little dangerous to skip. Chapter 6 introduces the C# type system, including a few concepts that are key to building types throughout the book. Also, Chapter 16 is sort of a continuation on the type system, describing value and reference types. It's important to understand the topics covered in those chapters. Those chapters cover some of the fundamental ways that C# is different from these other languages, so don't skip them.

For Busy People

One of the best parts about this book is that you don't need to read it all. Yes, that's right. It's not all mandatory reading to get started with C#. You could easily get away with only reading a part of this book, and still understand C#. In fact, not only understand it, but be able to make just about any program you can dream up. This is especially true if you already know a similar programming language.

At a minimum, you should start at the beginning and read through Chapter 18. That covers the basics of programming, all the way up to and including an introduction to making your own classes. (And if you're already a programmer, you should be able to fly through those introductory chapters quickly.)

The rest of the book could theoretically be skipped, though if you try to use someone else's code, you're probably going to be in for some surprises.

Once you've gone through those 18 chapters, you can then come back and read the rest of the book in more or less any order that you want, as you have extra time.

For Beginners

If you've never done any programming before, be warned: learning a programming language can be hard work. The concepts in the first 18 chapters of this book are the most important to understand. Take whatever time is necessary to really feel like you understand what you're seeing in these chapters. This gets you all of the basics, and gets you up to a point where you can make your own types using classes. Like with the *For Busy People* section above, Chapter 18 is the critical point that you've got to get to, in order to really understand C#. At that point, you can probably make any program that you can think of, though the rest of the book will cover additional tools and tricks that will allow you to do this more easily and more efficiently.

After reading through these chapters, skim through the rest of the book, so that you're aware of what else C# has. That's an important step if you're a beginner. It will familiarize you with what C# has to offer, and when you either see it in someone else's code or have a need for it, you'll know exactly where to come back to. A lot of these additional details will make the most sense when you have an actual need for it in a program of your own creation. After a few weeks or a few months, when you've had a chance to make some programs on your own, come back and go through the rest of the book in depth.

I Genuinely Want Your Feedback

Writing a book is a huge task, and no one has ever finished a huge task perfectly. There's the possibility of mistakes, plenty of chances to inadvertently leave you confused, or leaving out important details. I was tempted to keep this book safe on my hard drive, and never give it out to the world, because then those limitations wouldn't matter. But alas, my wife wouldn't let me follow Gandalf's advice and "keep it secret; keep it safe," and so now here it is in your hands.

If you ever find any problems with this book, big or small, or if you have any suggestions for improving it, I'd really like to know. After all, books are a lot like software, and there's always the opportunity for future versions that improve upon the current one. Also, if you have positive things to say about the book, I'd love to hear about that too. There's nothing quite like hearing that your hard work has helped somebody.

To give feedback of any kind, please visit **starboundsoftware.com/books/c-sharp/feedback**.

This Book Comes with Online Content

On my website, I have a small amount of additional content that you might find useful. For starters, as people submit feedback, like I described in the last section, I will post corrections and clarifications as needed on this book's errata page: **starboundsoftware.com/books/c-sharp/errata**.

Also on my site, I will post my own answers for all of the *Try It Out!* sections found throughout this book. This is helpful, because if you get stuck, or just want something to compare your answers with, you can visit this book's site and see a solution. To see these answers, go to: **starboundsoftware.com/books/c-sharp/try-it-out/**.

Additional information or resources may be found at **starboundsoftware.com/books/c-sharp**.

Part 1

Getting Started

The world of C# programming lies in front of you, waiting to be explored. In Part 1 of this book, within just a few short chapters, we'll do the following:

- Get a quick introduction to what C# is (Chapter 1).
- Get set up to start making C# programs (Chapter 2).
- Write our first program (Chapter 3).
- Dig into the fundamental parts of C# programming (Chapters 3 and 4).

The C# Programming Language

In a Nutshell

- Describes the general idea of programming, and goes into more details about why C# is a good language.
- Describes the core of what the .NET Framework is.
- Outlines how the compiler turns your C# source code into IL code, then JIT compiled by the CLR as the program is running.

I'm going to start off this book with a very brief introduction to C#. If you're already a programmer, and you've read the Wikipedia pages on C# and the .NET Framework, skip ahead to the next chapter.

On the other hand, if you're new to programming in general, or you're still a little vague on what exactly C# or the .NET Framework is, then this is the place for you.

I should point out that we'll get into a lot of detail about how the .NET Framework functions, and what it gives you as a programmer in Chapter 40. This chapter just provides a quick overview of the basics.

What Exactly is C#?

Computers only understand binary: 1's and 0's. All of the information they keep track of is ultimately nothing more than a glorified pile of bits. All of the things they do boil down to instructions, written with 1's and 0's.

But humans are notoriously bad at doing anything with a random pile of billions of 1's and 0's. So rather than doing that, we created programming languages, which are based on human languages (usually English) and structured in a way that allows you to give instructions to the computer. These instructions are called *source code*, and are made up of simple text files.

When the time is right, your source code will be handed off to a special program called a compiler, which is able to take it and turn it into the binary 1's and 0's that the computer understands, typically in the form of an .exe file. In this sense, you can think of the compiler as a translator from your source code to the binary machine instructions that the computer knows.

C# is one of the most popular of all of the programming languages available. There are literally thousands, maybe even tens of thousands of these languages, and each one is designed for their own purposes. C# is a simple general-purpose programming language, meaning you can use it to create pretty much anything, including desktop applications, server-side code for websites, and even video games.

C# provides an excellent balance between ease of use and power. There are other languages that provide less power and are easier to use (like Java) and others that provide more power, giving up some of its simplicity (like C++). Because of the balance it strikes, it is the perfect language for nearly everything that you will want to do, so it's a great language to learn, whether it's your first or your tenth.

What is the .NET Framework?

C# relies heavily on something called the .NET Framework. The .NET Framework is a large and powerful platform, which we'll discuss in detail in Chapter 40. (You can go read it as soon as you're done with this chapter, if you want.)

The .NET Framework primarily consists of two parts. The first part is the *Common Language Runtime*, often abbreviated as the *CLR*. The CLR is a virtual machine—a special type of software program that functions as though it is an actual computer. C# code is actually executed by the CLR, instead of being run by the actual physical computer. There are, of course, benefits and drawbacks to this kind of a setup, which we'll discuss in Chapter 40, but it turns out to be a good idea for everything except low level stuff, like an operating system or a device driver. C# is a bad choice for things like that. (Try C or C++ instead.)

I should also point out that while running your code through a virtual machine may make it a bit slower, in most cases this isn't enough to matter. In some situations, it could actually end up being *faster*. In other cases, you can call outside code that not running on a virtual machine. The bottom line is, don't stress too much about if the CLR will be fast enough.

Because C# code runs on the .NET Framework, the process your code goes through before executing is a little more complex than what I described earlier. Rather than just a single step to compile your source code to binary executable code, it goes through two steps. Here's what happens:

- The C# compiler turns your source code into *Common Intermediate Language* (*CIL* or *IL* for short).
- This IL code is packaged into an .exe file or .dll file, which can be shared with others.
- When it is time to run your program, the IL code in your .exe or .dll will be handed off to the CLR to run it.

- As your program is running, the CLR will look at the IL code and compile it into binary executable code that the computer it is running on understands. For each block of code you have, this will only happen once each time you run your program—the first time it needs it. This is called *Just-in-Time compiling*, or *JIT compiling*.

The other half of the .NET Framework is the Base Class Library, or BCL. The BCL is a massive library of code that you can reuse within your own programs, to accelerate the development of whatever you are working on. This library is much bigger than what you typically find packaged with a programming language (though Java also has a library of comparable size). It is impossible to cover all of this code in this or any book, though we'll cover the most important stuff. Before trying to build your own code for what seems like a common task, search the Internet to see if it already exists as a part of the BCL.

C# and .NET Versions

C# has gone through quite a bit of evolution over its history. The first release was in 2002, and established the bulk of the language features C# still has today.

A little over a year later, in 2003, C# 2.0 was released, adding in a lot of other big and powerful features, most of which will get quite a bit of attention in this book (generics, nullable types, delegates, static classes, etc.)

The next release, C# 3.0, expanded the language in a couple of very specific directions: LINQ and lambdas, both of which get their own chapters in this book.

The next two releases were somewhat smaller. C# 4.0 added the **dynamic** keyword, as well as named and optional method arguments. C# 5.0 added the **async** keyword, and better support for multi-threaded development.

In the C# 5 era, a new C# compiler was introduced: Roslyn. This compiler has a number of notable features: it's open source, it's written in C# (written in the language it's for), and it is available while your program is running (so you can compile additional code dynamically). Something about its construction also allows for people to more easily tweak and experiment with new features, which led to the features added in C# 6.0.

C# 6.0 adds a whole slew of little additions and tweaks across the language. While previous updates to the language could usually be summed up in a single bullet point or two, and are given their own chapters in this book, the new features in C# 6.0 are small and numerous. I try to point out what these new features are throughout this book, so that you are aware of them.

Alongside the C# language itself, both Visual Studio and the Base Class Library have both been evolving and growing. This book has been updated to work with Visual Studio 2015, .NET 4.6, and C# 6.0 at the time of publishing.

Future versions will, of course, arrive before long. Based on past experience, it's a pretty safe bet to say that everything you learn here in this book will still apply in future versions as well.

Installing Visual Studio

In a Nutshell

- To program in C#, we will need a program that allows us to write C# code and run it. That program is Microsoft Visual Studio.
- A variety of versions of Visual Studio exists, including the free Community Edition, as well as several higher tiers that offer additional features at a cost.
- You do not need to spend money to make C# programs.
- This chapter walks you through the various versions of Visual Studio to help you decide which one to use, but as you are getting started, you should consider the free Visual Studio 2015 Community Edition.

To make your own programs, people usually use a program called an *Integrated Development Environment* (IDE). An IDE combines together all of the tools you will commonly need to make software, including a special text editor designed for editing source code files, a compiler, and other various tools to help you manage the files in your project.

With C#, nearly everyone chooses to use some variation of Visual Studio, made by Microsoft. There are a few different levels of Visual Studio, ranging from the free Community edition, to the high-end Enterprise edition. In this chapter, I'll help guide you through the process of determining which one to choose.

As of the time of publication of this book, the latest version is the 2015 family. There will inevitably be future releases, but the bulk of what's described in this book should still largely apply in future versions. While new features have been added over time, the fundamentals of Visual Studio have remained the same for a very long time now.

There are three main flavors of Visual Studio 2015. Our first stop will be to look at the differences among these, and I'll point out one that is most likely your best choice, getting started. (It's free, don't worry!) I'll then tell you how to download Visual Studio and a little about the installation process. By the end of this chapter, you'll be up and running, ready to start doing some C# programming!

Versions of Visual Studio

Visual Studio 2015 comes in three editions: Community, Professional, and Enterprise. While I'm ultimately going to recommend the Community edition (it's free, and it still allows you to make and sell commercial applications with it) it is worth briefly considering the differences between the three.

From a raw feature standpoint, Community and Professional are essentially the same thing. Enterprise comes with some nice added bonuses, but at a significantly higher cost. These extra features generally are non-code-related, but instead deal with the surrounding issues, like team collaboration, testing, performance analysis, etc. While nice, these extra features are probably not a great use of your money as you're learning, unless you work for a company or attend school at a place that will buy it for you.

Microsoft will also push you heavily towards buying a subscription to the Microsoft Developer Network (MSDN) which gets you access to a lot of additional documentation and learning materials, as well as support. The reality is, between the free online documentation other web-based sources like stackoverflow.com, plus the option of buying a book or two for certain special topics, an MSDN subscription is usually not worth it when you're starting out.

Now that I've pushed you away from Enterprise and MSDN as a beginner, the only remaining question is what to actually use. And to answer that, we need to compare the Community Edition to the Professional Edition.

Community and Professional are essentially the same product with a different license. Microsoft wants to make Visual Studio available to everybody, but they still want to be able to bring in money for their efforts. With the current licensing model, they've managed to do that pretty well.

While Professional costs roughly $500 without an MSDN subscription, and about $1200 with one, Community is free. But the license prevents certain people (the ones with money) from using it. While the following interpretation is not legally binding by any means, the general interpretation of the Community license is essentially this:

You can use it to make software, both commercially and non-commercially, as long as you don't fit in one of the following categories:

- You have 5+ Visual Studio developers in your company. (If you're getting it for home, you don't count the place you work for. You have 1 developer.)
- You have 250+ computers or users in your company.
- You have a gross income of $1,000,000.

If any of the above apply, you don't qualify for the Community license, and you must buy Professional. But then again, if any of those apply to you, you've probably got the money to pay for Professional anyway.

There are a couple of exceptions to that:

- You're a student or educator, using it solely for educational uses.

- You're working solely on open source projects.

In short, for essentially everybody reading this book, you should be able to either use Community, or you're working somewhere that can afford to buy Professional or Enterprise.

This makes the decision simple: you will almost certainly want Visual Studio Community for now.

The Installation Process

Regardless of which version of Visual Studio you select, you should expect similar installation processes. You can find Visual Studio Community edition here:
https://www.visualstudio.com/products/visual-studio-community-vs

Once downloaded, start the installer and follow the steps you see. The installation process takes a while (even hours) depending on your Internet connection. Feel free to get it started and continue reading.

Once installed, it may place an icon on your desktop that you can use, but it will also put it in your Start Menu under All Programs. Look for it under Visual Studio 2015.

Visual Studio will ask you to sign in with a Microsoft account at some point. If you don't already have one, it's worth the effort. If you've got an MSDN subscription, signing in gives you access to MSDN. It also allows you to sync settings across different computers.

Following their steps to create an account is relatively painless, and while they require an email address, I've never gotten spam from them. Instead, I've just gotten updates on new versions of products coming out, which is usually welcome. So I recommend signing in with a Microsoft account right from the beginning, even if that means making an account from scratch.

Starting in the next chapter and throughout the rest of this book, we'll cover all sorts of useful tips and tricks on using Visual Studio. Towards the end of this book, in Part 5, we'll get into Visual Studio in a little more depth. Once you get through the next couple of chapters, you can jump ahead to that part whenever you're ready for it. You don't have to read through the book in order.

Throughout this book, I'm going to show you screenshots from Visual Studio 2015 Community. Depending on which version you are using, you may see some slight differences in the user interface. But all of the things that we talk about in this book will be available in all editions, and the steps to do things will be identical across the board.

> **Try It Out!**
> **Install Visual Studio.** Take the time now to choose a version of Visual Studio and install it, so that you're ready to begin making awesome programs in the next chapter.

Alternatives to Visual Studio

Almost everyone who writes programs in C# uses some version of Visual Studio. It is generally considered the best and most comprehensive option. However, there are other options available if you're looking for something different. When choosing an editor, please keep in mind that this book assumes you're using Visual Studio. Instructions on how to do various things may not apply in other IDEs.

- **Visual Studio Code (https://code.visualstudio.com/):** A free, extremely trimmed down version of Visual Studio itself, which also runs on Mac and Linux. While the full Visual Studio is useful for full development, it can be pretty heavy. If you're just interested in making some quick tweaks to some code, Visual Studio Code may be a great additional companion tool to the full version.
- **MonoDevelop (http://monodevelop.com/):** The most popular C# IDE aside from Visual Studio, and it's free (and open source). It is also cross-platform (runs on a Mac and Linux natively). This has most of the features that you get in Visual Studio, though plugins and extensions are rarer than what you find in the Visual Studio ecosystem.
- **Xamarin Studio (http://xamarin.com/studio):** A customized paid-for version of MonoDevelop with some added features for cross-platform mobile development, such as iOS and Android. This one isn't free.
- **SharpDevelop (http://www.icsharpcode.net/OpenSource/SD/Default.aspx):** A free, lighter weight IDE than Visual Studio, while still having the key features.

Hello World: Your First C# Program

In a Nutshell

- Start a new C# Console Application by going to **File > New > Project...**, choosing the Console Application template, and giving your project a name.
- Inside of the **Main** method, you can add code to write out stuff using a statement like **Console.WriteLine("Hello World!");**
- Compile and run your program with **F5** or **Ctrl + F5**.
- The template includes code that does the following:
  - **using** directives make it easy to access chunks of previously written code in the current program.
  - The **namespace** block puts all of the contained code into a single collection.
  - The code we actually write goes into the **Program** class in a method called **Main**, which the C# compiler recognizes as the starting point for a program.

In this chapter we'll make our very first C# program. Our first program needs to be one that simply prints out some variation of "Hello World!" or we'll make the programming gods mad. It's tradition to make your first program print out a simple message like this, whenever you learn a new language. It's simple, yet still gives us enough to see the basics of how the programming language works. Also, it gives us a chance to compile and run a program, with very little chance for introducing bugs.

So that's where we'll start. We'll create a new project and add in a single line to display "Hello World!" Once we've got that, we'll compile and run it, and you'll have your very first program!

After that, we'll take a minute and look at the code that you have written in more detail before moving on to more difficult, but infinitely more awesome stuff in the future!

Creating a New Project

Let's get started with our first C# program! Open up Visual Studio, which we installed in Chapter 2.

When the program first opens, you will see the Start Page come up. To create a new project, you can either select the **New Project...** button on the Start Page, or you can go up to the menu and choose **File > New > Project...** from the menu bar.

Once you have done this, a dialog will appear asking you to specify a project type and a name for the project. This dialog is shown below:

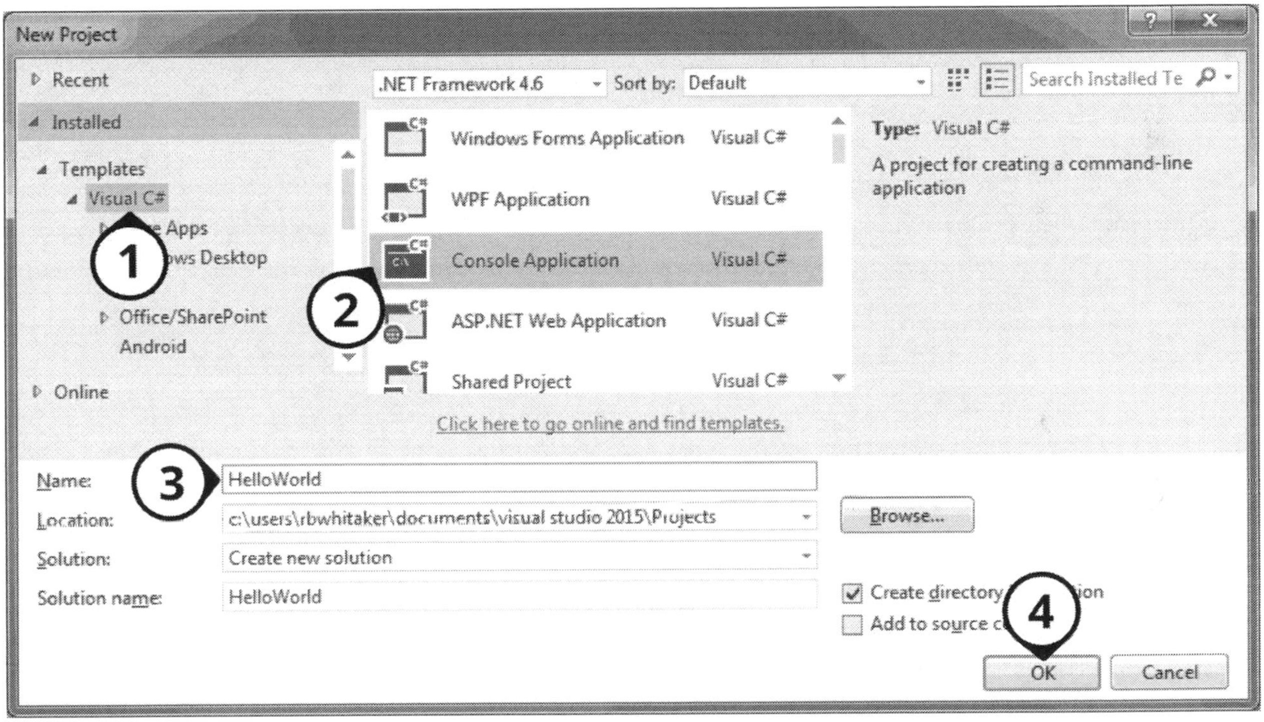

On the left side, you will see a few categories of templates to choose from. Depending on what version of Visual Studio you have installed and what plugins and extensions you have, you may see different categories here, but the one you'll want to select is the Visual C# category, which will list all C#-related templates that are installed.

Once that is selected, in the list in the top-center, find and select the **Console Application** template. The Console Application template is the simplest template and it is exactly where we want to start. For all of the stuff we will be doing in this book, this is the template to use.

As you finish up this book, if you want to start doing things like making programs with a graphical user interface (GUI), game development, smart phone app development, or web-based development, you will be able to put these other templates to good use.

At the bottom of the dialog, type in a name for your project. I've called mine "HelloWorld." Your project will be saved in a directory with this name. It doesn't really matter what you call a project, but you want to name it something intelligent, so you can find it later when you are looking at a list of all of your projects. By default, Visual Studio tries to call your programs "ConsoleApplication1" or "ConsoleApplication2." If you haven't chosen a good name, you won't know what each of these do.

By default, projects are saved under your Documents or My Documents directory (Documents/Visual Studio 2015/Projects/).

Finally, press the **OK** button to create your project! After you do this, you may need to wait for a little bit for Visual Studio to get everything set up for you.

A Brief Tour of Visual Studio

Once your project has loaded, it is worth a brief discussion of what you see before you. We'll look in depth at how Visual Studio works later on (Chapter 41) but it is worth a brief discussion right now.

By this point, you should be looking at a screen that looks something like this:

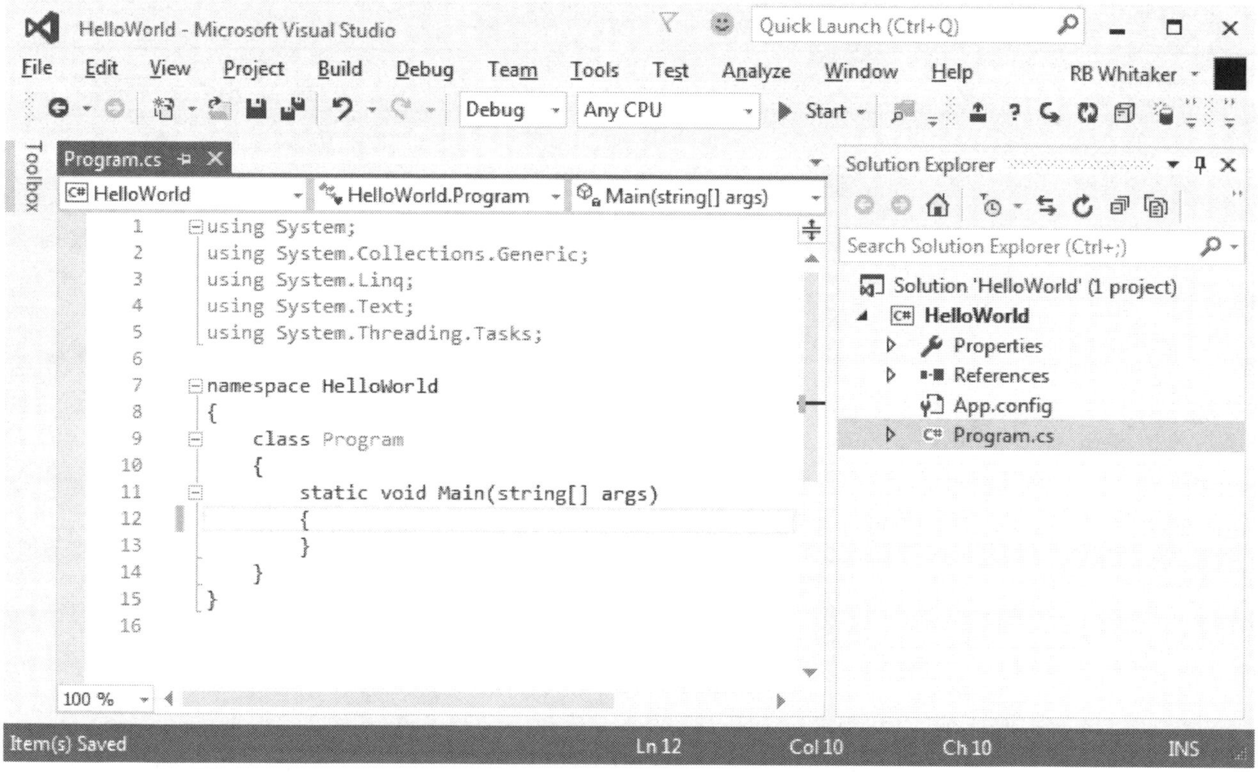

```
using System;
using System.Collections.Generic;
using System.Linq;
using System.Text;
using System.Threading.Tasks;

namespace HelloWorld
{
    class Program
    {
        static void Main(string[] args)
        {
        }
    }
}
```

Depending on which version of Visual Studio you installed, you may see some slight differences, but it should look pretty similar to this.

In the center should be some text that starts out with **using System;**. This is your program's source code! It is what you'll be working on. We'll discuss what it means, and how to modify it in a second. We'll spend most of our time in this window.

On the right side is the Solution Explorer. This shows you a big outline of all of the files contained in your project, including the main one that we'll be working with, called "Program.cs". The *.cs file extension means it is a text file that contains C# code. If you double click on any item in the Solution Explorer, it will open in the main editor window. The Solution Explorer is quite important, and we'll use it frequently.

As you work on your project, other windows may pop up as they are needed. Each of these can be closed by clicking on the 'X' in the upper right corner of the window.

If, by chance, you are missing a window that you feel you want, you can always open it by finding it on either the **View** menu or **View > Other Windows**. For right now, if you have the main editor window open with your Program.cs file in it, and the Solution Explorer, you should be good to go.

Building Blocks: Projects, Solutions, and Assemblies

As we get started, it is worth defining a few important terms that you'll be seeing spread throughout this book. In the world of C#, you'll commonly see the words *solution*, *project*, and *assembly*, and it is worth taking the time upfront to explain what they are, so that you aren't lost.

These three words describe the code that you're building in different ways. We'll start with a project. A *project* is simply a collection of source code and resource files that will all eventually get built into the same executable program. A project also has additional information telling the compiler how to build it.

When compiled, a project becomes an *assembly*. In nearly all cases, a single project will become a single assembly. An assembly shows up in the form of an EXE file or a DLL file. These two different extensions represent two different types of assemblies, and are built from two different types of projects (chosen in the project's settings).

A *process assembly* appears as an EXE file. It is a complete program, and has a starting point defined, which the computer knows to run when you start up the .exe file. A *library assembly* appears as a DLL file. A DLL file does not have a specific starting point defined. Instead, it contains code that other programs can access on the fly.

Throughout this book, we'll be primarily creating and working with projects that are set up to be process assemblies that compile to EXE files, but you can configure any project to be built as a library assembly (DLL) instead.

Finally, a *solution* will combine multiple projects together to accomplish a complete task or form a complete program. Solutions will also contain information about how the different projects should be connected to each other. While solutions can contain many projects, most simple programs (including nearly everything we do in this book) will only need one. Even many large programs can get away with only a single project.

Looking back at what we learned in the last section about the Solution Explorer, you'll see that the Solution Explorer is showing our entire solution as the very top item, which it is labeling "Solution 'HelloWorld' (1 project)." Immediately underneath that, we see the one project that our solution contains: "HelloWorld." Inside of the project are all of the settings and files that our project has, including the Program.cs file that contains source code that we'll soon start editing.

It's important to keep the solution and project separated in your head. They both have the same name and it can be a little confusing. Just remember the top node is the solution, and the one inside it is the project.

Modifying Your Project

We're now ready to make our program actually do something. In the center of your Visual Studio window, you should see the main text editor, containing text that should look identical to this:

```
using System;
using System.Collections.Generic;
```

```
using System.Linq;
using System.Text;
using System.Threading.Tasks;

namespace HelloWorld
{
    class Program
    {
        static void Main(string[] args)
        {
        }
    }
}
```

In a minute we'll discuss what all of that does, but for now let's go ahead and make our first change—adding something that will print out the message "Hello World!"

Right in the middle of that code, you'll see three lines that say **static void Main(string[] args)** then a starting curly brace ('{') and a closing curly brace ('}'). We want to add our new code right between the two curly braces.

Here's the line we want to add:

```
Console.WriteLine("Hello World!");
```

So now our program's full code should look like this:

```
using System;
using System.Collections.Generic;
using System.Linq;
using System.Text;
using System.Threading.Tasks;

namespace HelloWorld
{
    class Program
    {
        static void Main(string[] args)
        {
            Console.WriteLine("Hello World!");
        }
    }
}
```

We've completed our first C# program! Easy, huh?

Try It Out!

Hello World! It's impossible to understate how important it is to actually *do* the stuff outlined in this chapter. Simply reading text just doesn't cut it. In future chapters, most of these *Try It Out!* sections will contain extra things to do, beyond the things described in the actual body of the chapter. But for right now, it is very important that you simply go through the process explained in this chapter. The chapter itself is a *Try It Out!*

So follow through this chapter, one step at a time, and make sure you're understanding the concepts that come up, at least at a basic level.

Compiling and Running Your Project

Your computer doesn't magically understand what you've written. Instead, it understands special instructions that are composed of 1's and 0's called *binary*. Fortunately for us, Visual Studio includes a thing called a *compiler*. A compiler will take the C# code that we've written and turn it into binary that the computer understands.

So our next step is to compile our code and run it. Visual Studio will make this really easy for us.

To start this process, press **F5** or choose **Debug > Start Debugging** from the menu.

There! Did you see it? Your program flashed on the screen for a split second! (Hang on... we'll fix that in a second. Stick with me for a moment.)

We just ran our program in debug mode, which means that if something bad happens while your program is running, it won't simply crash. Instead, Visual Studio will notice the problem, stop in the middle of what's going on, and show you the problem that you are having, allowing you to debug it. We'll talk more about how to actually debug your code in Chapter 44.

So there you have it! You've made a program, compiled it, and executed it!

If it doesn't compile and execute, double check to make sure your code looks like the code above.

Help! My program is running, but disappearing before I can see it!

You likely just ran into this problem when you executed your program. You push **F5** and the program runs, a little black console window pops up for a split second before disappearing again, and you have no clue what happened.

There's a good reason for that. Your program ran out of things to do, so it finished and closed on its own. (It thinks it's so smart, closing on its own like that.)

But we're really going to want a way to make it so that *doesn't* happen. After all, we're left wondering if it even did what we told it to. There are two solutions to this, each of which has its own strengths and weaknesses.

Approach #1: When you run it *without* debugging, console programs like this will always pause before closing. So one option is to run it without debugging. This option is called Release Mode. We'll cover this in a little more depth later on, but the bottom line is that your program runs in a streamlined mode which is faster, but if something bad happens, your program will just die, without giving you the chance to debug it.

You can run in release mode by simply pressing **Ctrl + F5** (instead of just **F5**). Do this now, and you'll see that it prints out your "Hello World!" message, plus another message that says "Press any key to continue..." which does exactly what it says and waits for you before closing the program. You can also find this under **Debug > Start Without Debugging** on the menu.

But there's a distinct disadvantage to running in release mode. We're no longer running in debug mode, and so if something happens with your program while it is running, your application will crash and die. (Hey, just like all of the other "cool" programs out there!) Which brings us to an alternative approach:

Approach #2: Put another line of code in that makes the program wait before closing the program. You can do this by simply adding in the following line of code, right below where you put the **Console.WriteLine("Hello World!");** statement:

```
Console.ReadKey();
```

So your full code, if you use this approach, would look like this:

```
using System;
using System.Collections.Generic;
using System.Linq;
using System.Text;
using System.Threading.Tasks;

namespace HelloWorld
{
    class Program
    {
        static void Main(string[] args)
        {
            Console.WriteLine("Hello World!");
            Console.ReadKey();
        }
    }
}
```

Using this approach, there is one more line of code that you have to add to your program (in fact, every console application you make), which can be a little annoying. But at least with this approach, you can still run your program in debug mode, which you will soon discover is a really nice feature.

Fortunately, this is only going to be a problem when you write console apps. That's what we'll be doing in this book, but before long, you'll probably be making windows apps, games, or awesome C#-based websites, and this problem will go away on its own. They work in a different way, and this won't be an issue in those worlds.

Try It Out!

See Your Program Twice. I've described two approaches for actually seeing your program execute. Take a moment and try out each approach. This should only take a couple of minutes, and you'll get an idea of how these two different approaches work.

Also, try combining the two and see what you get. Can you figure out why you need to push a key twice to end the program?

A Closer Look at Your Program

Now that we've got our program running, let's take a minute and look at each of the lines of code in the program we've made. I'll try to explain what each one does, so that you'll have a basic understanding of everything in your simple Hello World program.

Using Directives

```
using System;
using System.Collections.Generic;
using System.Linq;
```

```
using System.Text;
using System.Threading.Tasks;
```

The first few lines of your program all start with the keyword **using**. A *keyword* is simply a reserved word, or a magic word that is a built in part of the C# programming language. It has special meaning to the C# compiler, which it uses to do something special. The **using** keyword tells the compiler that there is a whole other pile of code that someone made that we want to be able to access. (This is actually a bit of a simplification, and we'll sort out the details in Chapter 26.)

So when you see a statement like **using System;** you know that there is a whole pile of code out there named *System* that our code wants to use. Without this line, the C# compiler won't know where to find things and it won't be able to run your program. You can see that there are four **using** directives in your little program that are added by default. We can leave these exactly the way they are for the near future.

Namespaces, Classes, and Methods

Below the **using** directives, you'll see a collection of curly braces ('{' and '}') and you'll see the keywords **namespace**, **class**, and in the middle, the word **Main**. Namespaces, classes, and methods (which **Main** is an example of) are ways of grouping related code together at various levels. Namespaces are the largest grouping, classes are smaller, and methods are the smallest. We'll discuss each of these in great depth as we go through this book, but it is worth a brief introduction now. We'll start at the smallest and work our way up.

Methods are a way of consolidating a single task together in a reusable block of code. In other programming languages, methods are sometimes called functions, procedures, or subroutines. We'll get into a lot of detail about how to make and use methods as we go, but the bulk of our discussion about methods will be in Chapter 15, with some extra details in Chapter 27.

Right in the middle of the generated code, you'll see the following:

```
static void Main(string[] args)
{
}
```

This is a method, which happens to have the name **Main**. I won't get into the details about what everything else on that line does yet, but I want to point out that this particular setup for a method makes it so that C# knows it can be used as the starting point for your program. Since this is where our program starts, the computer will run any code we put in here. For the next few chapters, everything we do will be right in here.

You'll also notice that there are quite a few curly braces in our code. Curly braces mark the start and end of code blocks. Every starting curly brace ('{') will have a matching ending curly brace ('}') later on. In this particular part, the curly braces mark the start and end of the **Main** method. As we discuss classes and namespaces, you'll see that they also use curly braces to mark the points where they begin and end. From looking at the code, you can probably already see that these code blocks can contain other code blocks to form a sort of hierarchy.

When one thing is contained in another, it is said to be a *member* of it. So the **Program** class is a member of the namespace, and the **Main** method is a member of the **Program** class.

Classes are a way of grouping together a set of data and methods that work on that data into a single reusable package. Classes are the fundamental building block of object-oriented programming. We'll get into this in great detail in Part 3, especially Chapters 17 and 18.

In the generated code, you can see the beginning of the class, marked with:

```
class Program
{
```

And later on, after the **Main** method which is contained within the class, you'll see a matching closing curly brace:

```
}
```

Program is simply a name for the class. It could have been just about anything else. The fact that the **Main** method is contained in the **Program** class indicates that it belongs to the **Program** class.

Namespaces are the highest level grouping of code. Many smaller programs may only have a single namespace, while larger ones often divide the code into several namespaces based on the feature or component that the code is used in. We'll spend a little extra time detailing namespaces and **using** directives in Chapter 26.

Looking at the generated code, you'll see that our **Program** class is contained in a namespace called "HelloWorld":

```
namespace HelloWorld
{
    ...
}
```

Once again, the fact that the **Program** class appears within the **HelloWorld** namespace means that it belongs to that namespace, or is a member of it.

Whitespace Doesn't Matter

In C#, whitespace such as spaces, new lines, and tabs don't matter to the C# compiler. This means that technically, you could write every single program on only one line! But don't do that. That would be a pretty bad idea.

Instead, you should use whitespace to help make your code more readable, both for other people who may look at your code, or even yourself, a few weeks later, when you've forgotten what exactly your code was supposed to do.

I'll leave the decision about where to put whitespace up to you, but as an example, compare the following pieces of code that do the same thing:

```
static void Main(string
[] args) { Console
.WriteLine                                          (
             "Hello World!"                 );}
```

```
static void Main(string[] args)
{
    Console.WriteLine("Hello World!");
}
```

For the most part, in this book I'll use the style in the last block, because I feel it is the easiest to read.

Semicolons

You may have noticed that the lines of code we added all ended with semicolons (';').

This is often how C# knows it has reached the end of a statement. A *statement* is a single step or instruction that does something. We'll be using semicolons all over the place as we write C# code.

Try It Out!

Evil Computers. In the influential movie *2001: A Space Odyssey*, an evil computer named HAL 9000 takes over a Jupiter-bound spaceship, locking Dave, the movie's hero, out in space. As Dave tries to get back in, to the ship, he tells HAL to open the pod bay doors. HAL's response is "I'm sorry, Dave. I'm afraid I can't do that." Since we know not all computers are friendly and happy to help people, modify your Hello World program to say HAL 9000's famous words, instead of "Hello World!"

This chapter may have seemed long, and we haven't even accomplished very much. That's OK, though. We have to start somewhere, and this is where everyone starts. We have now made our first C# program, compiled it, and executed it! And just as important, we now have a basic understanding of the starter code that was generated for us. This really gets us off on the right foot. We're off to a great start, but there's so much more to learn!

4

Comments

Quick Start

- Comments are a way for you to add text for other people (and yourself) to read. Computers ignore comments entirely.
- Comments are made by putting two slashes (**//**) in front of the text.
- Multi-line comments can also be made by surrounding it with asterisks and slashes, like this: **/* this is a comment */**

In this short chapter we'll cover the basics of comments. We'll look at what they are, why you should use them, and how to do them. Many programmers (even many C# books) de-emphasize comments, or completely ignore them. I've decided to put them front and center, right at the beginning of the book—they really are that important.

What is a Comment?

At its core, a *comment* is text that is put somewhere for a human to read. Comments are ignored entirely by the computer.

Why Should I Use Comments?

I mentioned in the last chapter that whitespace should be used to help make your code more readable. Writing readable and understandable code is a running theme you'll see in this book. Writing code is actually far easier than reading it, or trying understanding what it does. And believe it or not, you'll actually spend far more time reading code than writing it. You will want to do whatever you can to make your code easier to read. Comments will go a very long way towards making your code more readable and understandable.

You should use comments to describe what you are doing so that when you come back to a piece of code that you wrote after several months (or even just days) you'll know what you were doing.

Writing comments—wait, let me clarify—writing *good* comments is a key part of writing good code. Comments can be used to explain tricky sections of code, or explain what things are supposed to do. They are a primary way for a programmer to communicate with another programmer who is looking at their code. The other programmer may even be on the other side of the world and working for a different company five years later!

Comments can explain what you are doing, as well as why you are doing it. This helps other programmers, including yourself, know what was going on in your mind at the time.

In fact, even if you know you're the only person who will ever see your code, you should still put comments in it. Do you remember what you ate for lunch a week ago today? Neither do I. Do you really think that you'll remember what your code was supposed to do a week after you write it?

Writing comments makes it so that you can quickly understand and remember what the code does, how it does it, why it does it, and you can even document why you did it one way and not another.

How to Make Comments in C#

There are three basic ways to make comments in C#. For now, we'll only really consider two of them, because the third applies only to things that we haven't looked at yet. We'll look at the third form of making comments in Chapter 15.

The first way to create a comment is to start a line with two slashes: **//**. Anything on the line following the two slashes will be ignored by the computer. In Visual Studio the comments change color—green, by default—to indicate that the rest of the line is a comment.

Below is an example of a comment:

```
// This is a comment, where I can describe what happens next...
Console.WriteLine("Hello World!");
```

Using this same thing, you can also start a comment at the end of a line of code, which will make it so the text after the slashes are ignored:

```
Console.WriteLine("Hello World!"); // This is also a comment.
```

A second method for creating comments is to use the slash and asterisk combined, surrounding the comment, like this:

```
Console.WriteLine("Hi!"); /* This is a comment that ends here... */
```

This can be used to make multi-line comments like this:

```
/* This is a multi-line comment.
   It spans multiple lines.
   Isn't it neat? */
```

Of course, you can do multi-line comments with the two slashes as well, it just has to be done like this:

```
//  This is a multi-line comment.
//  It spans multiple lines.
//  Isn't it neat?
```

In fact, most C# programmers will probably encourage you to use the single line comment version instead of the **/* */** version, though it is up to you.

The third method for creating comments is called XML Documentation Comments, which we'll discuss later, because they're used for things that we haven't discussed yet. For more information about XML Documentation Comments, see Chapter 15.

How to Make Good Comments

Commenting your code is easy; making *good* comments is a little trickier. I want to take some time and describe some basic principles to help you make comments that will be more effective.

My first rule for making good comments is to write the comments for a particular chunk of code as soon as you've got the piece more or less complete. A few days or a weekend away from the code and you may no longer really remember what you were doing with it. (Trust me, it happens!)

Second, write comments that add value to the code. Here's an example of a bad comment:

```
// Uses Console.WriteLine to print "Hello World!"
Console.WriteLine("Hello World!");
```

The code itself already says all of that. You might as well not even add it. Here's a better version:

```
// Printing "Hello World!" is a very common first program to make.
Console.WriteLine("Hello World!");
```

This helps to explain *why* we did this instead of something else.

Third, you don't need a comment for every single line of code, but it is helpful to have one for every section of related code. It's possible to have too many comments, but the dangers of over-commenting code matter a whole lot less than the dangers of under-commented (or *completely* uncommented code).

When you write comments, take the time put in anything that you or another programmer may want to know if they come back and look at the code later. This may include a human-readable description of what is happening, it may include describing the general method (or *algorithm*) you're using to accomplish a particular task, and it may explain why you're doing something. You may also find times where it will be useful to include why you aren't using a different approach, or to warn another programmer (or yourself!) that a particular chunk of code is tricky, and you shouldn't mess with it unless you really know what you're doing.

Having said all of this, don't take it to an extreme. Good comments don't make up for sloppy, ugly, or hard to read code. Meanwhile nice, clean, understandable code reduces the times that you need comments at all. (The code is the authority on what's happening, not the comments, after all.) Make the code as readable as possible first, then add just enough comments to fill in the gaps and paint the bigger picture.

When used appropriately, comments can be a programmer's best friend.

Try It Out!

Comment *ALL* the things! While it's overkill, in the name of putting together everything we've learned so far, go back to your Hello World program from the last chapter and add in comments for each part of the code, describing what each piece is for. This will be a good review of what the pieces of that simple program do, as well as give you a chance to play around with some comments. Try out both ways of making comments (**//** and **/* */**) to see what you like.

Part 2

The Basics

With a basic understanding of how to get started behind us, we're ready to dig in and look at the fundamentals of programming in C#.

It is in this part that our adventure really gets underway. We'll start learning about the world of C# programming, and learn about the key tools that we'll use to get things done.

In this section, we cover aspects of C# programming that are called "procedural programming." This means we'll be learning how to tell the computer, step-by-step, how to get things done.

We'll look at how to:

- Store data in variables (Chapter 5).
- Understand the type system (Chapter 6).
- Do basic math (Chapters 7 and 9).
- Get input from the user (Chapter 8).
- Make decisions (Chapter 10).
- Do things repeatedly (Chapter 12 and 13).
- Create enumerations (Chapter 14).
- Package related code together in a way that allows you to reuse it (Chapter 15).

Variables

In a Nutshell

- You can think of variables as boxes that store information.
- A variable has a name, a type, and a value that is contained in it.
- You declare (create) a variable by stating the type and name: **int number;**
- You can assign values to a variable with the assignment operator ('='): **number = 14;**
- When you declare a variable, you can initialize it as well: **int number = 14;**
- You can retrieve the value of a variable simply by using the variable's name in your code: **Console.WriteLine(number);**
- This chapter also gives guidelines for good variable names.

In this chapter, we're going to dig straight into the heart of one of the most important parts of programming in C#. We're going to discuss variables, which are how we keep track of information in our programs. We'll look at how you can create them, place different values in them, and use the value that is currently in a variable.

What is a Variable?

A core part of any programming language and any program you make is the ability to store information in memory while the program is running. For example, you might want to be able to store a player's score, or a person's name, so that you can refer back to it later or modify it.

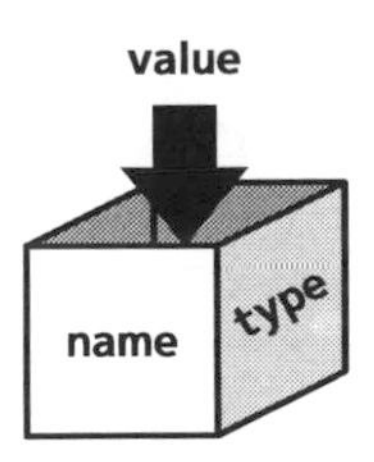

You may remember discussing variables in math classes, but these are a little different. In math, we talk about variables being an "unknown quantity" that you are supposed to solve for. Variables in math are a specific value that you just need to figure out.

In programming, a *variable* is a place in memory where you can store information. It's like a little box or bucket to put stuff in. At any point in time, you can look up the contents of the variable or rewrite the contents of the variable with new stuff. When you do this, the variable itself doesn't need to change, just the contents in the box.

Each variable is given a name and a type. The name is what you'll use in your program to access it by reading its contents or putting new stuff in it.

The variable's *type* indicates what kind of information you can put in it. C# has a large assortment of types that you can use, including a variety of integer types, floating point (real valued) types, characters, strings (text), Boolean (true/false), and a whole lot more.

In C#, types are a really big deal. Throughout this book, we'll spend a lot of time learning how to work with different types, change things from one type to another, and ultimately building our own types from the ground up.

In the next chapter, we'll get into types in great detail. For now, though, let's look at the basics of how to create a variable.

Creating Variables

Let's make our first variable. The process of creating a variable is called *declaring* a variable.

Let's start by going to Visual Studio and creating a brand new console project, just like we did with the Hello World project, back in Chapter 3. Inside of the **Main** method, add the following single line of code:

```
int score;
```

So your code should looks something like this:

```
using System;
using System.Collections.Generic;
using System.Linq;
using System.Text;
using System.Threading.Tasks;

namespace Variables
{
    class Program
    {
        static void Main(string[] args)
        {
            int score;
        }
    }
}
```

Congratulations! You've made your first variable! When you declare a variable, the computer knows that it will need to reserve a place in memory for this variable.

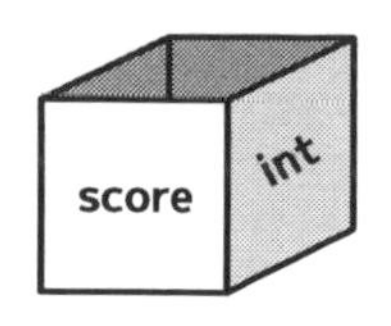

As you can see, when you declare a variable, you need to indicate the variable's name and type. This one line has both of those parts on it. The first part you see here is **int**. This is the variable's type. We'll look at the different types that are available in a minute, but for now all we need to know is that the **int** type is for storing integers. (In case you don't remember from math class, integers are whole numbers and their negatives, so 0, 1, 2, 3, 4, ..., and -1, -2, -3, -4, ...) Because we've made a variable that stores integers, we know we could put the number 100 in it, or -75946. But we could *not* store the number 1.3483 (it's not an integer), and we also could not store a word like "hamburger" (it's not an integer either), because the variable's type determines what kind of stuff we can put in it.

The second part of declaring a variable is giving it a name. It is important to remember that a variable's name is really for humans only. The computer doesn't care what it is called. (In fact, once you hand it off to the computer, it changes the name to a memory location anyway.) So you want to choose a name that makes sense to humans, and accurately describes what you're putting in it. In math, we often call variables by a single letter (like x), but in programming we can be more precise and call it something like **score** instead.

As always, C# statements end with a ';', which tells the computer that it has reached the end of the statement. After this line, we have created a new variable with the name **score** and with a type of **int** which we're now able to use!

Assigning Values to Variables

The next thing we want to do is put a value in your variable. This is called *assigning* a value to the variable, and it is done using the assignment operator: "=". The line of code below assigns the value 0 to the score variable we just created:

```
score = 0;
```

You can add this line of code right below the previous line we added.

If you're new to programming, it is worth pointing out that this use of the equals sign is different than what you might be used to from math. In math, the equals sign indicates that two things are in fact the same, even though they may be written in different forms. In C# and many other programming languages, it means we're going to take the stuff on the right-hand side of the equals sign and stick it in the variable that is named on the left.

Of course, you can assign any value to **score**:

```
score = 4;
score = 11;
score = -1564;
```

You can assign a value to a variable whenever you want, as long as it is after the variable has been declared. Of course, we haven't really learned very powerful tools for programming yet, so "whenever you want" doesn't mean a whole lot yet. (We'll get there soon, don't worry!)

When we create a variable, we often want to give it a value right away. (The C# compiler is not a big fan of you trying to see what's inside an empty variable box.) While you can declare a variable and assign it a value in separate steps, it is also possible to do both of them at the same time:

```
int theMeaningOfLife = 42;
```

In this statement, we created a variable called **theMeaningOfLife** with a type of **int**, and gave it a starting value of 42.

Retrieving the Contents of a Variable

As we just saw, we can put values in a variable by assigning the value to the variable using the assignment operator ('='). In a similar way, you can see and use the contents of a variable, simply by using the variable's name. When the computer is running your code and it sees a place that a variable's name is used, it will go to the variable, look up the contents inside, and use that value in its place.

Remember how we printed out the words "Hello World!" a couple of chapters ago? You could do the same kind of thing using a variable:

```
int number = 3;
Console.WriteLine(number); // Console.WriteLine prints lots of things, not just text.
```

When you access a variable, here's what the computer does:

1. Locates the variable that you asked for in memory.
2. Looks in the contents of the variable to see what value it contains.
3. Makes a **copy** of that value to use where it is needed.

The fact that it grabs a copy of the variable is important. For one, it means the variable keeps the value it had. Reading from a variable doesn't change the value of the variable. Two, whatever you do with the copy won't affect the original. (We'll learn more about how this works in Chapter 16, when we learn about value and reference types.)

As another example, let's look at some code that creates two different variables, and assigns the value of one to the other:

```
int a = 5;
int b = 2;

b = a;
a = -3;
```

With what you've learned so far about variables, what value will **a** and **b** have after the computer gets done running it?

> **Try It Out!**
>
> **Playing with Variables.** Take the little piece of code above and make a program out of it. Follow the same steps you did in Chapter 3 when we made the Hello World program, but instead of adding code to print out "Hello World!", add the lines above. Use **Console.WriteLine**, like we did before and print out the contents of the two variables. Before you run the code, think about what you expect to be printed out for the **a** and **b** variables. Go ahead and run the program. Is it what you expected?

Right at the beginning of those four lines, we create two variables, one named **a**, and one named **b**. Both can store integers, because we're using the **int** type. Like we saw before, we're also assigning the value 5 to **a**, and 2 to **b**. So after the first two lines, this is what we're looking at:

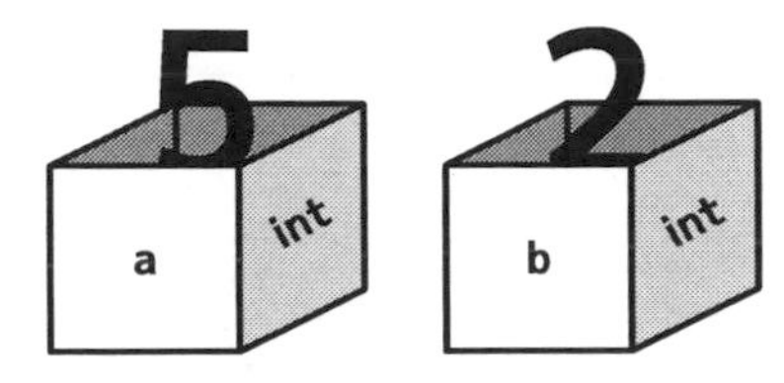

We then use the assignment operator to take the value inside of **a** and copy it to **b**:

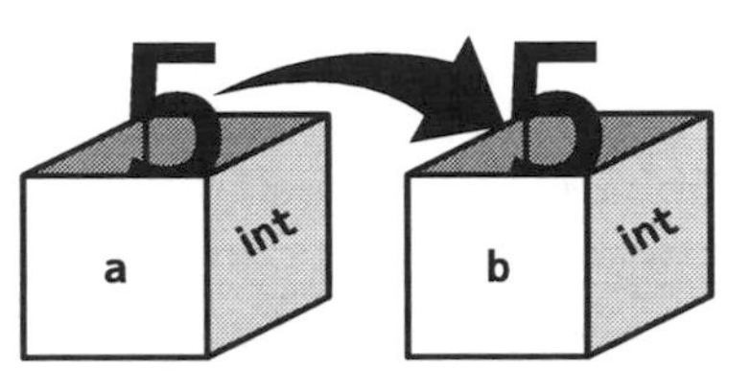

Finally, on the last line we assign a completely new value to **a**:

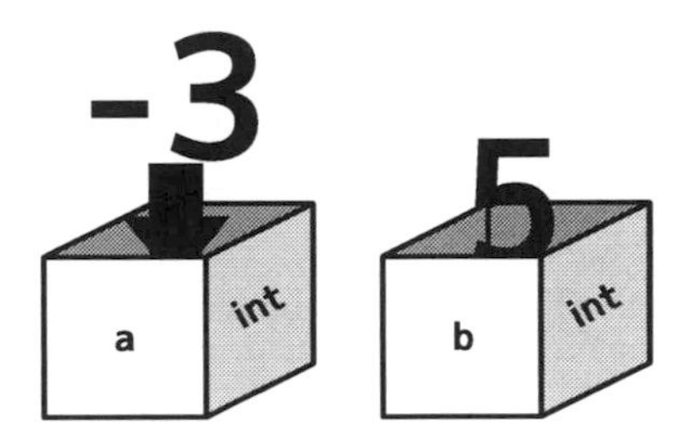

If we printed out **a** and **b**, we would see that **a** is -3 and **b** is 5 by the time this code is finished.

How Data is Stored

Before we move into a discussion about the C# type system, we need to understand a little about how information is stored on a computer. This is a key part of what drives the need for types in the first place. It is important to remember that computers only work with 1's and 0's. (Technically, they're tiny electric pulses or magnetic fields that can be in one of two states which we label 1 and 0, but you don't need to worry too much about that unless you want to make hardware.)

A single 1 or 0 is called a *bit*, and a grouping of eight of them is called a *byte*. If we do the math, this means that a single byte can store up to a total of 256 different states.

To get more states than this, we need to put multiple bytes together. For instance, two bytes can store 65,536 possible states. Four bytes can store over 4 billion states, and eight bytes combined can store over 18 quintillion possible states.

But we need a way to take all of these states and make sense out of them. This is what the type system is for. It defines how many bytes we need to store different things, and how those bits and bytes will be interpreted by the computer, and ultimately, the user.

For example, let's say we want to store letters. Modern systems tend to use two bytes to store letters. Programmers have assigned each letter a specific pattern of bits. For instance, we assign the capital letter 'A' to the bit pattern 00000000 01000001. 'B' is one up from that: 00000000 01000010. Because we're using two bytes, we have 65,536 different possibilities. That's enough to store every symbol in every language that is currently spoken on Earth, including many ancient languages, and still have room for a plethora of random extra characters.

For different types of data, we interpret the underlying bits and bytes in a different way. The **int** type that we were using earlier works like this. The **int** type uses four bytes. For brevity, in this discussion, I'm leaving off the first three bytes, which contain all zeros. The value 0 is represented with the bit pattern 00000000. The value 1 is represented with the bit pattern 00000001. 2 is represented with 00000010. 3 is 00000011. This is basically counting in a base two numbering system. (If you don't already know about counting in base two, often called *binary*, it is worth taking a detour to the Internet to check it out.)

Other types will use their bits and bytes in other ways. We won't get into the specifics about how they all work, as that's really beyond the scope of this book. I'll just point out that the way C# interprets bits and bytes fits in with the standard methods of storing information that nearly every programming language and computer use.

Good Variable Names

Before we wrap up here, I want to take a minute and explain some things about naming variables. There's a lot of debate in the programming world about the specifics of what makes a variable name good. Not everyone agrees. But I'm going to give you the rules I try to follow, which you'll discover is pretty typical, and not too far off from what most experienced programmers do.

The purpose of a variable name is to give a human-readable label for the box that you intend to put stuff in. It should make it obvious so that anyone who stumbles across the variable can instantly know what information it's supposed to contain.

It's easy to write code. It's hard to write code that you can actually go back and read and understand. Like comments, good variable names are an absolutely critical part of writing readable code, and it's not something that can be ignored.

Rule #1: Meet C#'s Requirements. C# has a few requirements for variable names. All variable names have to start with a letter (a-z or A-Z) or the underscore ('_') character, and can then contain any number of other letters, numbers, or the underscore character. You also cannot name a variable the same thing as one of the reserved keywords that C# defines. These keywords are highlighted in blue in Visual Studio, but includes things like **namespace**, **int**, and **public**. If you don't, your program won't compile, so this rule is mandatory.

Rule #2: Variable names should describe the stuff you intend on putting in it. If you are putting a player's score in it, call it **score**, or **playerScore**, or even **plrscr** if you have to, but don't call it **jambalaya**, **p**, or **monkey**. But speaking of **plrscr**...

Rule #3: Don't abbreviate or remove letters. Looking at the example of **plrscr**, you can tell that it resembles "player score." But if you didn't already know, you'd have to sit there and try to fill in the missing letters. Is it "plural scar," or "plastic scrabble"? Nope, it is "player score." You just have to sit there and study it. The one exception to this rule, and that is with common abbreviations or acronyms. HTML is fine.

Rule #4: A good name will usually be kind of long. In math, we usually use single letters for variable names. In programming, you usually need more than that to accurately describe what you're trying to do. In most cases, you'll probably have at least three letters. Often, it is 8 to 16. Don't be afraid if it gets longer than that. It's better to be descriptive than to "save letters."

Rule #5: If your variables end with a number, you probably need a better name. If you've got **count1** and **count2**, there's probably a better name for them. (Or perhaps an array, which we'll talk about later.)

Rule #6: "data", "text", "number", and "item" are usually not descriptive enough. For some reason, people seem to fall back to these all the time. They're OK, but they're just not very descriptive. It's best to come up with something more precise in any situation where you can.

Rule #7: Make the words of the variable name stand out from each other. This is so it is easier to read a variable name that is composed of multiple words. **playerScore** (with a capital 'S') and **player_score** are both more readable than **playerscore**. My personal preference is the first, but both work.

Try It Out!

Variables Quiz. Answer the following questions to check your understanding. When you're done, check your answers against the ones below. If you missed something, go back and review the section that talks about it.

1. Name the three things all variables have.
2. **True/False.** You can use a variable before it is declared.
3. How many times must a variable be declared?
4. Out of the following, which are legal variable names? answer, 1stValue, value1, $message, delete-me, delete_me, PI.

Answers: (1) name, type, value. **(2)** False. **(3)** 1. **(4)** answer,value1,delete_me,PI.

The C# Type System

In a Nutshell

- It is important to understand the C# type system. There are several built-in types that are discussed here, including:
  - Integral types (**int**, **short**, **long**, **byte**, **sbyte**, **uint**, **ushort**, **ulong**, and **char**) which store integers with various ranges.
  - Floating point types (**float**, **double**, **decimal**) that store floating point numbers with various levels of precision and range.
  - The **bool** type which stores truth values (**true**/**false**).
  - The **string** type which stores text.
  - Other types will be discussed in future chapters.

An Introduction to the Type System

Now that we understand a bit about how the computer stores information and how raw bits and bytes are interpreted, we can begin to look at the C# type system from a broader perspective.

In C#, there are tons of types that have been defined for you. In addition, as you go through this book, you'll see that C# gives you the option to create and use your own types.

In this introduction to types, we will cover all of the *primitive types*, or *built-in types*. These are types that the C# compiler knows a whole lot about. The language itself makes it easy to work with these types because they are so common. We're going to cover a lot of ground, so as you read this, remember that you don't need to master it all in one reading. Get a feel for what types are available, and come back for the details later.

The int Type

We already saw the **int** type in the last chapter, but it is worth a little more detail here to start off our discussion on types. The **int** type uses 4 bytes (32 bits) to keep track of integers. In the math world, integers can be any size we want. There's no limit to how big they can be. But on a computer, we need to be able to store them in bytes, so the **int** type has to be artificially limited. The **int** type can store numbers roughly in the range of -2 billion to +2 billion. In a practical sense, that's a bigger range than you will need for most things, but it *is* limited, and it is important to keep that in mind.

The byte, short, and long Types

Speaking of limitations in the size of the **int** type, there are three other types that also store integers, but use a different number of bytes. As a result, they have a different range that they can represent. These are the **byte**, **short**, and **long** types. The **byte** type is the smallest. It uses only one byte and can store values in the range 0 to 255. The **short** type is larger, but still smaller than **int**. It uses two bytes, and can store values in the range of about -32,000 to +32,000. The **long** type is the largest of the four, and uses 8 bytes. It can store numbers roughly in the range of -9 quintillion to +9 quintillion. That is an absolutely massive range, and it is only in very rare circumstances when that wouldn't be enough.

Type	Bytes	Range
byte	1	0 to 255
short	2	-32,768 to 32,767
int	4	-2,147,483,648 to 2,147,483,647
long	8	-9,223,372,036,854,775,808 to 9,223,372,036,854,775,807

So that's four different types that we've learned (**byte**, **short**, **int**, and **long**) which take up a different numbers of bytes (1, 2, 4, and 8, respectively). The larger they are, the bigger the number they can store. You can think of these different types as different sizes of boxes to store different sizes of numbers in.

It is worth a little bit of an example to show you how you create variables with these types. It is done exactly like earlier, when we created our first variable with the **int** type:

```
byte aSingleByte = 34;
aSingleByte = 17;

short aNumber = 5039;
aNumber = -4354;

long aVeryBigNumber = 395904282569;
aVeryBigNumber = 13;
```

The sbyte, ushort, uint, and ulong Types

We've already talked about four different types that all store integers of various sizes. Now, I'm going to introduce to you four additional types that are related to the ones we've already discussed.

Before introducing them to you, I should start by describing *signed types* vs. *unsigned types*. When I refer to a type as being signed, I mean it can include a + or - sign in front of it. In other words, the values of a signed type can be positive or negative. If you look back at the four types we already know about, you'll see that **short**, **int**, and **long** are all signed. You can store positive or negative numbers. The **byte** type, on the other hand, has no sign. (It is typically assumed that no sign is the equivalent of being positive, but that doesn't have to be the case.)

For each of the signed types we have looked at, we could have an alternate version where we shift the valid range to only include positive numbers. This would allow us to go twice as high, at the expense of not being able to use negative values at all. The **ushort**, **uint**, and **ulong** types do just this. The **ushort** type uses two bytes, just like the **short** type, but instead of going from -32,000 to +32,000, it goes from 0 to about 64,000. The same thing applies to the **uint** and **ulong** types. The exact ranges of these types will be shown a bit later.

Likewise, we can take the **byte** type, which is unsigned, and shift it so that it includes negative numbers as well. Instead of going from 0 to 255, we would go from -128 to +127. This gives us the signed byte type, **sbyte**.

Not surprisingly, we can create variables of these types just like we could with the **int** type:

```
ushort anUnsignedShortVariable = 59485; // This could not be stored in a normal short.
```

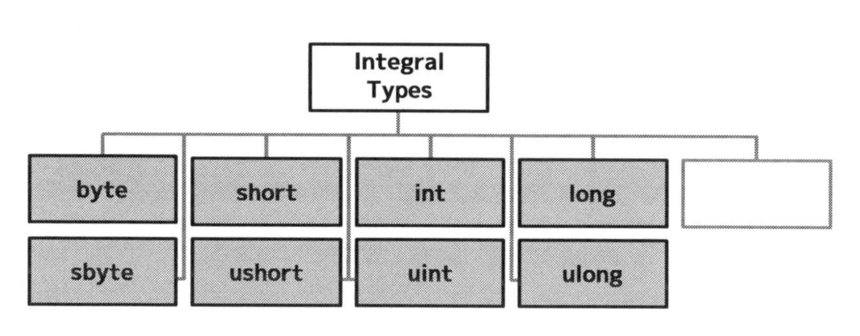

We now know of eight total types, all of which are designed to store only integers. These types, as a collection, are known as *integral types* (sometimes "integer types"). We can organize these types in the hierarchy diagram to the right. As we continue our discussion of types throughout this chapter and this book, we'll continue to expand this chart. You may notice that we're still missing one additional integral type, which we'll discuss shortly.

Each of the types we've discussed use a different number of bytes and store different ranges of values. That means each of them is best suited for storing different types of information. So how do you know what type to use? Here are three things to consider as you're choosing a type for a particular variable:

- Do you need signed values? (Can things be negative?) If so, you can immediately eliminate all of the unsigned types.
- How large might your values be? For example, if you are using a variable to store the population of a country, you know you'll need to use something bigger than a **byte** or a **short**.
- When in doubt, many programmers tend to default to the **int** type. This may be sort of a Goldilocks thing (not too big, not too small, but just about right). As a result, sometimes **int** is overused, but it may be a good starting point, at any rate.

The char Type

We have one more integral type to discuss before we move on. This last type is called the **char** type, and is used for storing single characters or letters, as opposed to numbers.

It is strange to think of characters as an integral type, which are all designed to store integers. Behind the scenes though, each letter can basically be assigned to a single number, which is why it gets lumped in among the integral types. However, when we use the **char** type, the computer will know to treat it as a letter instead, so we'll see it print out the letters instead of numbers. As we'll see later, the close association of the **char** type and the other integral types means that we can do most of the same things with any of these types.

The **char** type is two bytes big, giving it over 64,000 possible values. It's enough to store every letter in every language in the world (including some dead languages and some fictional languages like Klingon) with room to spare. The numeric value assigned to each letter follows the widely used Unicode system.

You can create variables with the **char** type in a way that is very similar to the other types we've seen, though you put the character you want to store inside single quotes:

```
char favoriteLetter = 'c';     // Because C is for cookie. That's good enough for me.
favoriteLetter = '&';
```

Later in this chapter, we'll talk about storing text of any length.

To summarize where we're at with the C# type system, we've learned 9 different integral types, including four for signed integers, four for unsigned integers, and one for characters:

Type	Bytes	Range
byte	1	0 to 255
short	2	-32,768 to 32,767
int	4	-2,147,483,648 to 2,147,483,647
long	8	-9,223,372,036,854,775,808 to 9,223,372,036,854,775,807
sbyte	1	-128 to +127
ushort	2	0 to 65,535
uint	4	0 to 4,294,967,295
ulong	8	0 to 18,446,744,073,709,551,615
char	2	U+0000 to U+ffff (All Unicode characters)

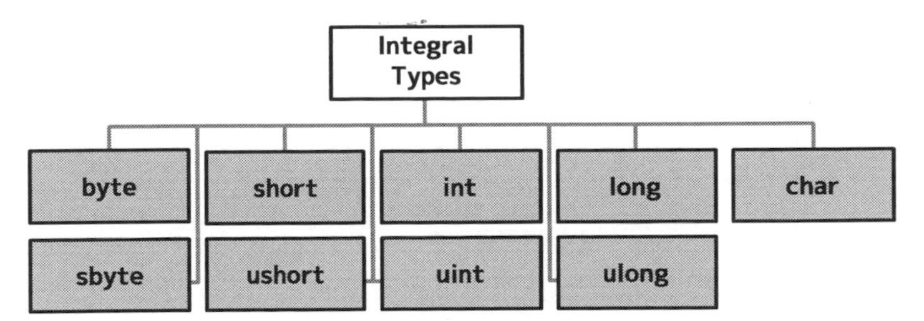

The float, double, and decimal Types

You're probably sick of integral types by now, so let's expand our discussion to include some additional types to store what are called *floating point numbers* ("real numbers" in the math world). Floating point numbers are able to store not just integer values, but also numbers with a decimal part. For example, the number 1.2 is not an integer and can't be stored by any of the integral types that we've talked about. But it *can* be stored in any of the three floating point types that C# provides.

We'll start off with the **float** type. The **float** type uses four bytes to store floating point numbers. With the **float** type, you get seven digits of precision. The largest values you can store are very large:

about 3.4 x 10^{38}. This is much higher than any of the integral types. But even these floating point types are limited in their own way. Because it only has seven digits of precision, the computer can't tell 1,000,000,000 and 1,000,000,001 apart. (Though it can tell 1000 and 1001 apart.) Of course, maybe you don't care about that kind of a difference either, because it's so small. It may be close enough for all practical purposes. Note that the integral values don't have this problem. Within the range that they can store, integral types are always exact.

There is also a limit to how small (as in, "close to zero") the **float** type can store, and that is about ±1.5 x 10^{-45}—a very tiny number.

The **double** type uses 8 bytes—twice as many as the float type (hence the name "double"). This has 15 or 16 digits of precision, and can store values as small as ±5 x 10^{-324} or as large as ±1.7 x 10^{308}. If you thought the **float** type had a huge range, think again. This dwarfs what the **float** type can hold, albeit at the expense of taking up twice as much space.

Creating variables with either of these types is pretty straightforward as well:

```
double pi = 3.14159265358979323846;
float anotherPi = 3.1415926f;
```

Take a close look at those two numbers. Obviously, the **double** version has more digits, which it can handle. But do you see what's hanging out there at the end of the number we've used for the **float**?

There's a letter 'f'.

In code, whenever we write out a value directly it is called a *literal*. When we say **a = 3;** the 3 is an integer literal. Here, when we put 3.1415926, it's a floating point literal. When we use a floating point literal like this, the compiler assumes it is the **double** type. But knowing that the **double** type is a much bigger "box" that uses 8 bytes, we can't just stuff that in a variable that is using the **float** type. It's too big. Sticking the letter 'f' on the end of a floating point literal tells the C# compiler to treat is as a literal of the **float** type, instead of the **double** type. Note that we could use lower case 'f' or capital 'F' and we would get the same result.

While we're on the subject, I want to point out that if we're working with an integer literal, we can put lower case 'l' or upper case 'L' on the end to indicate that it is a **long** integer literal. Though, I'd recommend that you always use the upper case version ('L') because lower case 'l' and the number '1' look too much alike. So you could do this if you want:

```
long aBigNumber = 39358593258529L;
```

You are also able to stick a 'u' or 'U' on the end of an integer literal to show that it is supposed to be an unsigned integer literal:

```
ulong bigOne = 2985825802805280508UL;
```

Anyway, back to the **float** and **double** types. Both of these types follow standards that have been around for decades (see the IEEE 754 floating point arithmetic standard) and are a native part of many computers' circuitry and programming languages.

C# provides another type, though, that doesn't conform to "ancient" standards. It's the **decimal** type. This type was built with the idea of using it for calculations involving money. Because it doesn't have the same hardware-level support that **float** and **double** have, doing math with it is substantially slower, but it doesn't have the same issues with losing accuracy that **float** and **double** have. This type allows for the range ±1.0 x 10^{-28} up to ±7.9 x 10^{28}, but has 28 or 29 significant digits. Essentially, a smaller range overall, but much higher precision.

You can create variables using the **decimal** type in a way that is very similar to all of the other types that we've talked about. Similar to the **float** type, you'll need to put an 'm' or 'M' at the end of any literals that you want to have be treated as the **decimal** type:

```
decimal number = 1.495m;
number = 14.4m;
```

Type	Bytes	Range	Digits of Precision
float	4	±1.0e-45 to ±3.4e38	7
double	8	±5e-324 to ±1.7e308	15-16
decimal	16	±1.0 × 10e-28 to ±7.9e28	28-29

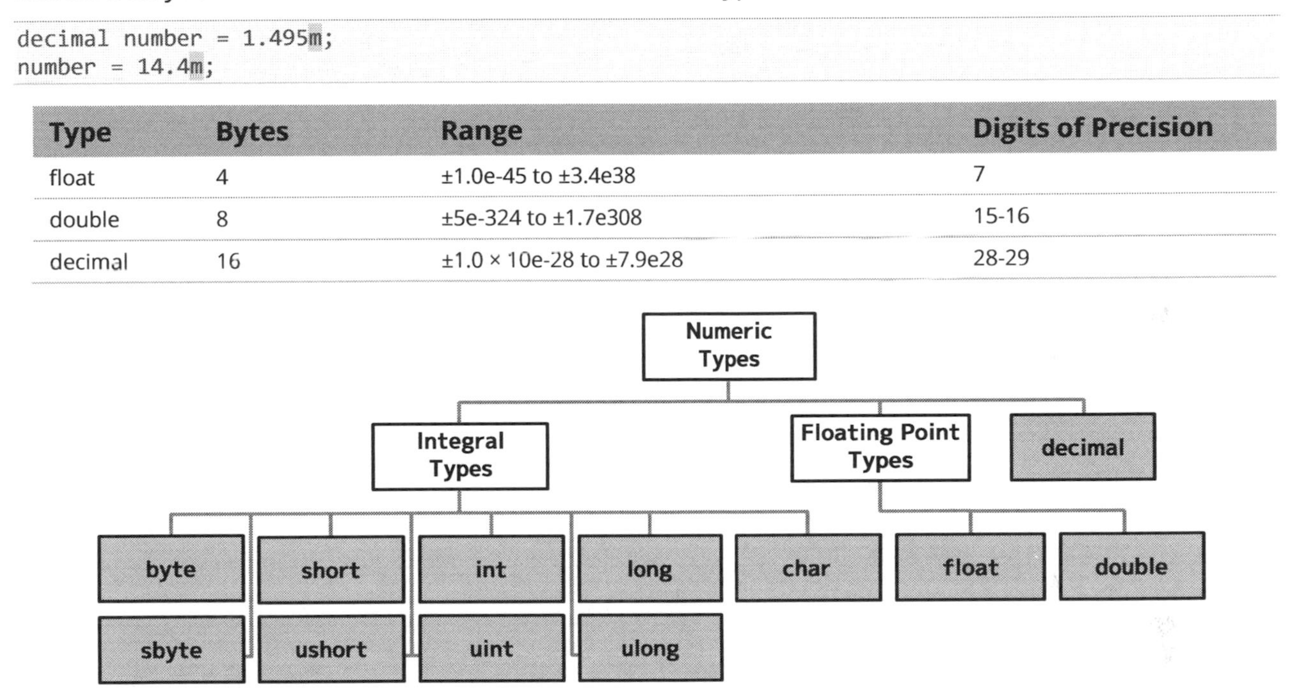

The bool Type

We'll now finally move away from numbers and discuss a completely different kind of type. The **bool** type is used to store Boolean or "truth" values. They can either be **true** or **false**. Boolean values are named after Boolean logic, which in turn was named after its inventor, George Boole. At first glance, this type may not seem very useful, but in a few chapters we'll begin to talk about decision making (Chapter 10) and these will become very important.

You can create or assign values to a variable with the **bool** type like this:

```
bool itWorked = true;
itWorked = false;
```

Note that both **true** and **false** are considered Boolean literals.

In many other languages, the **bool** type is treated like a special case of an integral type, where 0 represents false and 1 represents true. (In fact, in many of those languages, it is more like 0 represents false and anything else represents true.) While the C# **bool** type uses 0 and 1 to store true and false behind the scenes, the two things are treated very different from each other. For example, you cannot assign 1 to a variable with the **bool** type, or **true** to an integer.

While a **bool** type only keeps track of two states, and could theoretically be stored by a single bit, the **bool** type uses up a whole byte, simply because single bytes are the smallest unit of memory that you can access or reference at a time.

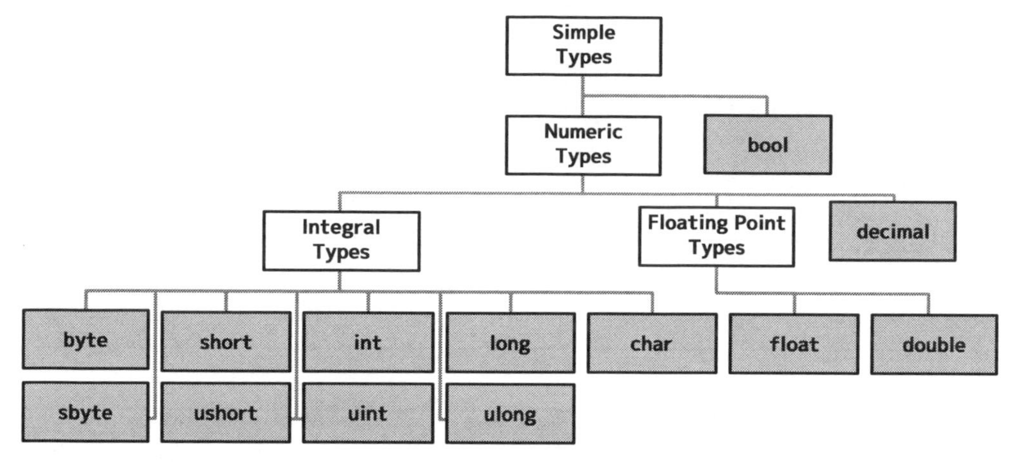

The string Type

The last type we are going to talk about here is the **string** type. Behind the scenes, this type is actually very different from all of the other types that we've discussed so far, but we won't get into the details about why or how it is different until Chapter 16.

The **string** type is used to store text of any length. The name "string" comes from the field of formal languages, where a string is defined as a sequence of symbols chosen from a set of specific symbols. In our case, that set of symbols is the set of all Unicode characters, which we discussed with the **char** type.

To create or assign values to **string** type variables, you follow the same pattern as all of the other types we've seen. String literals are marked with double quotes:

```
string message = "Hello World!";
message = "Purple Monkey Dishwasher";
```

When you see this, you'll probably think back to the Hello World program we made, because we saw a very similar thing there. It turns out, when we used the line **Console.WriteLine("Hello World!");** we were using a **string** there as well!

Try It Out!

Variables Everywhere. Create a new console application and in the **Main** method, create one variable of every type we've discussed here. That's 13 different variables. Assign each of them a value. Print out the value of each variable.

In this section, we've covered a lot of types, and I know it can be a bit overwhelming. Don't worry too much. This chapter will always be here as a reference for you to come back to later.

There are a lot more to types than we've discussed here (beyond the built-in types) but this has given us a good starting point. As we continue to learn C#, we'll learn even more about the C# type system. The diagram on the next page shows all of the types we've discussed so far. You can see that we're still missing a few pieces, but we'll pick those up as we go.

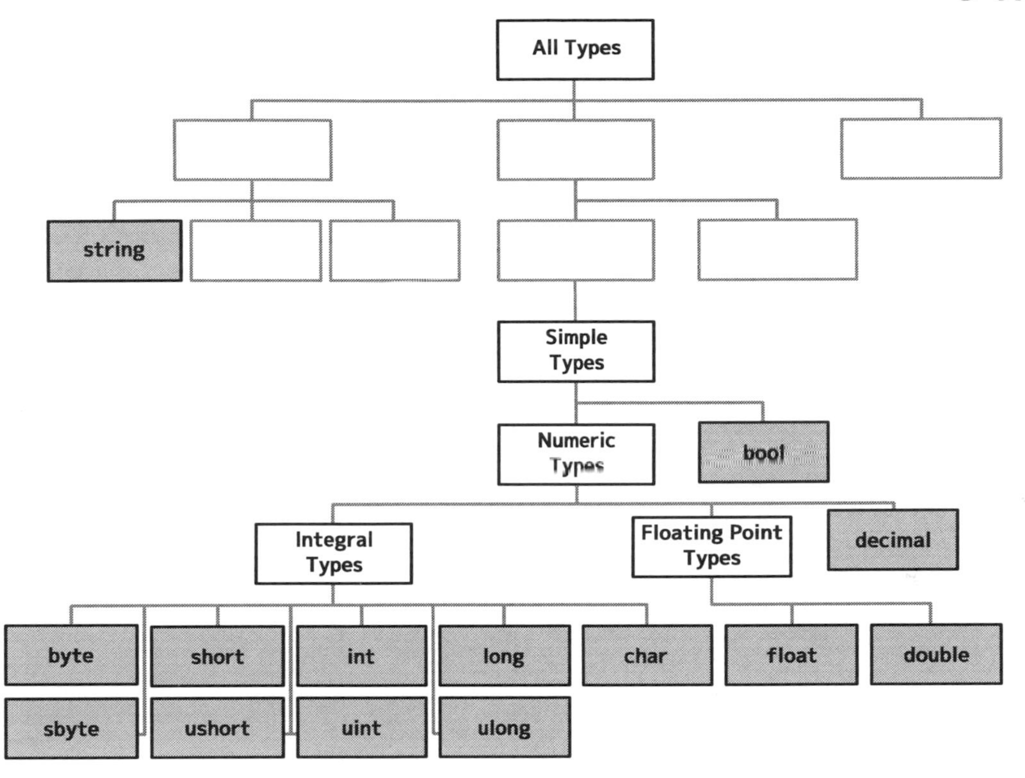

Try It Out!

Types Quiz. Answer the following questions to check your understanding. When you're done, check your answers against the ones below. If you missed something, go back and review the section that talks about it.

1. **True/False.** The **int** type can store any possible number.
2. **True/False.** The **char** type is an integral type.
3. Order the following by how large of numbers they can store, from smallest to largest: **short**, **long**, **int**, **byte**.
4. **True/False.** The **byte** type is a signed type.
5. Which can store higher numbers, **int** or **uint**?
6. What three types can store floating point numbers?
7. Which of the options in question 6 can store the largest numbers?
8. Which of the options in question 6 is considered the most precise?
9. In the following, should **myVariable** have the type **int**, **char**, or **string**, based on the literal value on the right hand side? **[type] myVariable = "8";**
10. What type is used to store true or false values?

Answers: (1) false. **(2)** true. **(3)** byte, short, int, long. **(4)** false. **(5)** uint. **(6)** float, double, decimal. **(7)** double. **(8)** decimal. **(9)** Double quotes indicate a string literal. **(10)** bool.

7

Basic Math

In a Nutshell

- Arithmetic in C# is very similar to virtually every other programming language out there.
- Addition: **a = 3 + 4;**
- Subtraction: **b = 7 - 2;**
- Multiplication: **c = 4 * 3;**
- Division: **d = 21 / 7;**
- The remainder operator ('%') gets the remainder of a division operation like this: **e = 22 % 7;** In this case, the variable **e** will contain the value 1.
- You can combine multiple operations into one, and use other variables too: **f = 3 * b - 4;**
- Order of operations matters and follows standard math rules, with multiplication and division happening before addition and subtraction, in a left to right order. Parentheses come even before multiplication and division, so you can use parentheses to make something happen before something else: **m = (y2 - y1) / (x2 - x1);**
- Compound assignment operators (**+=**, **-=**, **/=**, ***=**, and **%=**) do the desired operation and assign it to the variable on the left. So for instance **a += 3;** is the same as **a = a + 3;** in effect, adding 3 to the value of **a**.
- Chapter 9 covers additional math stuff.

With a basic understanding of variables behind us, we're ready to get in to some deeper material: math. This math is all pretty easy. The kind of stuff you learned in elementary school. Basic arithmetic. Computers love math. In fact, it is basically all they can do. (Well, that and push bytes around.) But they do it incredibly fast.

In this chapter, we'll cover how to do basic arithmetic in C#. We'll start with addition and subtraction, move up through multiplication and division, a lesser known operator called the remainder operator, positive and negative numbers, order of operations, and conclude with some compound assignment operators, which do math *and* assign a value all at once.

Operations and Operators

Let's start with a quick introduction to operations and operators. Whether you know it or not, you've been working with operations since you first learned how to add. In math, an operation is a calculation that takes two (usually) numbers and does something with them to get a result. Addition is an example of an operation, as is multiplication. Operations usually have two pieces to them: an operator (like the '+' sign) and operands. *Operand* is just a fancy name for the thing that the operator works on. So if we look at the operation 2 + 3, the numbers 2 and 3 are operands.

Interestingly, not all operators require two operands. Those that do are called *binary operators*. These are probably the ones you will be most familiar with. We'll also see some that only need one operand. These are called *unary operators*. C# also has an operator that requires three operands. It's a *ternary operator*. Because of what it does, it won't make sense to discuss it yet. We'll look at it in Chapter 10.

Addition, Subtraction, Multiplication, and Division

Doing the basic operations of addition, subtraction, multiplication, and division should be pretty straightforward. Let's start with a simple math problem that you probably did in second grade. The following code adds the numbers 3 and 4 and stores it into the variable called **a**:

```
int a = 3 + 4;
```

The same thing works for subtraction:

```
int b = 5 - 2;
```

I should also point out that while both of the above examples show doing math on the same line as a variable declaration (which is why we see the variable's type listed), all of these operations that we'll look at can be done anywhere, not just when a variable is first declared. So this would work too:

```
int a;              // Declaring the variable a.
a = 9 - 2;          // Assigning a value to a, using some math.
a = 3 + 3;          // Another assignment.

int b = 3 + 1;  // Declaring b and assigning a value to b all at once.
b = 1 + 2;          // Assigning a second value to b.
```

Remember that a variable must be declared before you can use it, as shown in the above code. You can also use any previously existing variables in your math. You don't just have to use numbers:

```
int a = 1;
int b = a + 4;
int c = a - b;
```

Don't rush by that last part too quickly. The fact that you can use variables on the right hand side of an equation is how you'll be able to calculate one thing based on another, step-by-step. As your program is running, the computer will figure out the result of the calculation on the right hand side of the equals sign and stick that value in the variable on the left.

You can also chain many operations together on one line:

```
int result = 1 + 2 - 3 + 4 - 5 + a - b + c - d;
```

Multiplication and division work in the exact same way, though there's a few tricks we need to watch out for if we do division with integers. We'll look at that more in Chapter 9, but for now, we'll just use the **float** or **double** type for division instead of integers. The sign for multiplication in C# is the asterisk (*****) and the sign for division in C# is the forward slash (**/**).

```
float totalCost = 22.54f;
float tipPercent = 0.18f;                    // Remember, this is the same as 18%
float tipAmount = totalCost * tipPercent;
```

```
double moneyMadeFromGame = 100000; // Note that you DO NOT put commas in the number.
                                   // C# won't know what to do with it.
double totalProgrammers = 4;
double moneyPerPerson = moneyMadeFromGame / totalProgrammers; // We're rich!
```

Below is a more complicated example, which calculates the area of a circle based on its radius.

```
// The formula for the area of a circle is pi * r ^ 2
float radius = 4;
float pi = 3.1415926536f;  // The 'f' makes it a float instead of a double.
float area = pi * radius * radius;

// Using the + operator with strings results in "concatenation".
Console.WriteLine("The area of the circle is " + area + ".");
```

In the above code, we have variables for the radius of a circle, the value of the number π, and the area, which we calculate using a standard formula from geometry class. We then display the results.

Notice that we can use the '+' operator with strings (text) and numbers. This is a cool feature of C#. It knows that you can't technically add a string and a number in a mathematical sense, but there is an intelligent way of handling it. It knows that it can turn the number into a string (the text representation of the number) and then it knows of a way to add or combine strings together: concatenation. It sticks one piece of text at the end of the other to create a new one. If you run this code, you will see that the result is the computer prints out, "The area of the circle is 50.26548."

By the way, back in Chapter 3, we defined what a statement was. At this point, it is worth defining another related term: *expression*. An expression is something that evaluates to a single value. Something like **3 + 4** is an expression, because it can get turned into the value 7. **(3 * 2 - 17 + 8 / 4.0)** is also an expression. It's quite a bit more complex, but it can still be evaluated to a single value. You'll see as we go through this book that expressions and statements can get very large.

Try It Out!

Area of a Triangle. Following the example above for calculating the area of a circle, create your own program that calculates the area of a triangle. The formula for this is:

$$A = \frac{1}{2}bh$$

A is the area of the triangle, *b* is the base of the triangle, and *h* is the height of the triangle.

You will want to create a variable in your program for each variable in the equation. Print out your results. Run through a few different values for *b* and *h* to make sure that it is working. For instance, check to make sure that if *b* is 5 and *h* is 6, the result is 15, and that if *b* is 1.5 and *h* is 4, you get 3.

The Remainder Operator

Do you remember in elementary school, not too long after you first learned division, how you always did things like: "23 divided by 7 is 3 remainder 2"? Well in programming, calculating the remainder has its own operator. This operator is called the *remainder operator*, though it is sometimes called the *modulo operator*, or *mod* for short. This operator *only* gives you back the remainder, not the division result. The sign for this operator is the percent sign (**'%'**). So it is important to remember that in C#, **'%'** does not mean "percent," it means "get the remainder of."

In Depth

Remainders Refresher. I know some people haven't done much with remainders since elementary school, so as a refresher, remainders work like this. For 23 % 7, we know that 7 goes in to 23 a total of three whole times. But since 7 * 3 is 21, there will be two left over. In this case, 2 is our remainder.)

Say you have 23 apples, and 7 people want the apples, this means everyone gets 3, and there are 2 remaining, or left over. This little example would look like this in C#:

```
int totalApples = 23;
int people = 7;
int remainingApples = totalApples % people; // this will be 2.
```

At first glance, the remainder operator seems like an operator that's not very useful, but if you know how to use it, you will find lots of ways to use it. For example, the remainder operator can be used to determine if a value is a multiple of another value. If it is, then the remainder operator will give us 0.

```
int remainder = 20 % 4; // This will be 0, which tells us 20 is a multiple of 4.
```

Try It Out!

Remainders. Create a simple program to write out the results of a division operation. Create two integer variables called **a** and **b**. Create a third integer variable called **quotient** (the result of a division) that stores the result of the division of **a** and **b**, and another integer variable called **remainder** that stores the remainder (using the % operator). Write out the results using **Console.WriteLine** or **Console.Write** to write out the results in the following form: if a = 17 and b = 4, print the following:

17/4 is 4 remainder 1

For bonus points check your work. Create another variable and store in it **b * quotient + remainder**. Print this value out as well. This value should be the same as **a** in all cases.

Edit your code to try multiple values for **a** and **b** to ensure it's working like it should.

Unary "+" and "-" Operators

Let's now turn our attention to a couple of unary operators. Remember, these only have one operand that they work on. You've already sort of seen these operators, but it is worth going into them a little bit more. The two operators that we want to talk about here are the unary "+" and unary "-" operators. (We're using the same sign as addition and subtraction, but it's technically a different operator.) You have probably seen these before in math. They indicate whether the number right after them is positive or negative.

```
// These are the unary '+' and unary '-' operators. They only work on
// the number right after them.
int a = +3;
int b = -44;
```

```
// These are the binary '+' and binary '-' operators, also called addition
// and subtraction that work on two numbers around them.
int a = 3 + 4;
int b = 2 - 66;
```

The unary '+' and unary '-' simply indicate the sign of the number (+ or -) that they are in front of, producing a new value from the result. (So **b = -a;** takes the value in **a**, creates a negative version of it, and stores it in **b**.)

Order of Operations and Parentheses

Perhaps you remember order of operations from math classes. C# also uses this, and it follows the same rules as math. As a refresher, with any mathematical equation or formula, you do the things in parentheses first, then powers and exponents (though C# doesn't have a built-in power operator), followed by multiplication and division, followed by addition and subtraction. C# uses the same rules.

Like in math, you can use parentheses to show that a particular part of the equation should be done first. For instance:

```
// Some simple code for the area of a trapezoid
// (http://en.wikipedia.org/wiki/Trapezoid)
```

```
double side1 = 5.5;
double side2 = 3.25;
double height = 4.6;

double areaOfTrapezoid = (side1 + side2) / 2 * height;
```

In math, if you want an especially complicated formula with lots of parentheses, you'll often use square brackets ('[' and ']') and curly braces ('{' and '}') as "more powerful" parentheses. In C#, like most programming languages, that's not how it's done. Instead, you put in multiple sets of nested parentheses like below. Be careful with multiple sets of parentheses to ensure everything lines up correctly.

```
// This isn't a real formula for anything. I'm just making it up for an example.
double a =3.2;
double b = -4.3;
double c = 42;
double d = -91;
double e = 4.343;

double result = (((a + b) * (c - 4)) + d) * e;
```

Why the '=' Sign Doesn't Mean Equals

I mentioned in Chapter 5 that the '=' sign doesn't mean that the two things equal each other. Not directly. Instead, it is an *assignment* operator, meaning that the stuff on the right gets calculated and put into the variable on the left side of the '=' sign.

Watch how this works:

```
int a = 5;
a = a + 1; // the variable a will have a value of 6 after running this line.
```

If you've done a lot of math, you may notice how strange that looks. From a math standpoint, it is impossible for a variable to equal itself plus 1. It's nonsense.

But from a programming standpoint, it makes complete sense, because what we are really saying is, "take the value currently in **a** (5), add one to it (to get 6), and assign that value back into the variable **a**." So after running that line, we've taken **a** and added one to it, replacing the old value of 5.

It's important to remember that the '=' sign indicates assignment, rather than equality.

Compound Assignment Operators

It turns out that what I just described (taking the current value of a variable, making a change to it, and updating the variable with the new value) is very common. So much so that there is a whole set of other operators that make it easy to do this. These operators are called *compound assignment operators*, because they perform a math function *and* assign a value all at once.

For instance, here's the compound assignment operator that does addition:

```
int a = 5;
a += 3;             // This is the same as a = a + 3;
```

There are also equivalent ones for subtraction, multiplication, division, and the remainder operator:

```
int b = 7;
```

```
b -= 3;    // This is the same as b = b - 3; At this point b would be 4.
b *= 5;    // This is the same as b = b * 5; At this point b would be 20.
b /= 4;    // This is the same as b = b / 4;  At this point b would be 5.
b %= 2;    // This is the same as b = b % 2;  At this point b would be 1.
```

We've covered a lot of math, and there's still more to come. But before we dig into even more math, we're going to take a little break in the next chapter and do something a little different. We'll look at how to allow the user to type in stuff for you to use in your program. We'll come back and discuss more math related things in Chapter 9.

User Input

In a Nutshell

- You can get input from the user with **Console.ReadLine();**
- Convert it to the type you need with various methods from the **Convert** class. For example, **Convert.ToInt32** converts to the **int** type, **Convert.ToDouble** converts to the **double** type, etc.

In this short chapter, we'll take a look at two additional pieces of code that we'll need to create a simple program that has some actual value.

We'll first learn how to get input from the user through the console window, then how to convert between the data types that we learned about in Chapter 6, so that you can change string-based user input to anything else.

User Input from the Console

We can get input from the user by using a statement like the following:

```
string whatTheUserTyped = Console.ReadLine();
```

This should look a lot like **Console.WriteLine**, which been using up to this point. Both **ReadLine** and **WriteLine** are methods. Methods are little blocks of reusable code. We'll get into methods in detail later (Chapter 15). For now, all we need to know is that we can use them to make things happen—read and write lines of text in this particular case. **Console.ReadLine()** will grab an entire line of text from the user (up until the <Enter> key is pressed) and put it in that **whatTheUserTyped** variable.

Converting Types

For many things, though, simply grabbing the user's input and sticking it in a **string** variable doesn't get it into the right data type for what we need. For instance, what if we need the user to type in a number? When we put it into a **string**, we can't do anything with it. We'll need to convert it to the right type of data before we can do anything with it. So to convert to the right data type, we're going to use another method that converts any of the built-in data types to another. For example, to convert this to an integer, we do this:

```
int aNumber = Convert.ToInt32(whatTheUserTyped);
```

There are similar methods for each of the other data types. For instance, **ToInt16** is for the type **short**, **ToBoolean** is for the type **bool**, **ToDouble** is for **double**, and so on. The one for the **float** type is tricky: **ToSingle**. Contrasted with the **double** type, the **float** type is said to have "single" precision.

These **Convert.ToWhatever** methods to convert things besides just strings. Any of the built-in types can be converted to any of the other built-in types. For instance, you could do the following to change a **double** to a **float**, or a **bool** to an **int** using the code below:

```
bool b = false;
int i = Convert.ToInt32(b);

double d = 3.4;
float f = Convert.ToSingle(d);
```

A Complete Sample Program

We now know enough to make a program that does something real: getting information from the user, doing basic stuff with it, and printing the results. Our next step is to actually *make* a real program. Even if you've been skipping the *Try It Out!* sections, take a few minutes and do the one below. While most of these *Try It Out!* sections have example code online, as I described in the Introduction, I've actually put my sample code for this particular problem here in the book as well.

Spend some time and actually write the program below. When you've got it complete, or if you get hung up on it, go down and compare what you've got with my code below. (It's worth mentioning that just because yours is different from the code below, doesn't mean it is wrong. There are always more than one way to get things done.)

Try It Out!

Cylinders: A Complete Program. We're going to create a simple program that will tell the user interesting information about a cylinder. A cylinder is usually defined by a height (**h**) and radius of the base (**r**). Create a program that allows the user to type in these two values. We'll do some math with those values to figure out the volume of that cylinder, as well as the surface area. (Don't worry about the formulas; I'll provide them to you.) The program will then output the results to the user.

The formula for the volume of a cylinder is this:

$$V = \pi r^2 h$$

The formula for the surface area of a cylinder is this:

$$SA = 2\pi r(r + h)$$

In both of these equations, **r** is the radius of the cylinder, **h** is the height of the cylinder, and of course, $\pi \approx 3.1415926$.

Once you've given the program above a shot, take a look at my code to accomplish this task below:

```
using System;
using System.Collections.Generic;
using System.Linq;
using System.Text;
using System.Threading.Tasks;

namespace CylinderCalculator
{
    class Program
    {
        static void Main(string[] args)
        {
            // Print a greeting message. After all, why not?
            Console.WriteLine("Welcome to Cylinder Calculator 1.0!");

            // Read in the cylinder's radius from the user
            Console.Write("Enter the cylinder's radius: ");
            string radiusAsAString = Console.ReadLine();
            double radius = Convert.ToDouble(radiusAsAString);

            // Read in the cylinder's height from the user
            Console.Write("Enter the cylinder's height: ");
            string heightAsAString = Console.ReadLine();
            double height = Convert.ToDouble(heightAsAString);

            double pi = 3.141592654; // We'll learn a better way to do PI in the next chapter.

            // These are two standard formulas for volume and surface area
            // of a cylinder.
            // You can find them on Wikipedia:
            //   http://en.wikipedia.org/wiki/Cylinder_(geometry)
            double volume = pi * radius * radius * height;
            double surfaceArea = 2 * pi * radius * (radius + height);

            // Now we output the results
            Console.WriteLine("The cylinder's volume is: " + volume +
                " cubic units.");
            Console.WriteLine("The cylinder's surface area is: " +
                surfaceArea + " square units.");

            // Wait for the user to respond before closing...
            Console.ReadKey();
        }
    }
}
```

Let's look at this a little at a time. The first part, with the **using** directives, is all template code that we discussed in Chapter 3.

The first thing we add is a simple welcome message to the user.

```
// Print a greeting message. After all, why not?
Console.WriteLine("Welcome to Cylinder Calculator 1.0!");
```

Again, this is basically identical to the Hello World program we did earlier. No surprises here.

Next, we prompt the user to enter the cylinder's radius and height and turn them into the right data type using the things we learned in this chapter:

```
// Read in the cylinder's radius from the user
Console.Write("Enter the cylinder's radius: ");
string radiusAsAString = Console.ReadLine();
double radius = Convert.ToDouble(radiusAsAString);

// Read in the cylinder's height from the user
Console.Write("Enter the cylinder's height: ");
string heightAsAString = Console.ReadLine();
double height = Convert.ToDouble(heightAsAString);
```

The first line is just simple output. We then read in the user's text and store it in a **string**. The third line uses more stuff we learned in this chapter and turns it into the correct data type. For this, I use the **double** type, but **float** or **decimal** would accomplish the same task with no meaningful difference. (You would just use **Convert.ToSingle** or **Convert.ToDecimal** instead.) We repeat the process with the cylinder's height.

We then define the value of PI and use the math operations from the last chapter to calculate the volume and surface area of the cylinder.

```
double pi = 3.141592654; // We'll learn a better way to do PI in the next chapter...

// These are two standard formulas for volume and surface area of a cylinder.
// You can find them on Wikipedia: http://en.wikipedia.org/wiki/Cylinder_(geometry)
double volume = pi * radius * radius * height;
double surfaceArea = 2 * pi * radius * (radius + height);
```

Finally, we output the results to the user, again using stuff from our first Hello World program:

```
// Now we output the results
Console.WriteLine("The cylinder's volume is: " + volume + " cubic units.");
Console.WriteLine("The cylinder's surface area is: " + surfaceArea +
        " square units.");
```

And finally, we end the program by waiting for the user to press any key:

```
// Wait for the user to respond before closing...
Console.ReadKey();
```

And there we have it! Our first *useful* C# program!

Escape Characters

When working with strings, we often run into things called escape characters. *Escape characters* (or *escape sequences*) are special sequence of characters that are supposed to have a different interpretation than what you actually see. These escape characters are designed to make it easy to represent characters that don't normally appear on the keyboard, or that the C# compiler will try to interpret as a part of the language, instead of just a normal character.

For example, how would you use **Console.WriteLine** to print out an actual quotation mark character? Saying **Console.WriteLine(""");** won't work. The problem is, the first quotation mark indicates the beginning of the string, and the C# compiler will interpret the middle one as the end of the string. Then it ends up with a final quotation mark that it can't understand.

To handle this, we can take advantage of escape characters. In C#, you start an escape sequence with the backslash character ('\'). Inside of a string literal, you can use a backslash, followed by a quotation mark, to get it to recognize it as a quotation mark that belongs inside of the string, as opposed to one that indicates the end of the string:

```
Console.WriteLine("\"");
```

It is worth pointing out that even though **\"** is visually two characters, it is treated as only one.

C# has lots of different escape characters that you'll find useful. For example, **\t** is the tab character, and **\n** is the newline character. You can see how this works in the following example:

```
Console.WriteLine("Text\non\nmore\nthan\none\nline.");
```

This will print out:

```
Text
on
more
than
one
line.
```

So what happens if you want to print out an actual backslash character? By default, the computer will try to take \ and combine it with the next letter to represent an escape character, since it marks the beginning of an escape sequence. To print out the backslash character, you will need to use two backslashes:

```
Console.WriteLine("C:\\Users\\RB\\Desktop\\MyFile.txt");
```

This will print out:

```
C:\Users\RB\Desktop\MyFile.txt
```

If you find that all of these extra slashes are kind of annoying, you can put the '@' symbol before the text (called a *verbatim string literal*) which tells the computer to ignore escape characters in the string. The following line is equivalent to the previous one:

```
Console.WriteLine(@"C:\Users\RB\Desktop\MyFile.txt");
```

String Interpolation

While we're on the subject of user input and output, there's a new feature of C# 6.0 that is worth talking about here. That is string formatting and the new string interpolation feature.

It's pretty common when you're going to present something to the user to mix in some text with some code. Earlier in this chapter, we did this by concatenating the strings and code together with a "+":

```
Console.WriteLine("The cylinder's volume is: " + volume + " cubic units.");
```

C# 6.0 introduces a new feature called string interpolation. This feature lets you write this same thing in a much more concise and readable way:

```
Console.WriteLine($"The cylinder's volume is: {volume} cubic units.");
```

String interpolation requires two steps. First, before the string, place a "$". This tells the C# compiler that this isn't just a normal string. It's one that has code embedded within it that needs to be evaluated. Within the string, when you want to put some code, you simply surround it with curly braces.

This translates to the same code as our first pass, but this is usually more readable, so it's usually preferred. The curly braces can contain any valid C# expression. It's not limited to just a single variable. That means you can do things like perform some simple math, or call a method, etc. You'll even get Visual Studio auto-completion and syntax highlighting inside the curly braces.

There's obviously a practical limit to how much you can stuff inside the curly braces before it becomes unreadable. At some point, the logic gets complicated and long enough that you lose track of how it's being formatted. At that point, it's better to pull the logic out onto a separate line and store it in a local variable.

More Math

In a Nutshell

- Division with integer types uses integer division. This means you won't get fractional or decimal results, just the integer part of the result.
- Casting can convert one type to another type. Many types are implicitly cast from narrow types to wider types. Wider types can be explicitly cast to certain narrower types using the type you want in parentheses in front: **float a = (float)3.4445;**
- Division by zero causes an error (Chapter 29) to be thrown for integer types, and results in infinity for floating point types.
- **NaN**, **PositiveInfinity**, and **NegativeInfinity** are defined for the **float** and **double** types (**float.NaN** or **double.PositiveInfinity**, for example).
- **MaxValue** and **MinValue** are defined for essentially all of the numeric types, which indicate the maximum and minimum values that a particular type can contain.
- π and *e* are defined in the **Math** class, and can be accessed like this: **float area = Math.PI * radius * radius;**
- Mathematical operations can result in numbers that go beyond the range of the type the value is being stored in. For integral types, this results in truncation (and wrapping around) while for floating point types, it results in **PositiveInfinity** or **NegativeInfinity**.
- You can use the increment and decrement operators (**++** and **--**) to add one or subtract one from a variable. For example, after the following, **a** will contain a value of 4: **int a = 3; a++;**

Here we are, back at more math. There's so much math that computers can do—in fact, it is *all* computers can do. This will be our last chapter that focuses purely on math. We'll soon move on to much cooler things.

But there are still quite a few things we need to discuss. The things we're going to talk about in this chapter are not very closely related to each other. Each section is its own thing. This means that if you already know and understand one topic, you can just jump down to the next.

Here are the basic things we're going to discuss in this chapter. We'll start by talking about doing division with integers ("integer division") and an interesting problem that comes up when we do that. We'll then talk about converting one type of data to another (called *typecasting* or simply *casting*). Next we'll talk about dividing by zero, and what happens in C# when you attempt this. We'll then talk about a few cool special numbers in C#, like infinity, NaN, the number *e*, and π. Then we'll take a look at overflow and underflow, and finish up with a cool and frequently used feature: incrementing and decrementing.

It's probably not required to read all of this if you're in a rush, but don't skip the section on casting or the section on incrementing and decrementing. You'll regret it if you do, I promise!

There's a lot to learn here. Don't worry if you don't catch it all your first time. You can always come back later on, and in fact, having some experience behind you will probably make it make that much more sense, too.

Integer Division

Let's start this section off with an experiment to help illustrate what's going on. In your head, figure out what 7 / 2 is. Got it? No really, don't just skip this. Get an answer in your head. Got it for real this time? Good. Your answer was probably 3.5, or maybe 3 remainder 1. Both are correct.

Now go into C#, and write the following, and see what ends up in **result**:

```
int a = 7;
int b = 2;
int result = a / b;
Console.WriteLine(result);
```

The computer thinks it is 3! Not 3.5, or 3 remainder 1. What we're doing isn't the normal division you learned in elementary school, but rather, a thing called *integer division*. In integer division, there's no such thing as fractional numbers.

Integer division works by taking only the integer part of the result, leaving off any fractional or decimal part. This is true no matter how close it is to the next higher number. For example, 99 / 100 is 0, even though from a math standpoint, it is 0.99, which is really close to 1.

Integer division is at work when you do division with any of the integral data types, but it does *not* apply with floating point types.

For comparison purposes, check out the following piece of code:

```
int a = 7;
int b = 2;
int c = a / b; // results in 3. Uses integer division.

float d = 7.0f;
float e = 2.0f;
float f = d / e; // results in 3.5. Uses "regular" floating point division.
```

This gets especially tricky when you mix different data types like this:

```
int a = 7;
int b = 2;
float c = 4;
float d = 3;
float result = a / b + c / d;
```

Here, the **a / b** part becomes 3 (like before), but the **c / d** part does floating point division (the normal division) and gives us 1.33333. Adding the two gives us 4.33333.

It is important to keep in mind that integer types will always try to do integer division. It can easily mess you up, but there are also plenty of times you can use it to your advantage. You just need to remember that it is happening.

If you don't want integer division, you'll need to convert the values to floating point values, as we'll discuss in the next section.

Working with Different Types and Casting

Typically, when you do math with two things that have the same type (adding two **int**s for example) the result is the same type as the things you started with. But what if you do math with two different types? For example, what if you add an **int** with a **long**? The computer basically only knows how to do math on two things with the same type. So to get around this problem, types can be changed to different types on the fly to allow the operation to be done. This conversion is called *typecasting* or simply *casting*.

There are two types of casting in C#. One is *implicit casting*, meaning it happens for you, without you having to say so, while the other is *explicit casting*, meaning you have to indicate that you want to do it, or it won't happen.

Generally speaking, implicit casting happens for you whenever you go from a *narrower* type to a *wider* type. This is called a *widening conversion*, and it is the kind of conversion that doesn't result in any loss of information. To help explain, remember that an **int** uses 32 bits, while a **long** uses 64 bits. Because of this, the **long** type can hold all values that the **int** type can handle plus a whole lot more. Because of this, we say that the **int** type is narrower, and the **long** type is wider. C# will happily cast the narrower **int** to the wider **long** when it sees a need, without having to be told. (It will implicitly cast from **int** to **long**.)

So for instance, you can do the following:

```
int a = 4049;
long b = 284404039;
long sum = a + b;
```

When you do the adding, it will take the value in **a** and turn it in to a **long** and add it to **b**.

This is a widening conversion. There's no risk of losing information because the wider type (like a **long**) can always contain any value of the narrower type (like the **int**). Because there's no risk of losing data, the conversion is safe to do, and will typically happen for you.

Floating point types are considered wider than integral types, so when there's a need, C# will convert integral types to floating point types. For example:

```
int a = 7;
float b = 2; // this converts the integer value 2 to the floating point value 2.0

// The value inside of 'a' will get converted to the float type before it is used in
// the division here. Interestingly, since the division is done with two float types,
// we will use floating point division instead of integer division.
float result = a / b;
```

On the other hand, there are also times that we will want to change from a wider type to a narrower type. This is done with an explicit cast, meaning we need to actually state, "Hey, I want to turn this type into another type." Explicit casts usually turn a value from a wider type into a narrower type, and because of that, it could mean a loss of data.

To do an explicit cast, you simply put the type you want to convert to in parentheses in front of the value you want to convert. For instance, look at this example, which turns a **long** into an **int**:

```
long a = 3;
int b = (int)a;
```

Casting doesn't just magically convert anything to anything else. Not all types can be converted to other types. It has its limitations. The compiler will give you an error if you are trying to do an explicit cast to something that it can't do. I should probably remind you that if you're trying to cast from one thing to another, and the compiler won't allow it, there's a possibility that one of the **Convert.ToWhatever()** methods discussed in Chapter 8 will do what you need.

Casting is considered an operator (the conversion operator) like addition or multiplication, and fits into the order of operations. It will be done after parentheses, but before the arithmetic operators (+, -, *, /, and %). For example, if you have the following code:

```
float result = (float)(3.0/5.0) + 1;
```

The following will happen:

1. The stuff in parentheses will be done first, taking 3.0 and dividing it by 5.0 to get 0.6 as a **double**.
2. The conversion/casting will be done next, turning the 0.6 as a **double** into a 0.6 as a **float**.
3. In preparation for doing addition with a 0.6 as a **float**, and a 1 as an integer, the 1 will be implicitly cast to a **float**.
4. Addition will be done with the 0.6 as a **float** and the 1.0 as a **float**, resulting in 1.6 as a **float**.
5. The 1.6 as a **float** will be assigned back into the **result** variable.

Try It Out!

Casting and Order of Operations. Using the above as an example, outline the process that will be done with the following statements:

- double a = 1.0 + 1 + 1.0f;
- int x = (int)(7 + 3.0 / 4.0 * 2);
- Console.WriteLine((1 + 1) / 2 * 3);

If this is confusing, try putting the code into Visual Studio and running it. You can try breaking out parts of it into different lines and debug it if you want. (Chapter 44 discusses debugging in depth.)

Division by Zero

You probably remember from math class that you can't divide by zero. It doesn't make sense mathematically. Bad things happen. You rip holes in the fabric of space-time, sucking you into a vortex of eternal doom, where you're forced to turn trees into toothpicks using nothing but one of those plastic sporks that you get at fast food restaurants.

So let's take a brief second and discuss what happens when you divide by zero in C#. If you divide by zero with integral types, a strange thing happens. An exception is "thrown." That's a phrase we'll come back to in more detail later when we talk about exceptions (Chapter 29), but for now, it is enough to know that an exception is simply an error. Your program will die, then and there if you are running without debugging. On the other hand, if you are running *with* debugging, Visual Studio will activate right at the line that the exception occurred at, allowing you to attempt to fix the problem. (For more information on what to do if this happens, see Chapter 44.)

Interestingly, if you are using a floating point type like **double** or **float**, it doesn't crash. Instead, you'll get the resulting value of **Infinity**. Integer types don't define a value that means **Infinity**, so they don't have this option. We'll look at **Infinity** and other special numbers in the next section.

Infinity, NaN, e, π, MinValue, and MaxValue

Infinity

There are a few special values that are worth discussing. Let's start off by looking at infinity. Both the **double** and the **float** type define special values to represent positive and negative infinity (+∞ and -∞). In the math world, doing stuff with infinity often results in some rather unintuitive situations. For example, **∞ + 1** is still ∞, as is subtracting 1: **∞ - 1 = ∞**.

To use these directly, you can do something like the following:

```
double a = double.PositiveInfinity;
float b = float.PositiveInfinity;
```

NaN (Not a Number)

NaN is another special value with the meaning "not a number." Like infinity, this can come up when you do something crazy, like **∞/∞**. This can be accessed like this:

```
double a = double.NaN;
float b = float.NaN;
```

E and π

Now let's move on to **e** and **π**. These are two special numbers that may be used frequently (or not) depending on what you are working on. Regardless of how often you might use them, they're worth knowing about. You can always do what we did a few chapters back, and create a variable to store those values, but why do that when there's already a pre-defined variable that does the same thing?

To use these, we'll use the **Math** class. (We're still going to talk about classes a lot more, starting in Chapter 17.) You can do this with the following code:

```
double radius = 3;
double area = Math.PI * radius * radius;

// You'll likely find more uses for pi than e.
double eSquared = Math.E * Math.E;
```

We don't need to create our own **pi** variable anymore, because it has already been done for us!

MinValue and MaxValue

Finally, let's talk quickly about **MinValue** and **MaxValue**. Most of the numeric types define a **MinValue** and **MaxValue** inside of them. These can be used to see the minimum or maximum value that the type can hold. You access these like NaN and infinity for floating point types:

```
int maximum = int.MaxValue;
int minimum = int.MinValue;
```

Overflow and Underflow

Think about this. A **short** can have a maximum value of up to 32767. So what if we do this?

```
short a = 30000;
short b = 30000;
short sum = (short)(a + b); // The sum will be too big to fit into a short.
                            // What happens?
```

First of all, I should point out that when you do math with the **byte** or **short** type, it will automatically convert them to the **int** type. So in the code above, I've had to do an explicit cast to get it back to a **short**.

Try out that code, and print out the **sum** variable at the end and see what you get.

Mathematically speaking, it should be 60000, but the computer gives a value of -5536.

When a mathematical operation causes something to go beyond the allowed range for that type, we get a thing called *overflow*. What happens is worth paying attention to. For integer types (**byte**, **short**, **int**, and **long**), the most significant bits (which overflowed) get dropped. This is especially strange, because the computer then interprets it as wrapping around. This is why we end up with a negative value in our example. You can easily see this happening if you start with the maximum value for a particular type (for example, **short.MaxValue**) and adding 1 to it. You'll end up at the minimum value.

For the floating point types, things happen differently. Because they have **PositiveInfinity** and **NegativeInfinity** defined, instead of wrapping around, they become infinity.

Another similar condition called *underflow* that can occur sometimes with floating point types. Imagine you have a very large floating point number. Something like 1,000,000,000,000,000,000,000. A **float** can store that number. Now let's say you have another very small number, like 0.00000000000000001. You can store that as a **float** as well. However, if you add the two together, a **float** cannot store the result: 1,000,000,000,000,000,000,000.00000000000000001.

A **float** just simply cannot be that big *and* that precise at the same time. Instead, the addition would result in 1,000,000,000,000,000,000,000,000 again. This is perhaps close enough for most things. But still, potentially valuable information is lost. This happens because floating point types have a certain number of digits of accuracy. Starting at the first digit that matters, it can only keep track of so many other digits before it just can't keep track of it anymore.

Underflow is not nearly as common as overflow, and it's entirely possible you may never encounter a time where it matters.

Incrementing and Decrementing

Let's talk about one final awesome feature of C#.

Perhaps you've noticed that lots and lots of times in this book, I've tried to add 1 to a value. We've already seen two ways of doing this:

```
int a = 3;
a = a + 1;  // Normal addition operator, and then assign it to the variable.
```

And:

```
int a = 3;
a += 1;     // The compound addition/assignment operator.
```

Here's yet another way of adding 1 to a value:

```
int a = 3;
a++;
```

This is called *incrementing*, and "++" is called the *increment operator*. In a few chapters, we'll see tons of places to use this.

As its counterpart, you can use "--" to subtract one from a number. This is called *decrementing*, and "--" is called the *decrement operator*.

Incidentally, the "++" in the name of the well-known C++ programming language comes from this very feature. C++ was designed to be the "next step" or "one step beyond" the programming language C, hence the name C++. (C, C++, and Java all have the increment operator too.)

One other thing worth mentioning with the increment and decrement operators is that you can write it in one of two ways. You can write **a++;**, or you can also write **++a;**. (Likewise, you can write **a--;** and **--a;**.) With the ++ at the end, it is called *postfix notation*, and with it at the beginning, it is called *prefix notation*.

There's a subtle difference between prefix and postfix notation. To understand the difference, it is important to realize that this operation, like many others, "returns a value." This is a phrase that we'll frequently see later on, but it is a concept we have already been working with. Take the addition operator, for instance. When we do **2 + 1**, the math happens and we're left with a result (**3**).

As it happens, using the increment or decrement operators will also return a result, aside from just modifying the variable involved. With the more common postfix notation (**a++;**) the *original* value of **a** is returned. With the less common prefix notation (**++a;**) the new, incremented value is returned. This might seem a little confusing, so here's an example:

```
int a = 3;
int b = ++a; // Both 'a' and 'b' will now be 4.

int c = 3;
int d = c++; // The original value of 3 is assigned to 'd', while 'c' is now 4.
```

a++ means give me the value in **a**, then increment **a**, while **++a** means increment **a** first, then give me the resulting value.

Using this operator to both modify a variable and return a result can be confusing and often less readable. Even experienced programmers often don't readily remember the subtle differences between postfix and prefix notations, and have to look it up when they run into spots where it might matter. It is more common to split up the logic across two lines, which sidesteps the confusion:

```
int a = 3;
a++;
int b = a;

int c = 3;
int d = c;
c++;
```

10

Decision Making

In a Nutshell

- Decision making is the ability for a program to make choices and do different things based on different states of the program.
- The **if** statement is the cornerstone of decision making in C#, and can optionally include **else** statements and **else-if** statements to create more sophisticated behavior:

```
if(condition)
{
    // ...
}
else if(another condition)
{
    // ...
}
else
{
    // ...
}
```

- In an **if** statement's condition block, you can use the operators **==**, **!=**, **<**, **>**, **<=**, and **>=** to check if something is "equal to", "not equal to", "less than", "greater than", "less than or equal to", or "greater than or equal to" the second side.
- The **!** operator reverses Boolean types.
- The operators **&&** (*and* operator) and **||** (*or* operator) allow you to check multiple things in an **if** statement.
- **if** statements can be nested.

Any program that does any *real* work will have to make decisions. These decisions are made based on whether a particular condition holds true or not. This allows the computer to run sections of code only when certain conditions are met, or the program is in a particular state. The central piece of decision making is a special C# statement called an **if** statement. We'll start by looking at the **if** statement, along with a few related statements. Then we'll look at a variety of ways to compare two values. We'll then look at a few other special operators called logical operators (not nearly as scary as it sounds, I promise!) that help us make more sophisticated conditions.

The 'if' Statement

Imagine a simple scenario where a teacher is assigning grades and wants to know what grade a student should get, based on their test score.

Grading systems vary throughout the world, but a pretty typical one is to give out the letters A, B, C, D, and F, where A represents a very high level of competence, F represents a failure, and B, C, and D represent varying levels in between.

The basic process is for the teacher to take a look at the student's score, and if it's 90 or higher, the student gets an 'A'. If they're lower than that, but still at 80 or higher, then they get a 'B', and so on, down to a failing grade. In order to figure out what grade a student should get, you need to make decisions based on certain things. We do things *conditionally*—meaning it only happens some of the time. We only give *some* students A's, while others get B's, and others fail.

As we start talking about decision making, let's go through the process of building a program that will determine a student's grade, based on their score.

Let's start at the beginning. In C#, decision making will always start with an **if** statement. A simple **if** statement can check if two things are equal, like the code below:

```
if(score == 100)
{
    // Any code between the curly braces gets executed by the program only when
    // the condition in the parentheses is true.
    Console.WriteLine("Perfect score!");
}
```

That should be fairly straightforward. There are a few parts to a basic **if** statement. We start off by using the **if** keyword. Then in parentheses, we put the condition we're checking for. In a minute, we'll discuss some additional cool and useful ways to make all sorts of conditions. For now, though, we simply use the **==** operator to check for equality. We then use the curly braces ('{' and '}') to show a code block that should only be run when the condition is true—the student's score was 100, in this particular case.

So to put this in the context of a complete program, here is what this might look like in your code:

```
using System;
using System.Collections.Generic;
using System.Linq;
using System.Text;
using System.Threading.Tasks;

namespace DecisionMaking
{
    class Program
    {
        static void Main(string[] args)
        {
            // Everything up here, before the if-statement will
            // always be executed.
```

```
            int score;

            Console.WriteLine("Enter your score: ");
            string scoreAsText = Console.ReadLine();
            score = Convert.ToInt32(scoreAsText);

            if (score == 100)
            {
                // The stuff here inside of the if-statement's code
                // block only gets executed when the condition is met--
                // in this case, if the score is 100.
                Console.WriteLine("Perfect score! You win!");
            }

            // Everything down here, after the if-statement will
            // also always be executed.

            Console.ReadKey();
        }
    }
}
```

Remember back in Chapter 7, that I said the '=' operator doesn't mean equals, like we find in math? Neither does the equality operator ("==") that we see here, though it is much closer. In math, to say that two things are equal is an assertion that the two are exactly equivalent to each other, just written in different forms. The equality operator in C# is a little different, in that it is a check, or a query to see if two things currently have the same value. If they do, the comparison evaluates to **true**, and if not, the comparison evaluates to **false**.

The 'else' Statement

So what if you want to do something in one case, but otherwise, do something else? For instance, what if you want to print out "You win!" if the student got 100, but "You lose!" if they didn't? That's easy to do, using an **else** block immediately after the **if** block:

```
if(score == 100)
{
    // This code gets executed when the condition is met.
    Console.WriteLine("Perfect score! You win!");
}
else
{
    // This code gets executed when it is not.
    Console.WriteLine("Not a perfect score. You lose.");
}
```

The thing to remember here (and use it to your advantage) is that one or the other block of code will be executed, but not both. If it does one, it won't do the other, but it will for sure do one of them.

'else if' Statements

You can also get more creative with **if** and **else**, stringing together many of them:

```
if(score == 100)
{
    Console.WriteLine("Perfect score! You win!");
}
```

```
else if(score == 99)
{
    Console.WriteLine("Missed it by THAT much."); // Get Smart reference, anyone?
}
else if(score == 0)
{
    Console.WriteLine("You must have been TRYING to get that bad of a score.");
}
else
{
    Console.WriteLine("Ah, come on! That's just boring.");
    Console.WriteLine("Seriously. Next time pick a more interesting score.");
}
```

In this code, one (and only one) of the blocks will get executed, based on the score.

I should point out that while most of the **if** and **else** blocks that we've looked at so far have had only one statement in them, that's not a requirement. You can have as much code inside of each set of curly braces as you want.

Curly Braces Not Always Needed

So far, our examples have always had a block of code after the **if** or **else** statement, surrounded by curly braces. If you have exactly one statement, then you don't actually need the curly braces. So we could have written the previous code like this:

```
if(score == 100)
    Console.WriteLine("Perfect score! You win!");
else if(score == 99)
    Console.WriteLine("Missed it by THAT much."); // Get Smart reference, anyone?
else if(score == 0)
    Console.WriteLine("You must have been TRYING to get that bad of a score.");
else
{
    Console.WriteLine("Ah, come on! That's just boring.");
    Console.WriteLine("Seriously. Next time pick a more interesting score.");
}
```

As you can see from that last part, if you have multiple statements, you will always need the curly braces.

In a lot of cases, if it's just a single line, it can be more readable to leave off the curly braces, which have a tendency to add a lot of mental weight to the code. So the code feels lighter without them.

It's important to remember one maintenance problem when you leave off the curly braces. There's a danger that as you make changes to your code later on, you might accidentally add in the new lines and *forget* to add the curly braces. This completely change the way the code is executed, as shown here:

```
if(score == 100)
    Console.WriteLine("Perfect score! You win!");
    Console.WriteLine("That's awesome!");          // Inadvertently executed every time.
```

Remember, whitespace doesn't matter in C#, so even though that last statement *looks* like it belongs in the **if** statement, it actually isn't in there. The middle line will only be executed when the condition is met, but the last line will always be executed, because it isn't in the **if** statement.

Usually as long as you're careful when you go to modify **if** statements, and add the curly braces back in when you need multiple lines, the shorter syntax with no curly braces is shorter and easier to read. But if you keep finding yourself making mistakes because you forgot to add the curly braces in, it might be worth just having a personal habit of always adding curly braces no matter what.

Relational Operators: ==, !=, <, >, <=, >=

Let's take a look at some better and more powerful ways to specify conditions. So far, we've only used the == operator to check if two things are exactly equal, but there are many others.

The **==** operator that we just saw is one of many *relational operators*, which is a fancy way of saying "an operator that compares two things." There are several others.

For instance, the **!=** operator checks to see if two things are *not* equal to each other. You use this in the same way that we use ==, but it does the opposite:

```
if(score != 100)
{
    // This code will be executed, as long as the score is not 100.
}
```

Then there's the **>** and **<** operators, which determine if something is greater than or less than something else. These work just like they do in math:

```
if(score > 90)
{
    // This will only be executed if the score is more than 90.
    Console.WriteLine("You got an 'A'!");
}

if(score < 60)
{
    // This will only be executed if the score is less than 60.
    Console.WriteLine("You got an 'F'. Sorry.");
}
```

Of course, you may have noticed that the code above isn't exactly what we set out to do in the original problem statement. We wanted to give an A if they scored *at least* 90. In the above code, 90 doesn't result in an A, because 90 is not greater than 90.

Which brings us to the last two relational operators **>=** and **<=**. These two mean "greater than or equal to" and "less than or equal to" respectively.

```
if(score >= 90)
{
    // This will only be executed if the score is 90 or higher...
    // Subtly different than the above example, because it also picks up 90.
    Console.WriteLine("You got an 'A'!");
}
```

In math, we would use the symbols ≤ and ≥, but since those aren't on the keyboard, C# and most other programming languages will use <= and >= instead.

This gives us everything we need to write the full program we set out to do:

```
using System;
using System.Collections.Generic;
```

```
using System.Linq;
using System.Text;
using System.Threading.Tasks;

namespace DecisionMaking
{
    class Program
    {
        static void Main(string[] args)
        {
            int score;

            Console.Write("Enter your score: ");
            string scoreAsText = Console.ReadLine();
            score = Convert.ToInt32(scoreAsText);

            // This if-statement is separate from the rest of them. Not because of the blank
            // line between this statement and the next block, but because that starts all
            // over with a brand new if-statement.
            if (score == 100)
                Console.WriteLine("Perfect score! You win!");

            // This checks each condition in turn, until it finds the first one that
            // is true, at which point, it executes the chosen code block, then jumps down
            // to after the end of the whole if/else code.
            if (score >= 90)
                Console.WriteLine("You got an A.");
            else if (score >= 80)
                Console.WriteLine("You got a B.");
            else if (score >= 70)
                Console.WriteLine("You got a C.");
            else if (score >= 60)
                Console.WriteLine("You got a D.");
            else
                Console.WriteLine("You got an F.");

            Console.ReadKey();
        }
    }
}
```

Using 'bool' in Decision Making

Remember when we first discussed data types in Chapter 6 that I told you about the **bool** type, and that it would turn out to be useful? We're there now. You can use variables with the **bool** type in making decisions; in fact, it is very common to do so. For instance, check out this block of code, which determines if a player has enough points to pass the level:

```
int score = 45; // Ideally, this would be set up in a way that the score changes as a
                // player progresses through a game.

int pointsNeededToPass = 100;

bool levelComplete;

if(score >= pointsNeededToPass)
    levelComplete = true;
else
```

```
    levelComplete = false;

if(levelComplete)
{
    // We'll be able to do more here later, as we learn more C#
    Console.WriteLine("You've beaten the level!");
}
```

Note that relational operators *return* or "give back" a **bool** value—meaning you can use a relational operator like **==** or **>** to directly assign a value to a **bool**:

```
int score = 45; // Ideally, this would be set up in a way that the score changes as a
                // player progresses through a game.

int pointsNeededToPass = 100;

// The parentheses below are optional.
bool levelComplete = (score >= pointsNeededToPass);

if(levelComplete)
{
    // We'll be able to do more here later, as we learn more C#
   Console.WriteLine("You've beaten the level!");
}
```

Try It Out!

Even and Odd. In Chapter 7, we talked about the remainder operator (%). This can be used to determine if a number is even or odd. If you take a number and divide it by 2 and the remainder is 0, the number is even. If the remainder is 1, the number is odd.

Write a program that asks the user for a number and displays whether the number is even or odd.

The ! Operator

A second ago, we looked at the **!=** operator, which sees if two things are *not* equal. There's another related way that we'll see the '!' sign, and that is to *negate* conditions. The **!** operator just returns the opposite value of whatever it is used on. So for instance:

```
bool levelComplete = (score >= pointsNeededToPass);

if(!levelComplete) // If the level is NOT complete...
    Console.WriteLine("You haven't won yet. Better keep trying...");
```

You can also combine this with all of the conditional operators that we've talked about, but the **!** operator has higher precedence than the relational operators in the order of operations, meaning it happens first, before comparisons. If you want to do the comparison first and then negate it, you have to use parentheses. To illustrate:

```
if(!(score > oldHighScore))
{
}

// That's effectively the same as:
if(score <= oldHighScore)
```

```
{
}
```

Conditional Operators: && and || (And and Or)

There are a lot of ways your conditions could become more complicated. For instance, imagine you are making a game where the player controls a spaceship that has both shields and armor, and the player only dies when *both* shields and armor are gone. You will need to check against both things, rather than just one.

Here's where conditional operators come in to play. C# has an *and* operator, which looks like this: **&&**, and an *or* operator that looks like this: **||**. (The '|' key is above the **<Enter>** key on most keyboards and requires pushing **<Shift>**.) You can use these to check multiple things at once:

```
int shields = 50;
int armor = 20;

if(shields <= 0 && armor <= 0)
    Console.WriteLine("You're dead.");
```

This could be read as, "if **shields** is less than or equal to zero, and **armor** is less than or equal to zero." With the **&&** operator, both parts of the condition must be true in order for the whole expression to be true.

The **||** operator works in a similar way, though if either one is true, then the whole thing becomes true.

```
int shields = 50;
int armor = 20;

if(shields > 0 || armor > 0)
    Console.WriteLine("You're still alive! Keep going!");
```

One thing worth mentioning is that with either of these, the computer will do *lazy evaluation*. This means that it won't check the second part unless it needs to. So in the example above, the computer will always check to see if **shields** is greater than 0, but it only bothers checking if **armor** is greater than 0 if **shields** is less than or equal to 0.

You can combine lots of these together, and along with parentheses, make some pretty crazy conditions. Anything you want is possible, though readability may quickly become an issue. Which leads us to the next section, which might provide an alternative approach that may be more readable.

Nesting If Statements

You can also put **if** statements (and **if-else** statements) inside of other **if** statements. This is called *nesting* them. For example:

```
if(shields <= 0)
{
    if(armor <= 0)
        Console.WriteLine("Your shields and armor are both zero! You're dead!");
    else
        Console.WriteLine("Shields are gone, but armor is keeping you alive!");
}
```

```
else
{
    Console.WriteLine("You still have shields left. The world is safe.");
}
```

It doesn't have to end there, either. You could nest **if** statements inside of **if** statements inside of **if** statements inside of even more **if** statements. However, the fact that it is doable doesn't necessarily make it a good idea. Lots of nesting means the code is less readable. Use it when there's a need for it (and there will be) but you don't need to overdo it.

Try It Out!

Positive or Negative? One thing that a lot of people have trouble with is doing multiplication when negative numbers are involved. They typically run into trouble when trying to figure out if the result should be positive or negative.

You're going to write a program to help them! But there's a catch. In this example, you're banned from actually doing the multiplication to figure out the answer. (It would be all too easy to take two numbers and multiply them, and then check if the result is greater than or less than zero.) Instead, you will follow the same logic that a human has to follow to get your answer.

When you're multiplying two numbers together, if the two numbers have the same sign (both positive or both negative) the result is positive. If they have different signs, the result is negative.

Write a program that asks the user for two numbers and then, using the rule above, prints out whether the result should be positive or negative.

The Conditional Operator ?:

Now that we know a little about logic, I want to bring up another operator in C#. Remember when we first talked about operators in Chapter 7, how we discussed unary and binary operators? Unary operators only work on one thing, (for example, the negative sign) while binary operators work on two things (addition or subtraction). I'm going to introduce a new operator, which is a ternary operator. This means it operates on *three* parts. This might seem kind of strange to you. It probably ought to. It's not a very normal thing to see in math or programming (though this specific operator is in many languages).

This operator is called the conditional operator, and it works quite a bit like an **if-else** statement. It uses the **?** and **:** characters like this:

```
(boolean condition) ? value if true : value if false
```

In a practical example, it might look like this:

```
Console.WriteLine((score > 70) ? "You passed!" : "You failed.");
```

Everything that you do with the conditional operator could have been done without it. But it is a nice shorthand way to do things that can make things very readable. On the other hand, it can also reduce readability, so it isn't *always* a better choice. Just an alternative that you may find helpful.

Switch Statements

> **In a Nutshell**
> - **switch** statements are an alternative to **if** statements, especially if they involve a lot of "else if X... else if Y... else if Z..." kind of stuff.
> - You can use the following types in a switch statement: **bool**, **string**, and all integral types.
> - You can also use enumerations. (See Chapter 13.)
> - Unlike C++ and Java, there's no implicit fall-through, meaning you can't leave out the **break** statements and allow code to go from one case block to the next, with the exception of putting multiple case labels together (with no code at all between them).

In the previous chapter, we looked at basic **if** statements and decision making. Before moving on to the next real topic, we want to take a moment and discuss a statement in C# that is very similar to **if** statements. These statements are called **switch** statements. In this case, think of a switch like a railroad switch, which determines which track a train will end up travelling down.

We'll take a look at when we'd want to use **switch** statements, how to do them, and then wrap up with a couple of extra details about **switch** statements.

Anything you can do with a **switch** statement can also be done with an **if** statement. (Theoretically, you could skip this entire chapter if you want, but you'll eventually run into **switch** statements in the "real world," and it will be nice to know what they are about.)

I should point out that while there may be some small performance tradeoffs with using **switch** vs. **if-else** (**switch** is usually considered as fast or faster, depending on the situation) but it is usually not enough to overtake readability concerns. It is usually in your best interest to pick whichever version makes for the most readable code.

The Basics of Switch Statements

It is pretty common to have a variable and want to do something different depending on the value of that variable. For instance, let's say we have a menu with five choices (1-5). The user types in their choice, which we store in a variable. We want to do something different depending on which value they chose.

Using only the tools that we already know, we'd probably think about a sophisticated **if-else** statement. Or rather, an **if/else-if/else-if/else-if/else-if/else** statement. Something like this:

```
int menuChoice = 3;

if (menuChoice == 1)
{
    Console.WriteLine("You chose option #1.");
}
else if (menuChoice == 2)
{
    Console.WriteLine("You chose option #2. I like that one too!");
}
else if (menuChoice == 3)
{
    Console.WriteLine("I can't believe you chose option #3.");
}
else if (menuChoice == 4)
{
    Console.WriteLine("You can do better than 4....");
}
else if (menuChoice == 5)
{
    Console.WriteLine("5? Really? That's what you went with?");
}
else
{
    Console.WriteLine("Hey! That wasn't even an option!");
}
```

That gets the job done, but this could also be done with a **switch** statement.

To make a **switch** statement, we'll use the **switch** keyword, and a **case** keyword for each of the various "cases" or options that we have. We'll also use the **break** and **default** keywords, but we'll talk about those more in a second. The **if** statement we had above would look like this as a **switch** statement:

```
int menuChoice = 3;

switch (menuChoice)
{
    case 1:
        Console.WriteLine("You chose option #1");
        break;
    case 2:
        Console.WriteLine("You chose option #2. I like that one too!");
        break;
    case 3:
        Console.WriteLine("I can't believe you chose option #3.");
        break;
    case 4:
```

```
        Console.WriteLine("You can do better than 4....");
        break;
    case 5:
        Console.WriteLine("5? Really? That's what you went with?");
        break;
    default:
        Console.WriteLine("Hey! That wasn't even an option!");
        break;
}
```

As you can see, we start out with the **switch** statement, and in parentheses, we put the variable we are going to base our "switch" on.

Then, we have a sequence of **case** statements (sometimes called **case** labels), which indicate that if the variable matches the value of the **case** statement, the contained block of code is where the flow of execution will go.

It is important to note that the flow of execution will go into exactly one of the **case** labels, so you will never end up in a situation where more than one **case** block gets executed.

At the end of each **case** block, you must put the **break** keyword, which sends the flow of execution back outside of the entire **switch** statement, and down to the next part of the code.

Notice, too, that we can have a **default** label, which indicates where the flow of execution should go if none of the other case labels are a match. The **default** label doesn't *need* to be the last one, but typically that's where people put it, and it is good practice to do so. The **default** block works as a sort of catch-all for anything other than the specific situations of the other **case** labels. Note that this is like the final **else** block in our original, giant **if** statement.

Types Allowed with Switch Statements

In our above example, we see that you can use the **int** type in a **switch** statement. Not all types can be used in a **switch**, but many can. Here's the list of types that can be used: **bool**, **string**, and any of the integral types (including **char**).

We can also use enumerations in a switch statement, though we haven't discussed those yet. We will later on in Chapter 14.

No Implicit Fall-Through

If you are coming from the C++ or Java world into the C# world, you may be aware of a little trick that you can do where one case block "falls through" to the next block if you leave off the **break** statement.

So for instance, you can do this in C++ and Java:

```
// This doesn't work in C#
switch (menuChoice)
{
    case 1:
        Console.WriteLine("You chose 1.");
    case 2:
        Console.WriteLine("You chose 2. Or maybe you chose 1.");
        break;
}
```

Because there's no **break** in the **case 1** block, the code there would be executed, and then continue on down into the case 2 block, until you hit a **break** statement. This isn't allowed in C#. Every **case** block needs a **break** statement.

The reason for requiring this is that people *accidentally* ended up doing this far more often than they intentionally do it. They leave off the **break** statement by accident, resulting in a bug that is usually somewhat tricky to resolve. To prevent this, C# won't even allow you to leave off the **break** statement and fall through.

However, there is one situation where you *can* do this: multiple case blocks with no code in between it:

```
// This does work in C#
switch (menuChoice)
{
    case 1:
    case 2:
        Console.WriteLine("You chose option 1 or 2.");
        break;
}
```

This allows you to do the same thing, regardless of whether the value is 1 or 2. And in this case, while **case 1** doesn't have a **break** statement, it also doesn't need it, because we're simply saying, "case 1 and 2 are the same thing." You can write the same thing with an **if** statement as well, using the && operator.

Try It Out!

Making a Calculator. We're going to try making a program that is a calculator. This simple program is going to bring in much of everything else we've learned so far, so be ready for a challenge!

The program that we'll make is going to be a simple calculator. We're going to ask the user to type in two numbers and then type in a math operation to perform on the two numbers.

Use a switch statement to handle the different operations in different ways.

Allow the user to type in '+' for addition, '-' for subtraction, '*' for multiplication, '/' for division, and '%' for remainder.

For bonus points, allow the user to type in '^' for a power. (You can compute this using **Math.Pow**.) For example, you can do x^2 by using the code below:

```
double result = Math.Pow(x, 2);
```

Print out the results for the user to see.

Looping

In a Nutshell

- **while** loops, **do-while** loops, and **for** loops all allow you to repeat things in various ways.

```
while(condition) { /* ... */ }

do { /*... */ } while(condition);

for(initialization; condition; update) { /* ... */ }
```

- You can break out of a loop at any time with the **break** keyword and advance to the next iteration of the loop with the **continue** keyword.
- The **foreach** loop is not discussed here, but will be in the next chapter.

In this chapter we'll discuss yet another very powerful feature in the C# language: loops. Loops allow you to repeat sections of code multiple times. We'll discuss three types of loops in this chapter, and cover a fourth type in the next chapter when we discuss arrays. (It will make a lot more sense there.)

The While Loop

The first kind of loop we're going to talk about is the **while** loop. A **while** loop will repeat certain code over and over, as long as a certain condition is true. While loops are constructed in a way that looks a whole lot like an **if** statement:

```
while( condition )
{
    // this code is repeated until the condition is true...
}
```

Let's start with a really simple example that counts to ten:

```
int x = 1;
while(x <= 10)
{
```

```
    Console.WriteLine(x);
    x++;
}
```

Let's take a minute and look at what the computer does when it runs into this code. This starts with the variable **x** being 1. The program then checks the condition in the **while** loop and asks, "is **x** less than or equal to 10?" This is true, so the stuff inside the **while** loop gets executed. It writes out the value of **x**, which is 1, then increments **x**. (Remember that means it adds one to it.) When it hits the end curly brace, it has finished the while loop and goes back up to the beginning of the loop, checking the condition again. This time, however, **x** has changed. It is now 2. The condition is still true (2 is still less than or equal to 10) and the loop repeats again, printing out "2" and incrementing **x** to 3. This will happen 10 times total, until **x** gets incremented to 11. When the program checks to see if 11 is less than or equal to 10, the condition is no longer true, and the flow of execution jumps down past the end of the loop.

One thing to keep in mind is that it is easy to end up with a bug in your code that makes it so the condition of a loop is never met. Imagine (or better yet, try it out) that the **x++;** line wasn't there. The program would keep repeating the loop, and each time **x** would still be 1. The program would never end! This problem is actually common enough to be given its own name: an *infinite loop*. I promise you, by the time you're done making your first *real* program, you will be the proud writer of at least one infinite loop. It happens.

To fix the problem, you may need to pause your program's execution, see where it is getting stuck, close your program, fix the problem in your code, and restart. The process of debugging your program like this is discussed in Chapter 44.

While we're on the subject of infinite loops, I'm going to mention that sometimes people make them on purpose. It sounds strange, I know. But it is easy to do:

```
while(true)
{
    // Depending on what goes in here, you'll never end...
}
```

Depending on what you put inside the loop (like a **break** statement, which we'll discuss in detail later in this chapter) you can actually still get out of the loop.

I've occasionally heard an infinite loop that was done intentionally be called a "forever loop" instead. It's not a term all programmers will be familiar with, but there is some value in distinguishing intentional infinite loops from the accidental ones.

Moving on, here's a more complicated (and more useful) example which repeats over and over until the user enters a number between 0 and 10:

```
int playersNumber = -1;

while(playersNumber < 0 || playersNumber > 10)
{
    // This code will get repeated until the player types in a number between 0 and 10.

    Console.Write("Enter a number between 0 and 10: ");
    string playerResponse = Console.ReadLine();
    playersNumber = Convert.ToInt32(playerResponse);
}
```

One important thing to remember with a **while** loop is that it always checks the condition before even going into the loop. So if the condition is not met right from the get-go, it doesn't ever go in the loop. In this little example above, it is important that we initialize **playersNumber** to -1, because if we had started it at 0, the flow of execution would have jumped right over the **while** loop block, and the player would have never been able to choose a number.

Like with **if** statements, if you only have one statement inside of a **while** loop (or any of the other loops we'll discuss here) the curly braces are optional.

The Do-While Loop

The next type of loop we'll look at is a slight variation on the **while** loop. It is the **do-while** loop. Remember what I said, that a **while** loop will first check the condition to see if it is met, and if not, it could potentially skip the loop entirely? (Might I add that this isn't a bad thing; it's just something to remember. Most of the time, that's exactly how you want it.)

In contrast, the **do-while** loop will *always* be executed at least once. This is useful if you are trying to set up some stuff the first time through the loop, and you *know* it needs to be executed at least once.

Let's revisit that last example, because it is a prime candidate for a **do-while** loop. Remember, we needed to set up the player's number to -1, to *force* it to go through the loop at least once? Doing this as a **do-while** loop solves the need for that:

```
int playersNumber;

do
{
    Console.Write("Enter a number between 0 and 10: ");
    string playerResponse = Console.ReadLine();
    playersNumber = Convert.ToInt32(playerResponse);
}
while (playersNumber < 0 || playersNumber > 10);
```

To form a **do-while** loop, we put the **do** keyword at the start of the loop, and at the end, we put the **while** keyword. Also notice that you need a semicolon at the end of the **while** line. Everything else is the same, but this time, you don't need to initialize the **playersNumber** variable because we know that the loop will be executed at least once before the condition is checked.

The For Loop

Now let's take a look at a slightly different kind of loop: the **for** loop. **For** loops are very common in programming. They are an easy way of doing counting type loops, and we'll see how useful they are again in the next chapter with arrays. For loops are a bit more complicated to set up, because they require three components inside of the parentheses, whereas **while** and **do-while** loops have only required one. It is structured like this:

```
for(initial condition; condition to check; action at end of loop)
{
    //...
}
```

There are three parts separated by semicolons. The first part sets up the initial state, the second part is the condition (the same thing that was in the **while** and **do-while** loops), and the third part is an action that is performed at the end of the loop. An example will probably make this clearer, so let's do the counting to ten example again, this time as a **for** loop:

```
for(int x = 1; x <= 10; x++)
{
    Console.WriteLine(x);
}
```

Note that we can declare and initialize a variable right in the **for** loop like we do here, with **int x = 1;**.

One of the reasons why this kind of loop is so popular is because it separates the looping logic from what you're actually doing with the number. Rather than having the **x++** statement inside of the loop and declaring the variable before the loop, all of that stuff gets packed into the loop control mechanism, making the stuff you're actually trying to accomplish clearer.

One other cool thing to point out is that anything you can do with one type of loop you can do with the other two as well. Often, one type of loop is cleaner and easier to understand than the others, but they can all do the same stuff. If suddenly the C# world ran out of **while** keywords, and all you had left was **for**s, you'd be fine. Because of this, you should pick the looping mechanism that creates the code that is easiest to understand.

Breaking Out of Loops

Another cool thing you can do with a loop is *break* out of it any time you want. Sometimes, as you're working through a loop, you get to a point where you know there's no point in continuing with the loop. You can jump out of a loop whenever you want with the **break** keyword like this:

```
int numberThatCausesProblems = 54;

for(int x = 1; x <= 100; x++)
{
    Console.WriteLine(x);

    if(x == numberThatCausesProblems)
        break;
}
```

This code will only go until it hits 54, at which point the **if** statement catches it and sends it out of the loop. This is kind of a trivial example, but it is a good illustration of how you might use the **break** command. When we start the loop, we're fully expecting to get to 100, but then we realize at some point that there's a critical problem (or alternatively, we've found the result we were looking for) and we can ignore the rest of the loop.

By the way, this is the kind of thing that makes forever loops actually worth something:

```
while(true)
{
    Console.Write("What is thy bidding, my master? ");
    string input = Console.ReadLine();

    if(input == "quit" || input == "exit")
        break;
}
```

Before we leave the topic of forever loops, I should say that you rarely truly need to write one. We could have restructured the code above to not need it. (By moving the **input** variable outside of the loop and converting the **if** statement to be the condition in the while loop.) Occasionally, though, the code will be more readable this way, so it is good to know about.

Continuing to the Next Iteration of the Loop

Similar to the **break** command, there's another command that, rather than getting out of the loop altogether, jumps back to the start of the loop and checks the condition again. In other words, it continues on to the next iteration of the loop without finishing the current one.

This is done with the **continue** keyword:

```
for(int x = 1; x <= 10; x++)
{
    if(x == 3)
        continue;

    Console.WriteLine(x);
}
```

In this code sample, all of the numbers will get printed out with one exception. 3 gets skipped because of the **continue** statement. When it hits that point, it jumps back up, runs the **x++** part of the loop, checks the **x < 10** condition again, and continues on with the next cycle through the loop.

Nesting Loops

Like with **if** statements it is possible to nest loops. And to put **if** statements inside of loops and loops inside of **if** statements. You can go absolutely crazy with all of this control!

Let's go through a couple more complete examples.

I'm going to repeat a really simple example that I remember from when I first started learning to program. (Back then, it was C++, not C#, and we had to walk uphill both ways to school, and define our own **true** and **false**!) The task was to write a loop that would print out the following (you were only allowed to have the '*' character in your program once):

```
**********
**********
**********
**********
**********
```

The code below accomplishes this:

```
for(int row = 0; row < 5; row++)
{
    for(int column = 0; column < 10; column++)
    {
        Console.Write("*");
    }

    Console.WriteLine(); // This makes it wrap around to the beginning again.
}
```

Let's try one more, slightly harder one. If we want to do this:

```
*
**
***
****
*****
******
*******
********
*********
**********
```

The code would be:

```
for(int row = 0; row < 10; row++)
{
    for(int column = 0; column < row + 1; column++)
    {
        Console.Write("*");
    }

    Console.WriteLine();
}
```

Notice how tricky we were, using the **row** variable in the condition of the **for** loop with the **column** variable.

Oh, I should probably mention (because I'm sure you're wondering) that programmers seem to love 0-based indexing, meaning we love to start counting at 0. (There's actually a good reason for this, which we'll look at a little bit more when we look at arrays in the next chapter.)

So you'll see that I've done this frequently in these samples. I start with **row** and **column** at 0, and go up to the amount I want (10 in this case), but not including it. That does it 10 times. You could do it starting at 1, and use **row <= 10** instead, but what I wrote is more typical for your normal programmer.

Try It Out!

Print-a-Pyramid. Like the star pattern examples that we saw earlier, create a program that will print the following pattern:

```
    *
   ***
  *****
 *******
*********
```

If you find yourself getting stuck, try recreating the two examples that we just talked about in this chapter first. They're simpler, and you can compare your results with the code included above.

This can actually be a pretty challenging problem, so here is a hint to get you going. I used three total loops. One big one contains two smaller loops. The bigger loop goes from line to line. The first of the two inner loops prints the correct number of spaces, while the second inner loop prints out the correct number of stars.

Try It Out!

FizzBuzz. We now know everything we need to be able to do a popular test problem in programming: the FizzBuzz problem. This is a simple little toy problem that many people who claim to know a particular programming language still seem to struggle with.

If you can complete this problem, you're probably better off than half of the other people in the world who claim to know C#. It's actually a sad state of affairs, because the problem is so simple, but even still, it is a fact than many people who claim to know C# (or another language) can't even write this simple program in it. (Admittedly, they often don't have Visual Studio and a compiler to check their work like you will, which makes the problem a bit harder.)

The challenge is to print out all of the numbers from 1 to 100. Except if a number is a multiple of 3, print out the word "Fizz" instead. If the number is a multiple of 5, print out "Buzz". If a number is a multiple of both 3 and 5 (like 15 or 30) then print out "FizzBuzz".

A couple of things to remember that seem to derail people: Be sure you're going through the numbers 1-100, not 0-99, and you'll likely use the remainder operator (%) which can be used to determine if something is a multiple of another number.

Still to Come: Foreach

It is worth mentioning that we have one more type of loop to discuss, which we'll do in the next chapter: the **foreach** loop. I only bring that up because a full discussion of loops has to include this type of loop. But the **foreach** loop makes the most sense in conjunction with arrays, which we'll be talking about next.

13

Arrays

In a Nutshell

- Arrays store collections of related objects of the same type.
- To declare an array, you use the square brackets: **int[] numbers;**
- To create an array, you use the **new** keyword: **int[] scores = new int[10];**
- You can also create arrays using collection initializer syntax: **int[] scores = new int[] { 1, 2, 3, 4, 5 };**
- You can access and modify values in an array with square brackets as well: **int firstScore = scores[0];** and **scores[0] = 44;**
- Indexing of arrays is 0-based, so 0 refers to the first element in the array.
- You can create arrays of anything, including arrays of arrays: **int[][] grid;**
- You can also create multi-dimensional arrays: **int[,] grid = new int[5, 4];**
- You can use the **foreach** loop to easily loop through an array and do something with each item in the array: **foreach(int score in scores) { /* ... */ }**

In this chapter, we'll take a look at arrays. Arrays are powerful features of any programming language. They allow us to store collections or groups of related objects together. We'll talk about how to create them and use them, go through a few examples of how to work with them, how to make multi-dimensional arrays (what they call a *matrix* in the math world) and then wrap up with a discussion of the last type of loop that we didn't discuss in the previous chapter.

What is an Array?

An *array* is a way of keeping track of a group of many related things all together. Imagine that you have a high score board in a game, with 10 high scores on it. Using only the tools we know so far, you could create one variable for every score that you want to keep track of. Something like this:

```
int score1 = 100;
int score2 = 95;
```

```
int score3 = 92;
// Keep going to 10...
```

That's one way to do it. But what if you had 10,000 scores? All of a sudden, creating that many variables to store scores becomes overwhelming.

This brings us to arrays. Arrays are perfect for keeping track of things like this, since they could store 10 scores or 10,000 scores in a way that is both easy to create and easy to work with.

Creating Arrays

Declaring an array is very similar to declaring any other variable. You give it a type and a name, and you can initialize it at the same time if you want. You declare an array using square brackets (**[** and **]**).

```
int[] scores;
```

The square brackets here indicate that this is an array, not just a single **int**. (Note for C++ programmers: unlike C++, you cannot put the square brackets after the variable name (**int scores[];**. You must put it after the type as shown in the example.)

This would declare an array variable which can contain multiple **int**s. Like with all other types that we've discussed so far, this is basically just a named location in memory to stick stuff. To actually *create* a new array and place it in the variable, you will use the **new** keyword and specify the number of elements that you'll have in the array:

```
int[] scores = new int[10];
```

In the first example, we had not yet assigned a value to the array variable. We had just declared it, reserving a spot for it. In this second example, our array now exists and has room for 10 items in it.

Once we have set a size for an array, we can't change it. You can change the individual values in the array (we're still coming to that) but not the size. You can, however, create a new array with a different size, copy the values over into part of the new, larger array, and then assign that back in to your array variable. The variable itself doesn't care how large the array is specifically.

By the way, this is the first time we're seeing the **new** keyword, but it won't be the last. We use this keyword to create things that aren't just simple primitive data types. We'll see this come back in a big way later, when we start looking at classes in Chapter 17.

Arrays aren't limited to **int**s. You can make arrays of anything. For example, here's an array of **string**s:

```
string[] names = new string[10];
```

(For those coming from a C++ background, you may be wondering if there is a **delete** keyword, and when and how to clean things up. While C++ requires you to clean up memory that you are no longer using with **delete**, C# doesn't require or even allow this. C# uses managed memory, and uses garbage collection to clean up things that are no longer in use. We'll look at this in more detail in Chapter 16.)

Getting and Setting Values in Arrays

Now that we have an array declared and created, we can assign values to specific spots in the array. To access a specific spot in the array, we use the square brackets again, along with what is called a *subscript* or an *index*, which is a number that tells the computer which of the *elements* (things in an array) to access. For example, to put a value in the first spot of an array, we would use the following:

```
scores[0] = 99;
```

You can see here that the first spot in the array is number 0. It is very important to remember that array indexing is 0-based. If you forget about 0-based indexing, and start with 1, you'll run into what are called "off-by-one" errors, because you're on the wrong number.

You can access any value in an array by using the same technique:

```
int fourthScore = scores[3];
int eighthScore = scores[7];
```

If you try to access something beyond the end of the array (for example, **scores[54]** in an array with only 10 items in it), your program will crash, telling you the index was out of the bounds of the array.

Side Note

0-based Indexing. I know it seems kind of strange if you're new to programming, but there's a very good reason for 0-based indexing. The array itself has a specific spot in memory—a specific memory address. This address is called the base address.

For any particular array, the size of each slot in the array may be a different size based on the size of the type you're putting into the array. An array of **int**s will have slots that are 4 bytes big, since **int**s are 4 bytes. An array of **long**s will have slots that are 8 bytes big.

Because the computer knows the base memory address, and how big each slot is, it is easy for it to calculate the memory location of any particular index in the array using the simple formula:

$$address = base + index \times size$$

If we didn't use 0-based indexing (perhaps using 1-based indexing instead) the math here would become more complicated.

More Ways to Create Arrays

There are a couple of other ways of initializing an array that are worth discussing here. You can also create an array by giving it specific values right from the get-go, by putting the values you want inside of curly braces, separated by commas:

```
int[] scores = new int[10] { 100, 95, 92, 87, 55, 50, 48, 40, 35, 10 };
```

When you create an array this way, you don't even need to state the number of items that will be in the array (the '10' in the brackets, in this case). You can leave that out:

```
int[] scores = new int[] { 100, 95, 92, 87, 55, 50, 48, 40, 35, 10 };
```

Array Length

We can also tell how long an array is by using the **Length** property. We haven't talked about properties at all yet (we will cover them in depth in Chapter 19), but the code for doing this is useful enough and simple enough that it is worth taking the time to point out. For instance, the code below grabs an array's **Length** property and uses it to print out the length of an array:

```
int totalThingsInArray = scores.Length;
Console.WriteLine("There are " + totalThingsInArray + " things in the array.");
```

Some Examples with Arrays

Now that we know the basics of how arrays work, let's look at a couple of examples using arrays.

Minimum Value in an Array

In our first example, we'll calculate the minimum value in an array, using a **for** loop.

The basic process will require looking at each item in the array in turn. We'll create a variable to store the value that we know is the minimum that we've seen so far. As we go down the array, if we find one that is less than our current known minimum, we update the known minimum to the new one.

```
int[] array = new int[] { 4, 51, -7, 13, -99, 15, -8, 45, 90 };

int currentMinimum = Int32.MaxValue; // We start really high, so that any element
                                     // in the array will be lower than this.

for(int index = 0; index < array.Length; index++)
{
    if(array[index] < currentMinimum)
        currentMinimum = array[index];
}
// At this point, currentMinimum contains the minimum value in the array.
```

Average Value in an Array

Let's try a similar but different task: finding the average value in an array. We'll follow the same basic pattern as in the last example, but this time we're going to total up the numbers in the array. When we're done doing that, we'll divide by the total number of things in the array to get the average.

```
int[] array = new int[] { 4, 51, -7, 13, -99, 15, -8, 45, 90 };

int total = 0;

for (int index = 0; index < array.Length; index++)
    total += array[index];

float average = (float)total / array.Length;
```

Try It Out!

Copying an Array. Write code to create a copy of an array. First, start by creating an initial array. (You can use whatever type of data you want.) Let's start with 10 items. Declare an array variable and assign it a new (yes, that means we'll use the **new** keyword) array with 10 items in it. Use the things we've discussed to put some values in the array.

Now create a second array variable. Give it a new array with the same length as the first. Instead of using a number for this length, use the **Length** property to get the size of the original array.

Use a loop to read values from the original array and place them in the new array. Also, print out the contents of both arrays, to be sure everything copied correctly.

Arrays of Arrays and Multi-Dimensional Arrays

You can have arrays of anything. **int**s, **float**s, **bool**s. Even have arrays of arrays! This is one way that you can create a matrix. (A matrix is simply a grid or table of numbers with rows and columns.)

To create an array of arrays, one option is to use the following notation:

```
int[][] matrix = new int[4][];
matrix[0] = new int[4];
matrix[1] = new int[5];
matrix[2] = new int[2];
matrix[3] = new int[6];

matrix[2][1] = 7;
```

Notice that each of my arrays within the main array has a different length. You could, of course, make them all the same length (there's a better way to do this that we'll see in a second). When each array within a larger array has a different length, it is called a *jagged array*. If they're all the same length, it is often called a *square array* or a *rectangular array*.

There's another way to work with arrays of arrays, assuming you want a rectangular array (which is often the case). This is called a *multi-dimensional array*.

To do this, you put multiple indices inside of one set of square brackets like this:

```
int[,] matrix = new int[4, 4];
matrix[0, 0] = 1;
matrix[0, 1] = 0;
matrix[3, 3] = 1;
```

It is worth briefly describing how you might go about looking at each element in these more complicated arrays. For an array of arrays, or a jagged array, this might look like this:

```
int[][] matrix = new int[4][];
matrix[0] = new int[2];
matrix[1] = new int[6];
// Continue filling in values for the jagged array...

for(int row = 0; row < matrix.Length; row++)
{
    for(int column = 0; column < matrix[row].Length; column++)
    {
        Console.Write(matrix[row][column] + " "); // Each item in the row separated by spaces
    }
    Console.WriteLine(); // Rows separated by lines
}
```

Or with a multi-dimensional array:

```
int[,] matrix = new int[4,4];
// Fill in contents for multi-dimensional array

// Note: GetLength gives back the size of the multi-dimensional array for a specific index.
for(int row = 0; row < matrix.GetLength(0); row++)
{
    for(int column = 0; column < matrix.GetLength(1); column++)
    {
        Console.Write(matrix[row, column] + " ");
    }
    Console.WriteLine();
}
```

The 'foreach' Loop

To wrap up our discussion of arrays, let's go back to what was mentioned in the last chapter about loops. There's one final type of loop that works really well when you're doing stuff with arrays. This type of loop is called the **foreach** loop (as in, "do this particular task for each element in the array").

To use a **foreach** loop, you use the **foreach** keyword with an array, specifying the name of the variable to use inside of the loop:

```
int[] scores = new int[10];

foreach (int score in scores)
    Console.WriteLine("Someone had this score: " + score);
```

Inside of the loop, you can use the **score** variable. One key thing to note is that inside of the loop, you have no way of knowing what index you are currently at. (You don't know if you are on **scores[2]** or **scores[4]**.) In many cases, that's no big deal. You don't care what index you're at, you just want to do something with each item in the array.

If you need to know what index you're at, your best bet is to use a **for** loop instead:

```
int[] scores = new int[10];

for(int index = 0; index < scores.Length; index++)
{
    int score = scores[index];
    Console.WriteLine("Score #" + index + ": " + score);
}
```

Compared to a **for** loop, a **foreach** loop has a tendency to be more readable. There is not as much overhead clutter inside of the parentheses. Also, in many cases the first thing you do in many **for** loops is use the index to get the correct item out of the array. This is bypassed with a **foreach** loop.

But **foreach** loops are somewhat slower, simply as a result of their internal mechanics. If you're discovering that a particular **foreach** loop is taking way more time than you'd like, converting it to a **for** loop is an easy way to get a little extra speed out of it.

This leaves us with a final question: when can **foreach** loops be used? The short answer (which won't make much sense just yet) is that it can be used on anything that implements the **IEnumerable** interface. We haven't talked about interfaces at all yet, but virtually every container or collection across the .NET Framework uses it. That means that **foreach** can be used on just about everything that has multiple items in it.

> **Try It Out!**
>
> **Totaling and Averaging Revisited.** Earlier in this chapter, I presented code to go through an array and find the total or sum of all values in that array. I also presented code to average the values in an array. Using that code as a starting point, rewrite them to use **foreach** loops instead.
>
> Also, answer this question: why can't you use a **foreach** loop to copy one array to another, like we did in the last *Try It Out!* section? If you're struggling to find an answer, give it a try by coding it.

Enumerations

> **In a Nutshell**
>
> - Enumerations are a way of defining your own type of variable so that it has specific values.
> - Enumerations are defined like this: **public enum DaysOfWeek { Sunday, Monday, Tuesday, Wednesday, Thursday, Friday, Saturday };**
> - Enumerations are usually defined directly in a namespace, outside of any classes or other type definitions.
> - The fact that an enumeration can only take on the values that you explicitly listed, means that you won't end up with bad or meaningless data (unless you force it by casting).
> - You can assign your own numeric values to the items in an enumeration: **public enum DaysOfWeek { Sunday = 5, Monday = 6, Tuesday = 7, Wednesday = 8, Thursday = 9, Friday = 10, Saturday = 11 };**

We've been going at a pretty fast pace. We have had a lot to learn in every chapter. This one will be a bit of a break, though, as enumerations are pretty simple and easy.

Enumerations (or *enumeration types*, or *enums* for short) are a cool way to define your own type of variable. This is the first of many ways we'll see to create your own.

The word *enumeration* comes from the word *enumerate*, which means "to count off, one after the other," which is basically what we're going to be doing. We'll start by looking at the kinds of situations where enumerations might be useful. We'll then look at how to create your own enumeration and use it. We'll take a look at a few additional important features of enumerations before wrapping up the chapter.

The Basics of Enumerations

Let me start off with an example of when enumerations may be used. Let's say you are keeping track of something that has a known, small, specific set of possible values that you can use—for instance, the days in the week. One way you could do this is to assign numbers to each of these in your head. Sunday would be 1, Monday would be 2, Tuesday would be 3, and so on.

So your code could look something like this:

```
int dayOfWeek = 3;

if(dayOfWeek == 3)
{
    // Do something here that should only be done on Tuesday.
}
```

For the sake of readability, you might decide to create a variable to store each of these days:

```
int sunday = 1;
int monday = 2;
int tuesday = 3;
int wednesday = 4;
int thursday = 5;
int friday = 6;
int saturday = 7;

int dayOfWeek = tuesday;

if(dayOfWeek == tuesday)
{
    // Do something here because it is Tuesday.
}
```

This kind of situation is exactly what enumerations are for. You create an enumeration, and list the possible values it can have.

When we create an enumeration, we're defining a new type. We've discussed a wide variety of types before, but this is the first experience we'll have creating our own type. When we define a new type, we put them directly inside of the namespace. You use the **enum** keyword, give your enumeration a name, and list the values that your enumeration can have inside of curly braces:

```
public enum DaysOfWeek { Sunday, Monday, Tuesday, Wednesday, Thursday, Friday, Saturday };
```

Remember, enumerations are placed directly inside of the namespace (outside of other classes or types you might have, including the **Program** class that we've been using). This is shown in context below:

```
using System;
using System.Collections.Generic;
using System.Linq;
using System.Text;
using System.Threading.Tasks;

namespace Enumerations
{
    // You define enumerations directly in the namespace.
    enum DaysOfWeek { Sunday, Monday, Tuesday, Wednesday, Thursday, Friday, Saturday };

    public class Program
    {
        static void Main(string[] args)
        {
```

```
        }
    }
}
```

Alternatively, enumerations can be defined in their own file. We'll see how to do this when we learn about classes in Chapter 18.

Inside of your **Main** method, you can now create variables that have your new **DaysOfWeek** type, instead of just the built-in types that we've been working with. We've defined a brand new type! Creating a variable with your new type works just like any other variable:

```
DaysOfWeek today; // Indicate the type, and give it a name.
```

To assign it a value, you will use the '.' operator (usually read "dot operator"). We'll discuss the '.' operator more later, but for now what you need to know is that it is used for *member access*. This means that you use the '.' operator to use something that is a part of something else. The values are a part of the enumeration, so we use this operator.

```
today = DaysOfWeek.Tuesday;
```

This works like any other variable, even for things like comparison in an **if** statement:

```
DaysOfWeek today = DaysOfWeek.Sunday;

if(today == DaysOfWeek.Sunday)
{
    // ...
}
```

Why Enumerations are Useful

There's a very good reason to use enumerations. Let's go back to where we first started this chapter—doing days of the week with the **int** type. Remember how we had defined **dayOfWeek = 3;** which meant Tuesday? What if someone put **dayOfWeek = 17;**? As far as the computer knows, this should be valid. After all, **dayOfWeek** is just an **int** variable. Why not allow it to be 17? But for what we're doing, 17 doesn't make any sense. We are only using 1 through 7.

Enumerations force the computer (and programmers) to use only specific values, which you have defined. It prevents tons of errors, and makes your code more readable. For example, **dayOfWeek = DaysOfWeek.Sunday** is immediately clear what it means, while **dayOfWeek = 1** is not.

In addition, enumerations allow us to use the IntelliSense Visual Studio provides, which is another nice little bonus.

Underlying Types

Before moving on, it is worth looking at a couple of additional details about enumerations. Under the surface, C# is simply wrapping our enumeration around an *underlying type*. By default, this underlying type is the **int** type, though it is possible to choose a different integer type. So enumerations are, at their core, just numbers. (As I just mentioned though, it also makes sure that only the *right* numbers.) When you create an enumeration, it starts off giving them values, starting with 0, and going up by one for each new number. So in the enumeration that we created, Sunday has a value of 0, Monday is 1, and so on.

This means that you can cast to and from an **int** or other number like this:

```
int dayAsInt = (int)DaysOfWeek.Sunday;
DaysOfWeek today = (DaysOfWeek)dayAsInt;    // Both of these require an explicit cast.
```

By default, enumerations make sure that only *valid* values are used. However, if you use an explicit cast to convert an invalid number to an enumeration, you can break this:

```
DaysOfWeek today = (DaysOfWeek)17; // Legal, but a bad idea...
```

Because of this, you should be very cautious about casting to an enumeration type.

Assigning Numbers to Enumeration Values

When you create an enumeration, you can additionally assign it specific values if you want:

```
enum DaysOfWeek { Sunday = 5, Monday = 6, Tuesday = 7, Wednesday = 8,
                  Thursday = 9, Friday = 10, Saturday = 11 };
```

Try It Out!

Months of the Year. Using the **DaysOfWeek** enumeration as an example, create an enumeration to represent the months of the year. Assign them the values 1 through 12. Write a program to ask the user for a number between 1 and 12. Check to be sure that they gave you a value in the right range and use an explicit cast to convert the number to your month enumeration. Then, using a **switch** statement or **if** statement to print out the full name of the month they entered.

Methods

In a Nutshell

- Methods are a way of taking a chunk of code that does a specific task and put it together in a way that it is reusable.
- You can return stuff from a method by putting the return type in the method definition, and then using the **return** keyword inside of the method anywhere you want to return a value.
- You can pass parameters into a method by listing them inside of the parentheses after the method's name.
- You can make multiple methods with the same name (called method overloading) as long as their method signatures are different.
- Recursion is where you call a method from inside itself, and requires a base case to prevent the flow of execution from digging deeper and deeper into more method calls.

Think about a program like *Microsoft Word 2013* or a game like the latest *Assassin's Creed*. Think about how much code goes into creating something that big. They're enormous! Especially when compared to the little tiny programs that we've been working on so far.

How do you think you could manage all of that code? Where would you look if there was a bug?

As programmers, we've learned that when you have a large program, the best way to manage it is by breaking it down into smaller, more manageable pieces. This even gives us the ability to reuse these pieces. This basic concept of breaking things down into smaller pieces is called *divide and conquer*, and it is one we'll see over and over again.

C# provides us with a feature that allows us to break down our program into small manageable pieces. These small, reusable building blocks of code are called *methods*. Methods are sometimes called *functions*, *subroutines*, or *procedures* in other programming languages.

The idea of a method is to take a piece of your code and place it inside of a special block with a specific name, allowing you to run it from anywhere else in your program. We can more easily locate a bug or problem in a program because we can pinpoint it to a very specific method. In addition, we can reuse a method as many times as we want, which means we don't have to retype it, saving ourselves time.

In this chapter, we'll look at how to create a method and how to use it (called "calling" a method), as well as how to get data back from and send data to a method. We'll then look at creating multiple methods with the same name—a process called *overloading*.

We'll then take another look at the **Console** and **Convert** classes, which we've used in previous chapters, armed with the new knowledge we have about methods. We'll wrap up our discussion about methods with a quick look at a powerful (but sometimes confusing) trick that you can use with methods called *recursion*.

Creating a Method

Let's get started by creating our first method. If you start with a brand new project, you'll see yourself staring down at code that looks like this:

```
using System;
using System.Collections.Generic;
using System.Linq;
using System.Text;
using System.Threading.Tasks;

namespace Methods
{
    public class Program
    {
        static void Main(string[] args)
        {
        }
    }
}
```

If you remember from the Hello World program we made back in Chapter 3, this template code already defines a method (the **Main** method) for us. At the time, we only had a basic idea of what a method was, but we know more now. You can see from the template code that the **Main** method is contained inside of a class called **Program**. (We're still getting to classes.) When it comes down to it, we've already been using a method, without even really thinking about it!

Basically, all methods belong to a type. A class is one specific example of a type. Most of the types we create will be classes, but methods can belong to the handful of other types that we will look at over the course of this book. Because the method belongs to the class, it is a *member* of that class.

Whenever we create new methods, we will need to add them to a specific type. They can't just be floating around on their own. For now, we've already got the **Program** class where our **Main** method is, which is a perfect spot for our new methods.

So let's go ahead and make our first method. Let's make a method that simply prints out the numbers 1 through 10. This new method looks like the following:

```
static void CountToTen()
{
    for(int index = 1; index <= 10; index++)
        Console.WriteLine(index);
}
```

So for the moment, our full code should look like this:

```
using System;
using System.Collections.Generic;
using System.Linq;
using System.Text;
using System.Threading.Tasks;

namespace Methods
{
    public class Program
    {
        static void Main(string[] args)
        {
        }

        static void CountToTen()
        {
            for (int index = 1; index <= 10; index++)
                Console.WriteLine(index);
        }
    }
}
```

This code won't do anything yet. Remember, our program will run everything in the **Main** method by default, and we still have nothing in the **Main** method for the time being. Before we're done, we'll need to have our **Main** method "call" our newly created **CountToTen()** method, but we'll do that in a second. First, let's discuss the piece of code that we've just added.

There are just a few simple pieces to what we just added. The first piece is the **static** keyword. We'll talk about this a little more when we start creating our own classes. (See Chapter 18.) For now, we'll assume that it's just needed for what we're doing.

The second thing we added is the keyword **void**. We'll talk more about this in a second, when we discuss returning stuff from a method, but basically, the **void** keyword tells us our method isn't going to give us anything back. It's just going to do some task and be done.

The next part is the method's name: **CountToTen**. Like with variables, you can name your method all sorts of things. Giving methods descriptive names will make it easier to explain what it does. While it is not required, the standard convention is to have method names start with an upper case letter, while variables usually start with a lower case letter.

Next we have the parentheses. Right now, there's nothing inside of our parentheses, but we'll soon see that we can put stuff in here, allowing us to hand stuff off to a method.

Lastly, we have a set of curly braces where we put the *body* or *implementation* of the method. This is where we put the code that the method should do. Up until now, we have been putting all of our code in the body of the **Main** method. As of this chapter, we'll start putting our code in the body of different methods to organize our code into separate, reusable tasks.

We can see that in the body of our method, we have a simple **for** loop. Any of the code that we've written before can go inside of a method, including loops, **if** statements, math, and variable declarations.

It is also worth mentioning that the order that methods are arranged within a class doesn't matter. In the example above, we could have put **CountToTen** above **Main** and it wouldn't have made a difference. The C# compiler is smart enough to scan through your file and find all of the methods that exist in it. (For those of you with a C or C++ background, this is a major change. You no longer need to make function prototypes or anything like that.)

Calling a Method

Now that we've got our method created, we want to be able to use it. Doing this is really easy. Inside of your **Main** method, you do the following:

```
CountToTen();
```

So your complete code would look like this:

```
using System;
using System.Collections.Generic;
using System.Linq;
using System.Text;

namespace Methods
{
    public class Program
    {
        static void Main(string[] args)
        {
            CountToTen();
        }

        static void CountToTen()
        {
            for (int index = 1; index <= 10; index++)
                Console.WriteLine(index);
        }
    }
}
```

Now you can run your program (which automatically starts in the **Main** method) and see that it jumps over to the **CountToTen** method and does everything inside of it. When it reaches the end of a method, it goes back to where it came from. So looking at this code:

```
using System;
using System.Collections.Generic;
using System.Linq;
using System.Text;

namespace Methods
{
    public class Program
    {
        static void Main(string[] args)
        {
            Console.WriteLine("I'm about to go into a method.");
```

```
            DoSomethingAwesome();

            Console.WriteLine("I'm back from the method.");
        }

        static void DoSomethingAwesome()
        {
            Console.WriteLine("I'm inside of a method! Awesome!");
        }
    }
}
```

This will result in the following output:

```
I'm about to go into a method.
I'm inside of a method! Awesome!
I'm back from the method.
```

It is also worth mentioning that you can call a method from inside of any other method, so you can go crazy creating and calling methods all over in your program. (In fact, that's essentially what software engineering is! Creating and calling methods in a way that minimizes the craziness!)

Returning Stuff from a Method

Methods are created to do some specific task. Often, that task involves figuring something out, and we want to be able to know the results. A method has the option of giving something back. (How generous of them!) This is called *returning* something. (Even if it doesn't return anything, like the first method we wrote a second ago, people still talk about "returning from a method.")

So far, all of the methods we've looked at haven't returned anything. If you'll look back, you'll see that we've been putting the **void** keyword before our method name when we create it. This means that the method doesn't return anything. By changing that, our method can return something. To do this, instead of the **void** keyword, we simply place a different type there, like **int**, **float**, or **string**. This makes it so our method can (and must) return a value of that particular type.

Inside of a method, we use the **return** keyword to return a specific value. This is followed by the value we want to return. A very simple example of returning a value is this:

```
static float GetRandomNumber()
{
    return 4.385f; // Obviously very random.
}
```

You can see that we start off by stating the return type (**float** in this case, meaning we're going to return something of the type **float**), and inside of the method, we simply use the **return** keyword, along with the actual value we want to return. Obviously, this is not a very useful example, but it illustrates the point.

Let's go on to a more sophisticated example. So far in this book, there have been many times that we've done things like ask the user to type in a number. The code below creates a method that asks the user for a number between 1 and 10, and keeps asking until they finally give us something that works. At that point, we'll return the number the user gave us.

```
static int GetNumberFromUser()
{
    int usersNumber = 0;

    while(usersNumber < 1 || usersNumber > 10)
    {
        Console.Write("Enter a number between 1 and 10: ");
        string usersResponse = Console.ReadLine();
        usersNumber = Convert.ToInt32(usersResponse);
    }

    return usersNumber;
}
```

If your method returns nothing (**void**), you don't need a **return** statement. But if it *does* return something, we're required to have a **return** statement.

If a method returns something, we can grab the value that a method returns and do something with it, like put it in a variable or do some math with it. For instance, the code below will take the value that is returned by the **GetNumberFromUser** method and store it in a variable:

```
int usersNumber = GetNumberFromUser();
```

Does that look a little familiar? It should. We've been doing stuff like that all along! Now we finally understand what's going on though. For instance, when we've written things like **string usersResponse = Console.ReadLine();**, we're simply calling a method (**ReadLine**) that belongs to the **Console** class and getting the value that was returned by it!

While the **return** statement is often the last line in a method, it doesn't have to be. For instance, you can have an **if** statement right in the middle of a method that returns a value from a method only when a certain condition is met:

```
static int CalculatePlayerScore()
{
    int livesLeft = 3;
    int underlingsDestroyed = 17;
    int minionsDestroyed = 4;
    int bossesDestroyed = 1;

    // If the player is out of lives, they lose all of their points.
    if(livesLeft == 0)
        return 0;

    // Otherwise, the player gets 10 points for every underling destroyed, 100 points
    // for every minion destroyed, and 1000 points for every boss destroyed.
    return underlingsDestroyed * 10 +
           minionsDestroyed * 100 +
           bossesDestroyed * 1000;
}
```

If your method's return type is **void**, while you don't *need* the **return** keyword anywhere, you are allowed to put it wherever you want all by itself (since you aren't returning anything) like this:

```
static void DoSomething()
{
    int aNumber = 1;
```

```
    if(aNumber == 2)
        return;

    Console.WriteLine("This only gets printed if the 'return' statement wasn't executed.");
}
```

Note that as soon as a **return** statement is hit, the flow of execution *immediately* goes back to where the method was called from—nothing more gets executed in the method.

Sending Stuff to a Method

Sometimes, we want a method to *do* stuff with certain pieces of information. We can hand stuff off to a method by putting what we want it to work with inside of the parentheses. Earlier, we had made a method that was called **CountToTen**, which simply printed out the numbers 1 through 10. We can create an alternate method that requires you to supply a certain number to count to. The method could then be used to count to 10, 25, 1000, or anything else.

To do this, we need a way to hand off information for the method to use. This is done by putting a list of variables inside of the parentheses when we're defining the method. Our modified version of the **CountToTen** method, which works for any number, might look like this:

```
static void Count(int numberToCountTo)
{
    for(int current = 1; current <= numberToCountTo; current++)
        Console.WriteLine(current);
}
```

Where you define the method, you will put the type of the variable, and give it a name to use throughout the method. This particular kind of variable is called a *parameter*. So the variable **numberToCountTo** is a parameter.

When you call a method that has a parameter, you can "pass in" or hand off a value to the method by putting it in the parentheses as well:

```
Count(5);
Count(15);
```

This code, in conjunction with the example before it, will first print out the numbers 1 through 5, and then it will print out the numbers 1 through 15.

Passing in Multiple Parameters

You can also create a method that passes in multiple parameters. All of the parameters listed—everything inside of the parentheses—is called the *parameter list*. The code below is a simple example of passing two numbers into a **Multiply** method that multiplies them together, and returns the result:

```
static int Multiply(int a, int b)
{
    return a * b;
}
```

Having multiple parameters like this is extremely common. There's a technical limit to how many parameters you can have (just over 65,500) but the practical limit is far lower. Most programmers will start complaining about the number of parameters a method has well before you even reach 10.

If you need that many parameters, you should spend some time trying to come up with an approach that lets you break the task down into smaller tasks, where each task requires fewer pieces of data to work on.

Try It Out!

Reversing an Array. Let's make a program that uses methods to accomplish a task. Let's take an array and reverse the contents of it. For example, if you have 1, 2, 3, 4, 5, 6, 7, 8, 9, 10, it would become 10, 9, 8, 7, 6, 5, 4, 3, 2, 1.

To accomplish this, you'll create three methods: one to create the array, one to reverse the array, and one to print the array at the end.

Your **Main** method will look something like this:

```
static void Main(string[] args)
{
    int[] numbers = GenerateNumbers();
    Reverse(numbers);
    PrintNumbers(numbers);
}
```

The **GenerateNumbers** method should return an array of 10 numbers. (For bonus points, change the method to allow the desired length to be passed in, instead of just always being 10.)

The **PrintNumbers** method should simply use a **for** or **foreach** loop to go down the array, one at a time, and print out the items in it.

The **Reverse** method will be the hardest. Give it a try and see what you can make happen. If you get stuck, here's a couple of hints:

Hint #1: To swap two values, you will need to place the value of one variable in a temporary location to make the swap:

```
// Swapping a and b.
int a = 3;
int b = 5;

int temp = a;
a = b;
b = temp;
```

Hint #2: Getting the right indices to swap can be a challenge. Use a **for** loop, starting at 0 and going up to the length of the array / 2. The number you use in the **for** loop will be the index of the first number to swap, and the other one will be the length of the array minus the index minus 1. This is to account for the fact that the array is 0-based. So basically, you'll be swapping **array[index]** with **array[arrayLength – index – 1]**.

Method Overloading

One cool, but potentially confusing thing that you can do with methods is create multiple methods with the same name. This is called *method overloading*, or simply *overloading*.

The key is that while two methods can have the same name, they can't have the same *signature*. A method's signature is defined as the combination of the method name and the types and order of the parameters that get passed in. Note that this does *not* include the parameters' names, just their types.

As an example, to help illustrate what a method signature is, take the following example:

```
static int Multiply(int a, int b)
{
    return a * b;
}
```

This has a signature that looks like this: **Multiply(int, int)**.

You can only overload a method if you use a different signature. So for example, the following works:

```
static int Multiply(int a, int b)
{
    return a * b;
}

static int Multiply(int a, int b, int c)
{
    return a * b * c;
}
```

This works because the two multiply methods have a different number of parameters, and as a result, a different signature. We could also define a **Multiply** method that has no parameters (**int Multiply()**), or one parameter (**int Multiply(int)**), though in this particular case, I can't imagine how either of those two methods would do multiplication with zero or one thing. (Hey, it's just an example!) Likewise, you could define a **Multiply** method with eight or ten parameters, if there was a need for it.

Also, you can have the same *number* of parameters, if their types are different:

```
static int Multiply(int a, int b)
{
    return a * b;
}

static double Multiply(double a, double b)
{
    return a * b;
}
```

This works because the two **Multiply** methods each have their own signature (**Multiply(int, int)** and **Multiply(double, double)**).

The following *does not work*:

```
static int Multiply(int a, int b)
{
    return a * b;
}

static int Multiply(int c, int d) // This won't work. It has the same signature.
{
```

```
    return c * d;
}
```

The magic of method overloading is that you can have many methods that do very similar work on different types of data (like the **int** multiplication and the **double** multiplication above) without having to have a completely different method name. It makes things easier on anyone who uses the method. With different signatures, the C# compiler can easily determine which of the overloaded methods to use.

Before leaving the topic of method overloading, I should caution you to only use it when you are trying to do the exact same kind of thing with different kinds of data. If two methods have the same name, they should perform essentially the same task. If you know what one does, it should be obvious what the other does, even if it uses different data. If that's not the case, you should use a different method name altogether to avoid confusion.

Revisiting the Convert and Console Classes

With a basic understanding of how methods work in C#, it is worth a second discussion about some of the classes that we've already been using.

For instance, in the **Console** class, we've done things like this:

```
Console.WriteLine("Hello World!");
```

With our new knowledge of methods, we can see that there is a **Console** class, which has defined in it a method named **WriteLine** that we call. The **WriteLine** method has one parameter, which is a **string**. We pass in the string "Hello World!" and the method runs off and does its job.

This, though, is a perfect example of *why* we use methods. Because someone has already created a **WriteLine** method that writes stuff to the console window, we don't need to worry about *how* it does it. All we care about is that it does it, and that it is easy to use. We just use it and move on to the more interesting things our program needs to do.

We see similar things with the **Convert** class, which we have used several times as well:

```
int number = Convert.ToInt32("42");
```

The **Convert** class has a ton of methods that let you convert many types to other types. In the example above, we see a method called **ToInt32**, which takes a **string** as a parameter. This class uses method overloading extensively, because it has a **ToInt32** method that takes a **string**, one that takes a **double**, one that takes a **short**, and so on.

XML Documentation Comments

In Chapter 4, we introduced the idea of comments. I mentioned that there is one additional way to do comments that we couldn't really talk about back then. We're ready to discuss it here.

It is a good idea to add a comment by a method to describe what it does. This way, when you or somebody else wants to use it, they can easily figure out what it does, and how to make it work. This is especially important because methods are *designed* to be reused.

You can add comments to a method using any of the methods we discussed in Chapter 4, but there is one additional way that works really well if you are trying to describe what a method does. (By the way, this also applies to classes, structs, enumerations, and other things that we'll discuss shortly.) This additional way is called XML documentation comments.

Let's say you have a **Multiply** method like we were talking about a second ago:

```
public static int Multiply(int a, int b)
{
    return a * b;
}
```

To add XML documentation comments, go just above the method, and type three forward slashes: ///

As soon as you hit that third forward slash, Visual Studio will pop in a big comment that looks like this:

```
/// <summary>
///
/// </summary>
/// <param name="a"></param>
/// <param name="b"></param>
/// <returns></returns>
```

This is your XML documentation comment. In between the **<summary>** and **</summary>** lines, you can add a description of what the method does. In between the **param** tags, you can describe what each parameter of your method does, and in the **returns** tags, you can add what the return value should be. A completely filled in example of this might look like this:

```
/// <summary>
/// Takes two numbers and multiplies them together, returning the result.
/// </summary>
/// <param name="a">The first number to multiply</param>
/// <param name="b">The second number to multiply</param>
/// <returns>The product of the two input numbers</returns>
```

The nice thing about adding in an XML documentation comment is that we can then immediately start seeing this information appear in IntelliSense, as we will discuss in Chapter 41.

The Minimum You Need to Know About Recursion

There's one additional trick people use with methods that I think is worth bringing up here. It's kind of a complicated trick, so right now you don't need to master it. But it is worth mentioning, so you know what it is when it comes up, and so you can start thinking about how it might be useful.

The trick is called *recursion*, and it is where you call a method from inside itself. Here's a trivial example (which happens to break when you run it):

```
static void MethodThatUsesRecursion()
{
    MethodThatUsesRecursion();
}
```

This example is going to break on you because it just keeps going into deeper and deeper method calls. Eventually, it will run out of memory to make more method calls, and the program will crash.

The problem with this first example, is that there is no *base case*. Recursion always needs a state where it doesn't call itself again, and each time it calls itself, it should be getting closer and closer to that base case. If not, it will never get to a point where it is done, and can start returning from the method calls back up to the starting point.

One of the classic example of when you could use recursion is the mathematical factorial function. Perhaps you remember from math classes that factorial, written as an exclamation mark by a number, means to take it and multiply it by all smaller numbers. For example, **7!** means **7 * 6 * 5 * 4 * 3 * 2 * 1**.

72! would be **72 * 71 * 70 * 69 * ... * 3 * 2 * 1**. (With factorial, by the way, you really quickly run into the overflow issues that we were talking about back in Chapter 9.)

In this case, we know that **1!** is **1**. This can function as a base case. The factorial of every other number can be thought of as that number, multiplied by the factorial of the number smaller than it. **7!** is the same as **7 * 6!**.

This sets up a great opportunity for recursion:

```
static int Factorial(int number)
{
    // We establish our "base case" here. When we get to this point, we're done.
    if(number == 1)
        return 1;

    return number * Factorial(number - 1);
}
```

Try It Out!

The Fibonacci Sequence. If you're up for a good challenge involving recursion, try out this challenge. The Fibonacci sequence is a sequence of numbers where the first two numbers are 1 and 1, and every other number in the sequence after it is the sum of the two numbers before it. So the third number is 1 + 1, which is 2. The fourth number is the 2nd number plus the 3rd, which is 1 + 2. So the fourth number is 3. The 5th number is the 3rd number plus the 4th number: 2 + 3 = 5. This keeps going forever.

The first few numbers of the Fibonacci sequence are: 1, 1, 2, 3, 5, 8, 13, 21, 34, 55, ...

Because one number is defined by the numbers before it, this sets up a perfect opportunity for using recursion.

Your mission, should you choose to accept it, is to create a method called **Fibonacci**, which takes in a number and returns that number of the Fibonacci sequence. So if someone calls **Fibonacci(3)**, it would return the 3rd number in the Fibonacci sequence, which is 2. If someone calls **Fibonacci(8)**, it would return 21.

In your **Main** method, write code to loop through the first 10 numbers of the Fibonacci sequence and print them out.

Hint #1: Start with your base case. We know that if it is the 1st or 2nd number, the value will be 1.

Hint #2: For every other item, how is it defined in terms of the numbers before it? Can you come up with an equation or formula that calls the **Fibonacci** method again?

16

Value and Reference Types

In a Nutshell

- Your program's memory is divided into several sections including the stack and the heap.
- The stack is used to keep track of your program's state and local variables.
- The heap is used to store data that is accessible anytime from anywhere.
- The CLR manages memory in the heap for you, and cleans up unused memory using garbage collection.
- Value types store their data directly inside of the variable. Reference types store a reference to a location in the heap, where the rest of the data is stored.
- Value types have value semantics, which means that when you assign one variable the contents of another, the entire contents of the variable are copied, so you get a complete copy in the other variable. Reference types have reference semantics, which means when you assign one variable to another, only the reference is copied, and the two are referencing the same object. Changes to one affects the other.

We're going to change gears a little and spend a chapter in pure learning mode, instead of coding mode. We're going to discuss a few topics that are going to be important to understanding types and how classes work. This chapter is probably the most technical chapter in the book, but trust me when I say it is both important and useful.

The Stack and the Heap

When your program first starts, the operating system gives it a pile of memory for it to work with. Your program splits that up into several areas that it uses for different things, but most of it is used by two sections for storing data as your program runs. These two sections are the *stack* and the *heap*. The stack and the heap are used for slightly different things, and they're organized in different ways.

The stack is used to keep track of the current state of the program, including all of the local variables that you have. The stack behaves like a stack of containers. Whenever we enter a method, we place a new container on the top of the stack. These containers on the stack are called *frames*. All of the local variables (including parameters) for a particular method go in a single frame.

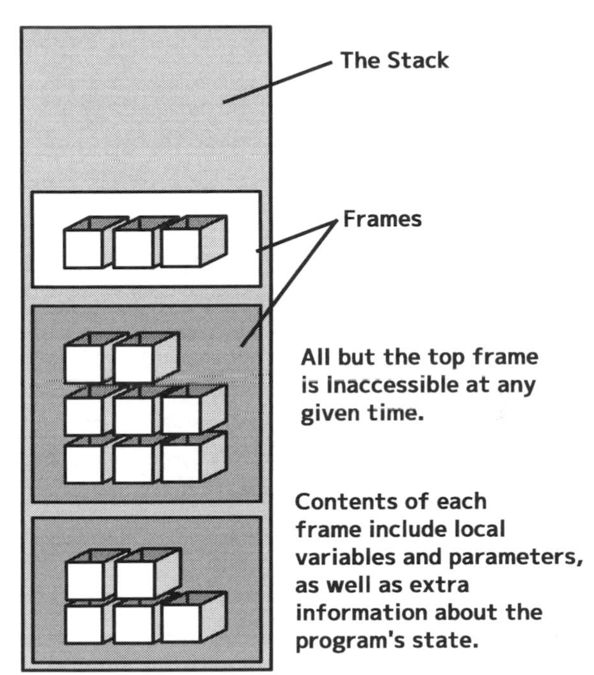

Additionally, a frame will include information to keep track of the program's current state of execution, like what line of code the program was at just before entering the new method. In fact, one of the primary purposes of the stack is to keep track of the program's current state of execution.

Like with a stack of containers, you can readily access the contents of the frame on the top of the stack, but you can't really get to the containers lower down. When we return from a method, the top frame is taken off the stack and discarded, and we go back down to the frame beneath it.

Interestingly, the debugger that is built into Visual Studio *can* inspect the entire stack, including the buried frames, and show them to you. (This is called a *stack trace*.)

The heap, on the other hand, is not concerned at all about keeping track of the program's state. Instead, it is only focused on storing data. The heap is organized in a way where it is easy to get access to things any time you need. It's easy for the program to grab some space in the heap and start using it to store the information it needs. There's not necessarily any logical organization to what gets stored where on the heap, so it is easy to think of it as just a pile of individual blocks of data.

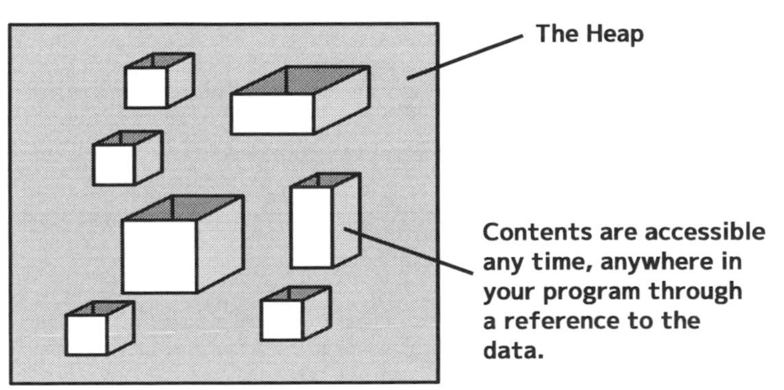

If you have a multi-threaded application (Chapter 37) each thread will have its own stack, but all threads in the program will share the same heap.

Memory Management and Garbage Collection

A program needs to manage the memory it is using. In particular, it needs to clean up old memory that is no longer being used. For the stack, this is easy. Any time it returns from a method, it knows it can throw the top frame away and reuse that space immediately.

The heap is a little more complicated. Back in the day, when you put stuff in the heap, you needed to remember it, and when you were done using it you had to tell the computer you were done with it. This would allow you to reuse that space again.

If you were putting something on the heap for just a moment, and then getting rid of it right after, this was easy to do. But this was not so simple when what you were putting on the heap needed to stay there a while before being cleaned up. It was very easy to forget that it was even there, and even if you remembered, it was usually all too easy to fail to free that memory correctly. You'd end up with garbage in the heap that you no longer wanted, but the computer didn't know it could reuse. This is called a memory leak, and if you had one (or many) then your program was eventually going to run out of space and die.

Fortunately for us, C# uses an approach where the .NET Framework manages your heap's memory for you. If you're starting to run low on memory for your program, a part of the .NET Framework called the *garbage collector* will run through and get rid of any old, unused stuff on your heap, freeing it up for other uses. If you've never had to manage your own program's memory, it may be hard to see this, but garbage collection is a huge deal, and it saves you lots of time, worry, and customer complaints.

References

Since data on the heap is sort of scattered about rather than structured like the stack is, you will need a way to keep track of the variable or data you are interested in.

Historically speaking, you would have used a thing called a *pointer*, which stored the memory address of the object you were interested in. The computer could use that memory address to go and look up the variable you wanted.

With C#, the heap's memory is managed, which means the CLR is going out of its way to take care of it for you. If the heap's memory wasn't managed, we couldn't do garbage collection.

Because the CLR is managing the memory on the heap for you, it may move things on the heap around as it sees a need. That means having an actual pointer to a location in memory is not the best idea, since it could get changed. Instead, C# uses a thing called a *reference*, which is like a pointer, but it is also managed for you. Even as things get moved around in memory, your reference will always direct your straight to the information you're interested in.

It is convenient to think of a reference as an arrow pointing you to a specific item or variable within the heap. However, references aren't truly arrows, but more of a unique identifier that tells the computer where to find the rest of the data within the heap.

Value Types and Reference Types

In C#, all types are divided up into two broad categories: value types and reference types. With our discussion on the stack vs. the heap, and what we now know about references, we're ready to dig into the difference between these two. This is one of the biggest stumbling blocks or misunderstandings for both new programmers and experienced programmers, so take the time you need to make sure you understand how these two difference categories of types work.

Value and reference types are one of the biggest things that make C# stand out from C++, Java, and other programming languages, which do things in a completely different way. If you're a programmer, it's time to pay extra attention, so that you understand this key difference.

Here is the key difference: when you have a variable that is a value type, the actual contents of that variable live where the variable lives. With a reference type, the contents of the variable live on the heap, and the variable itself contains only a reference to the actual content.

Go back and read that last paragraph again a time or two to make sure you understand it.

It turns out we have already been using both value and reference types. The **string** type is a reference type, and arrays are references as well (though the contents of the array may be reference or value types). All of the simple types that we've discussed before, as well as enumerations (Chapter 14) are all value types.

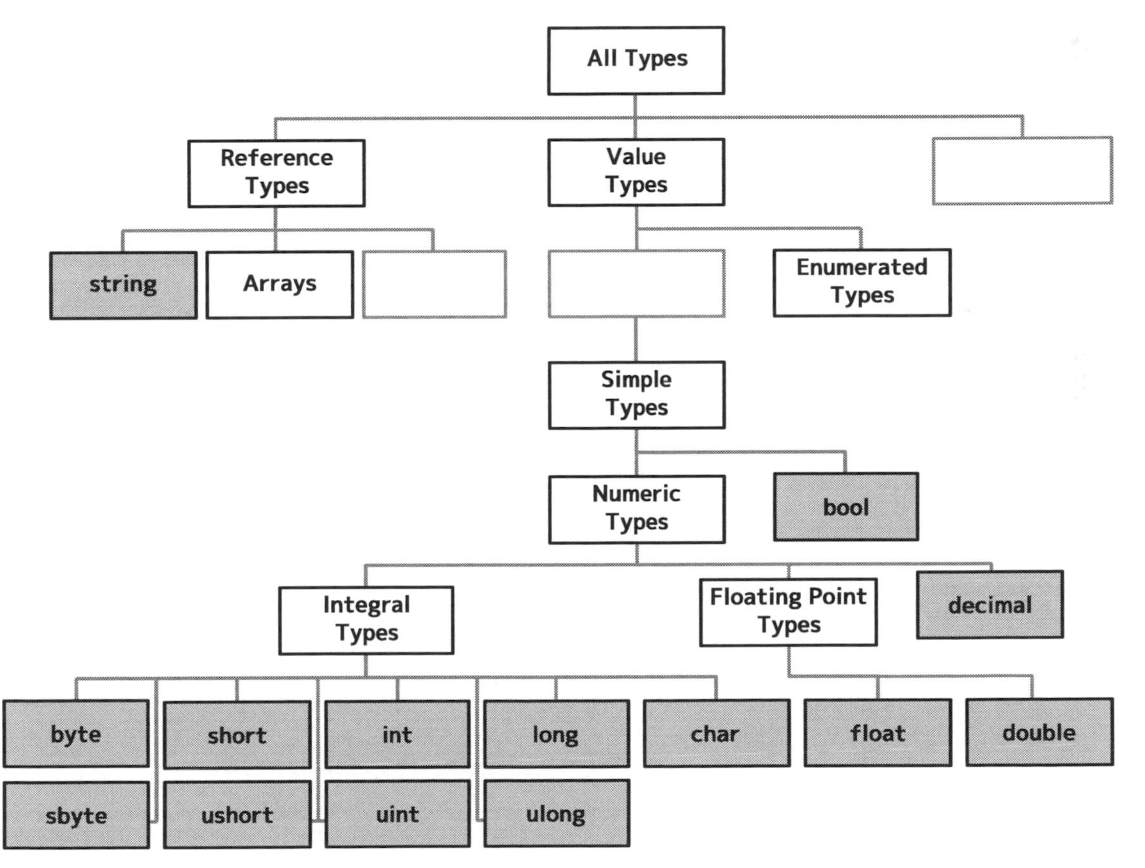

As we get started with Part 3 of this book, we're going to learn how you can create your own custom value or reference types, and finish filling in our type system diagram.

Let's now dig into this in a little more depth, starting with reference types. Look at the following code:

```
string message = "Hello!";
```

Since the **string** type is a reference type, the actual "Hello!" text will always be stored in the heap, and the **message** variable will simply contain a reference to that text. As a local variable or parameter, the **message** variable will be stored on the stack somewhere, and contain a reference to that actual text on the heap.

Depending on how things are set up, reference type variables can also live on the heap, so you could have references from one part of the heap pointing to other parts of the heap. This would be the case if you have an array of **string**s, both of which are reference types. For instance, look at the following:

```
string[] messages = new string[3] { "Hello", "World", "!" };
```

This might look like this:

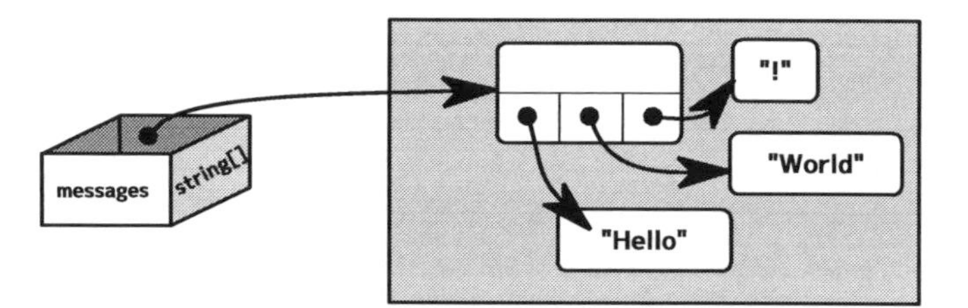

A local **messages** variable on the stack has a reference to the **string** array, but that actual array lives on the heap. Within that array, each item contains another reference to other parts of the heap—the **string**s that go in the array, in this case. Taking this further, it is entirely possible to have huge networks of references spread throughout the heap.

On the other hand, value types are stored entirely at the place that the variable exists. If they're local variables or parameters, this means they're stored entirely on the stack. We've seen this with nearly all of the built-in types that we've been working with so far, like **int** and **double**. The variable doesn't contain a reference, just the actual value.

This still holds true, even when the value type is defined on the heap. Value typed variables will always contain their data right there where the variable is at. Arrays give us a great example of how this works. Let's say you had an array of value types, like this**: int[] numbers = new int[3] { 2, 3, 4 };**. This would end up looking something like this:

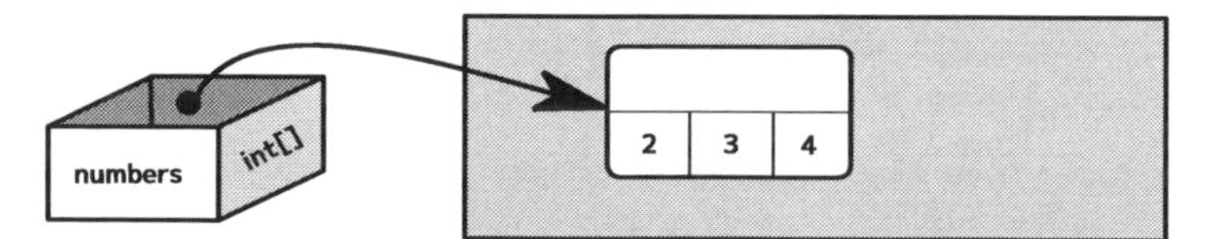

Note that the items in the array don't point to elsewhere like they did with our **string** array earlier. The numbers exist right where the variable exists.

That covers the basics of *how* value and reference types work. In a minute, we'll come back and take a look at what that means for us, when we talk about value and reference semantics. But first, a little bit of a detour to cover something else that we need to know about reference types.

Null: References to Nothing

One interesting thing about reference types is that you can have them reference nothing at all. This is called a *null reference*. Assigning a reference type a value of null can be done with the **null** keyword:

```
string message = null;
int[] numbers = new int[] { 2, 4, 6, 8 };
numbers = null;
```

Of course, you'll note that with the middle line, we create a new array, which is placed on the heap, and then we immediately reassign **numbers** to **null**. At this point, the **numbers** variable isn't referencing that array anymore, and since nothing else is referencing it either, it has become inaccessible, all alone in the heap. Because it is inaccessible, eventually it will be garbage collected.

If a reference type has a value of **null**, it means there's no data at the other end. If you inadvertently try to actually do something with it, your program is going to crash:

```
int[] numbers = null;
numbers[3] = 6;        // This crashes, since numbers doesn't reference anything.
```

To address this, you can do a simple check to see if a reference type is **null** first, if you think that could happen:

```
if(numbers != null)
    numbers[3] = 6;
```

Value types cannot be assigned a value of **null**.

Checking for null is incredibly common. While the approach above covers you in all cases, C# 6 introduced a new pair of operators that allow you to check for null using much more concise syntax. In fact, with the new syntax, checking for null with an if-statement is no longer the recommended approach in most cases. For more information on this new syntax, see the section called *Simple Null Checks: Null Propagation Operators* in Chapter 39.

Value and Reference Semantics

If the only difference between value and reference types was whether they stored their data "on site" or off elsewhere in the heap, I wouldn't normally worry about that kind of detail for an introductory book like this. However, there's more to it. Because of this difference, we get different behavior.

Reference types have what's called *reference semantics*, and value types have *value semantics* or *copy semantics*. Let's compare these two with an example, using the **int** type (a value type) and an array of **int**s (the array is a reference type, even though it contains things that are value types). Let's say you have the following code:

```
int a = 3;
int b = a;
b++;
```

We've created two variables with the **int** type. We place the value 3 inside **a**. We then assign the contents of **a** to **b**. Doing this means reading the contents of **a** (which is 3) and then put that value into **b**. Next, we modify **b** by incrementing it so **b** is now 4. But **a** is still the original value of 3. When we assign the contents of **a** to **b**, we made a new copy for **b**. Whatever happens with the **b** variable won't affect the original **a** variable.

Interestingly, the exact same thing happens with reference types, but because the variable contains only a reference instead of the actual value, we get different behavior entirely. Take a look at this example:

```
int[] a = new int[] { 1, 2, 3 };
int[] b = a;
b[0] = 17;
Console.WriteLine(a[0]); // This will print out 17.
```

When we create our array, the entire contents of the array are placed in the heap. While **a** may be stored on the stack, it will contain a reference to the array out in the heap. When we assign the value of **a** to the variable **b**, we read the value of the **a** variable, which is a reference, and copy it in the **b** variable. But copying a reference gives **b** a reference to the same object on the heap. At this point, both of our variables are referencing the same array in the heap. We then modify that array by going through the **b** variable, but since the two things are referencing the same thing on the heap, **a** has been modified as well.

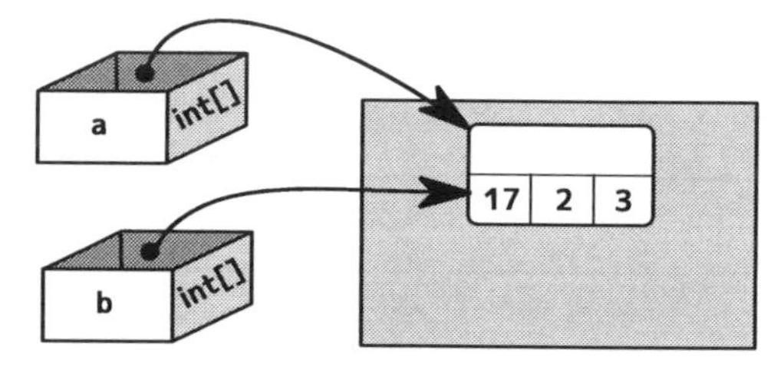

This same thing happens as you send things to methods as parameters. For value types, the value is copied over to the parameter. The copy that the method ends up with is completely different, and in the method, changing the variable won't affect the original, back in the calling method.

With reference types, while you're still technically passing a copy of the variable's contents, it is a copy of a reference, so inside of the method the parameter is still referencing the same thing on the heap. Making changes to it inside of the method affects the thing back in the calling method.

This is often done intentionally, handing an array or other reference type to a method with the specific task of modifying it in some way. But of course, if you hand a reference type to a method, expecting it to remain unchanged, and the method makes changes, you'll be in for a surprise.

This can be a useful thing or a liability, but it *is* a fact. You have to understand the differences between value types and reference types, or it will come back to haunt you repeatedly. Surprisingly, there are a lot of talented programmers who claim to know C# that haven't figured this heap vs. stack and value vs. reference type stuff yet. If you don't understand these differences, you will be constantly running into strange problems that just don't seem to make sense.

Try It Out!

Value and Reference Types Quiz. Answer the following questions to check your understanding. When you're done, check your answers against the ones below. If you missed something, go back and review the section that talks about it.

1. **True/False.** Anything on the stack can be accessed at any time.
2. **True/False.** The stack keeps track of local variables.
3. **True/False.** The contents of a value type can be placed in the heap.
4. **True/False.** The contents of a value type are *always* placed in the heap.
5. **True/False.** The contents of reference types are *always* placed in the heap.
6. **True/False.** The garbage collector cleans up old, unused space on both the heap and stack.
7. **True/False.** You can assign **null** to value types.
8. **True/False.** If **a** and **b** are reference types, and both reference the same object, modifying **a** will affect **b** as well.
9. **True/False.** If **a** and **b** are value types with the same value, modifying **a** will affect **b** as well.

Answers: (1) False. **(2)** True. **(3)** True. **(4)** False. **(5)** True. **(6)** False. **(7)** False. **(8)** True. **(9)** False.

Part 3

Object-Oriented Programming

C# is an object-oriented programming language, meaning that the code we write is typically organized into little blocks that are modeled after real-world objects. We define what kind of data those objects have and what those objects can do.

The object-oriented aspect of C# is a key part of the language. Without knowing how to create and use classes, our understanding of the language is far from complete.

Part 3 is all about designing and building your own types in C#. Building your own reusable types and assembling them together is how you'll be able to make truly amazing software.

We'll dig straight to the heart of object-oriented programming. We'll look at the following:

- Introduce what object-oriented programming is about (Chapter 17).
- Discuss how to make your own classes (Chapter 18) and structs (Chapter 20).
- Create easy access to data in your own custom types with properties (Chapter 19).
- Make classes that are special types of other classes with inheritance (Chapter 21).
- Making special types that can do the same tasks in their own way (Chapters 22 and 23).
- Make generic types (Chapters 24 and 25).

Classes and Objects

In a Nutshell

- Object-oriented programming is a way of programming where you create chunks of code that are modeled after real world objects.
- Classes in C# define what an entire "class" of objects can do (what all players can do, or what any car can do), including what kind of data it stores and the tasks that it can do.
- Classes are reference types.
- Creating a class is the most powerful way C# has of defining your own new types.
- You can create a new object with something like this: **Random random = new Random();**. This creates a new **Random** object, which is used to generate random numbers.

We have worked with a lot of types as we've been learning C#. All along, I've kept promising you that we'll learn how to make our own types, and the time has finally come for that!

To define our own types, we're going to use all of the other types we have learned about, combined with methods, to construct new types from scratch. We will then be able to use these types, just like we have done with the built-in types.

Before we start making our own types, we first want to start with a discussion about why we need to be able to create our own types, and how to go about designing our new types.

Modeling the Real World

Over the many, many years that people have been programming, we've come to learn that one of the best ways to structure code is to mimic the way the real world works. It helps to break things down into manageable pieces, and presents an intuitive way for programmers to work with their code.

Read that last paragraph again. It's important.

In the real world, there are objects. There are houses, cars, balls... you name it. Actually, let me be a little more specific. What we just listed were *types* of objects. Entire categories or "classes" of objects. I mean, we referred to balls in general, but there are also lots of *specific* balls. This ball and that ball, the red ball and the blue ball, and the basketball that was signed by LeBron James.

We know that all things that belong to a particular category or class will have certain attributes. We know all balls will have a size, and that they will have a color. Each ball may have a different size and a different color, but we know they'll all have a size and color. A ball's size and color defines its state.

We also know that all things of a particular category or class will have certain behaviors. Different things that they can do, or that can be done with them. For instance, balls can be thrown, or they can be popped. You can do this (or attempt to do this, anyway) with any and every ball.

In C#, as well as nearly every other object-oriented programming language, we can build new, special types, modeled on these ideas. In C#, we will create code, packaged together in what is called a **class**. You can think of a class as a blueprint for *objects* or *instances* of that type. While the class outlines how the entire category works, an object or instance is a single, specific one.

A class will outline three things about our new type:

1. The state or attributes that things of the new type can have, using variables (Chapter 5). These variables can be of any of the built-in types, or they can even be other types that you create.
2. The behaviors that things of the new type can do, or that can be done with them, using methods (Chapter 15).
3. The ways in which you can build or construct a new object or instance of the given type or class. This is done by using a special type of method called a constructor, which we'll learn more about in the next chapter.

Continuing with our ball example, we might create a **Ball** class that includes two variables, one for the ball's size, and one for the ball's color. The **Ball** class might also include a **Throw** and a **Pop** method. (In the next chapter, we'll look at how to actually create a class like this.)

Because a class combines a collection of variables and methods, you can also think of a class as simply a way of packaging related data and methods together into one single new cohesive piece.

It is important to keep the concepts of classes and objects separated in your head. They refer to two slightly different things, so they're easily confused. A class defines things about how the entire category of objects works, while an object or instance is a *specific* one. You might define a **Ball** class, but then using that class, you can create or build specific balls, each with their own size and color.

You can create a class for just about anything. We could make a **Ball** class, a **House** class, a **Car** class, or even our own **PurpleMonkeyDishwasher** class if we really want. Starting now, we're no longer going to be limited to just the built-in types like **int**, **string**, or **float**.

Working with an Existing Class

Rather than dive in to creating our own classes right away, let's start off by using a class that was created for us already. We'll use the **Random** class.

The **Random** class is a cool way to let you create random numbers. You can use this class for things like determining the outcome of a die roll, shuffling a deck, or making random events happen.

In Depth

Computers and Randomness. Generally speaking, random numbers on a computer aren't truly random. They are actually chosen using an algorithm that simply *appears* random. They need to be initialized, starting with a number chosen by the programmer. This number is called a seed. If you start with the same seed, you will get the same sequence of random numbers, over and over again.

Since the vast majority of the time, this is not something you want, programmers will often seed the random number generator with the current time, accurate to the millisecond. This makes it so that a different sequence is generated every time you run the program.

The **Random** class that we're using here will automatically seed the random number generation with the current time, but you can also specify a seed if you would like.

We can use the **Random** class in a way that is very similar to any other type:

```
Random random = new Random();
```

In this case, we want to use **Random** for our variable's type, and like usual, we give our variable a name (**random**). Like with any other variable, we can assign a value by using the assignment operator (=).

But after that, we'll see a couple of things that we haven't really seen. You can see the **new** keyword there. This tips us off to the fact that we're going to create a new instance of the **Random** class—an object with the **Random** type.

The **Random()** part might look like a method call to you, and it kind of is. In fact, it is a special kind of method called a *constructor*. A constructor is placed in a class and describes how to create or build a new object that has that type. This particular constructor has no parameters (hence the empty parentheses) but it is possible to have constructors with parameters, like any other method.

When that line is finished executing, our variable called **random** will contain a reference to a brand new object with the type **Random**.

Using an Object

Now that we've got a working object, we can do stuff with it. If you remember, an object has state, and it has behaviors. In the specific case of the **Random** object we just created, internally, it is keeping track of its state, but it doesn't expose any of that to us.

So while we can't modify a **Random** object's state, we can use its behaviors, defined by its methods. Other types, aside from the **Random** type, will often expose some of its state, or provide additional methods to allow us to change the object's state. (Like changing a ball's color from blue to red.)

We can access our object's behaviors through the methods it has defined in it. For starters, there's the **Next** method. We call this method by using the dot operator ("."):

```
Random random = new Random();
int nextRandomNumber = random.Next();
```

We've used the dot operator in the past, and it is probably time to really explain what it means. This operator is used for member access. In other words, to access something that belongs to something else. Here, we use it to say, "Look in **random** to find the method called **Next**." Things that belong to an object, like variables and methods, are called *members* of the type and are accessible via the "." operator.

The **Next** method is overloaded (Chapter 15), so there are multiple versions of the **Next** method, including one that lets us choose a range to use:

```
Random random = new Random();
int dieRoll = random.Next(6) + 1; // Add one, because Next(6) gives us 0 to 5.
```

The **Random** class also has a **NextDouble** method, which returns numbers between 0 and 1. This allows you to do things like random probabilities and so on.

Once you have created an object like this, you can use any of its methods in any order, at any time.

Try It Out!

Die Rolling. Tons of games use dice. And the **Random** class that we've been talking about gives us the ability to simulate die rolling. Many games give the player the task of rolling multiple six-sided dice and adding up the results.

We're going to write a program that makes life easier for the player of a game like this. Start the program off by asking the player to type in a number of dice to roll. Create a new **Random** object and roll that number of dice. Add the total up and print the result to the user. (You should only need one **Random** object for this.)

For bonus points, put the whole program in a loop and allow them to keep typing in numbers until they type "quit" or "exit".

The Power of Objects

Classes are a core part of object-oriented programming languages like C#. In fact, you could argue they are *the* core part. Classes and the objects they create are the fundamental building block of programming in these languages. In C#, nearly all of your code will belong to a particular class (or another custom-made type). In fact, all of the code we've been writing so far has been inside of a **Program** class! (Starting in the next chapter though, we're going to be building our own classes, and putting code elsewhere.)

The fact that classes and objects are modeled after real world objects means that you can build your code in a way that will intuitively make sense to anyone trying to use it, and that's worth a lot.

Classes provide us with reusable pieces of code. For example, the **Random** class was made by other people years ago, and yet, we were able to reuse it ourselves.

Classes provide a way to break a complex program into small pieces that are easily managed. Each class is able to manage its own information and the things it can do. Because things can be broken down into classes, it will be easy to work on our program in pieces. And when something goes wrong, it will be easier to track down where the problem lies.

Classes are Reference Types

It is critical to point out that all classes are reference types. This means that everything we learned in Chapter 16 about reference types applies to classes. For example, you can assign **null** to it:

```
Random random = null;
```

And two variables that are assigned the same reference *will* affect each other, because they use reference semantics:

```
Random random1 = new Random();
Random random2 = random1;
// random1 and random2 are actually referencing the same Random object now.
```

And like all reference types, this applies even as you pass the contents of a variable to a method as a parameter. If you pass an object to a method, the parameter inside of the method will reference the exact same object as whatever was passed in. This means changes to the object from inside of the method will affect anything that references the same object in the method that called it, for better or worse.

```
public static void Main(string[] args)
{
    Random random = new Random();
    DoSomething(random);
}

public static void DoSomething(Random r)
{
    // The variable 'r' references the same exact object as 'random' in Main.
    // The two variables are two different references that refer to the same object.
    Console.WriteLine(r.Next());
}
```

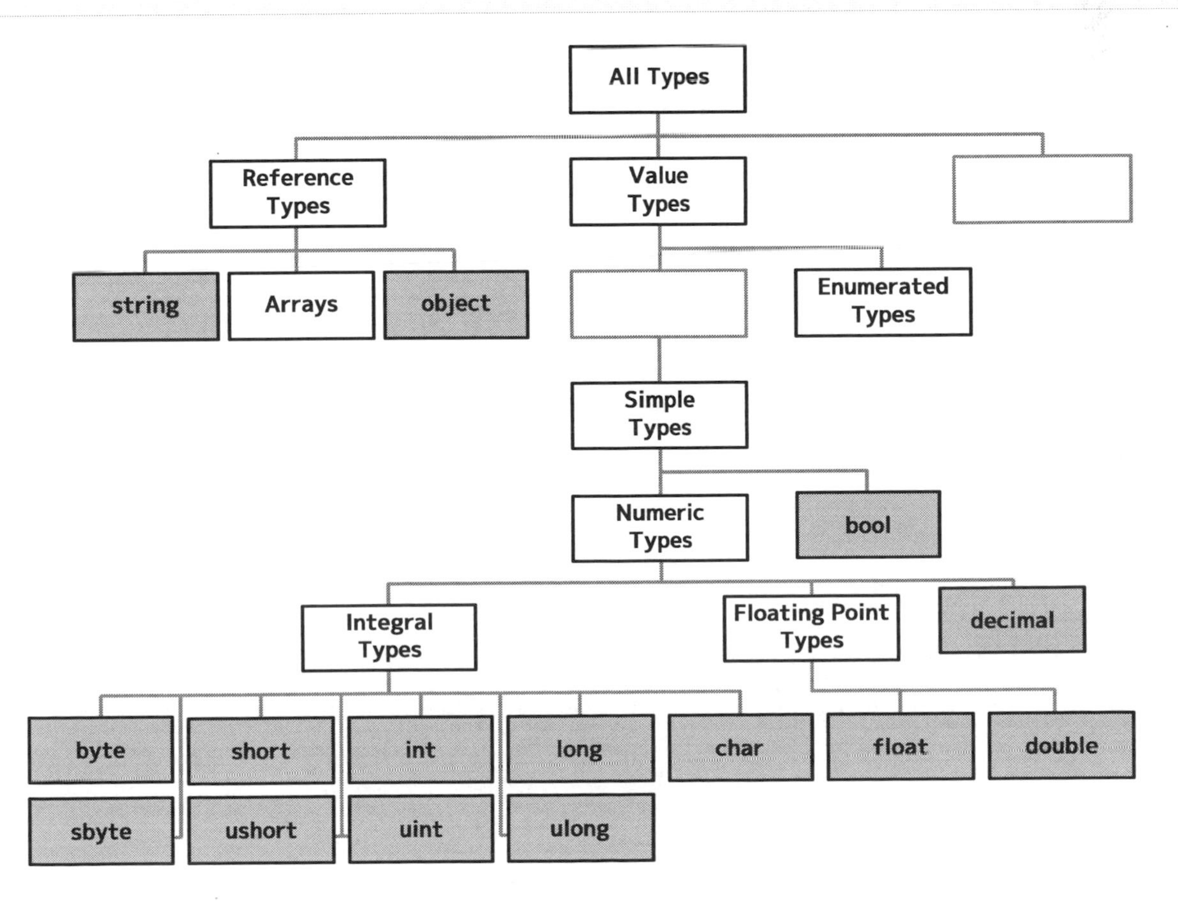

18

Making Your Own Classes

In a Nutshell

- Classes typically go in their own file (though C# doesn't require it).
- Anything that is a part of a class (or other type) is called a member. This includes instance variables, methods, and constructors.
- Variables can be added to a class—these are called instance variables.
- You can add as many constructors to a class as you want.
- You can also add methods to a class, which operate on the instance variables that the class has.
- The **private** keyword, attached to a method or instance variable, means that only stuff inside the class can see it.
- The **public** keyword, attached to a method or instance variable, means that anything (inside or outside of the class) can use it.
- This chapter also covers scope, name hiding, and the **this** keyword.

In the previous chapter we learned about classes, objects, and how to use them. Now, we'll move ahead and see how to create your own classes.

The ability to create your own classes is very important. As you write your own programs, you will be creating lots of your own classes. That's because building your own classes allows you to break your whole program into manageable pieces that you can work with.

In this chapter, we'll create a simple **Book** class, which we could use to represent some information about a book in a program. While we cover a lot of ground in this chapter, this is the end of the beginning. By the end of this chapter, you will know all the fundamentals to programming in C#! While there is more to cover, it is possible to write almost any program you want after this chapter.

Creating a New Class

We're now ready to create our very first class. Remember from the last chapter that a class works as a blueprint for what an entire class or category of things can do and keep track of.

While you are allowed to put multiple classes in a single file, it is a good idea to put each new class into its own file. This helps keep your files to a manageable size, and makes it easier to find things.

So our first step is to create a new file inside of our project. To do this, in the Solution Explorer, right-click on your project. The very top level item in the Solution Explorer is your *solution*, which is simply a collection of related projects. That's not the one you want. Your solution probably only has one project right now, and it likely has the same name as the solution. So in most cases, this means you'll be clicking on the second item in your Solution Explorer. After right-clicking, choose **Add > Class...**.

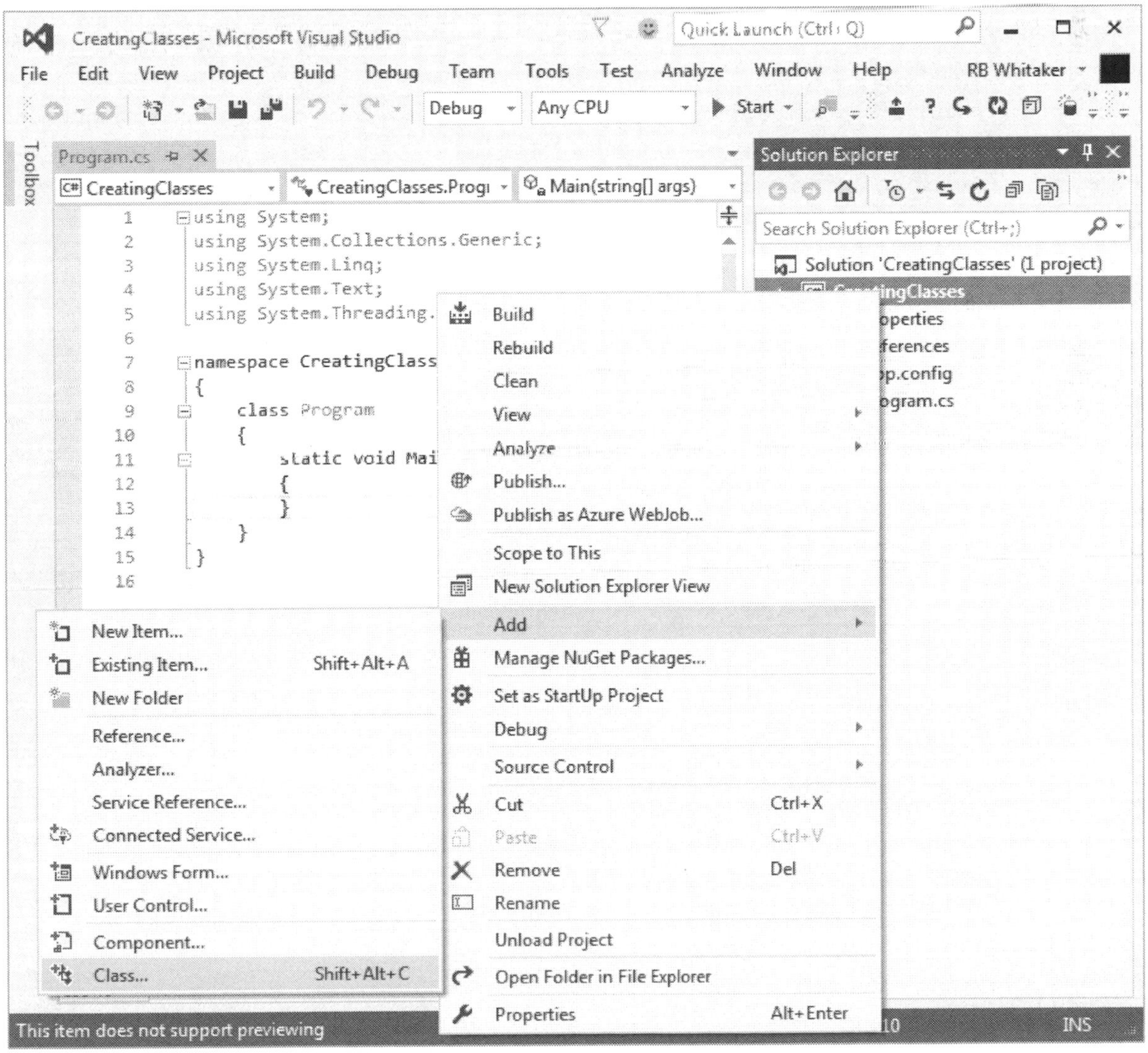

This will bring up the **Add New Item** dialog, with the Class template already selected. At the bottom, type in the name of your new class. I've called mine **Book.cs**, so my class will be called **Book**. Press the **Add** button, and your new file will be created.

When your new file is created, you'll see that Visual Studio generated some template code for you. **Book.cs** will contain code that looks something like this:

```
using System;
using System.Collections.Generic;
using System.Linq;
using System.Text;
using System.Threading.Tasks;

namespace CreatingClasses // Your namespace will likely be different here.
{
    class Book // Whatever name you chose will appear here as the class name.
    {
    }
}
```

This starts things off pretty close to where we want it to be. We've got our new class started!

The class that was generated for us is completely empty. Our task now will be to fill in the class with the things it needs. Anything that is a part of a class is called a *member* of the class. Remember that classes will outline three general categories of things: the data it has (as variables) the things it can do (methods) and ways to create or setup new instances of the class (constructors). We'll now go through the process of adding each of these to our new class.

Adding Instance Variables

We'll start by adding variables to our class, which are used to keep track of the information our class needs to store. Since each object or instance of the class will have its own set of data, these variables are called *instance variables*. If you have two different books, and each one is represented by its own **Book** object, they can each keep track of their own titles separately, each in its own set of instance variables.

For our simple book class, we're going to keep track of the book's title, author, number of pages, and the word count. Obviously, we could store a lot of other information about a book, like the publisher, the year of publishing, and even the full text of the book, but this is enough to illustrate the point.

To create an instance variable for these things, we'll add the following code inside of the class's curly braces ('{' and '}'):

```
private string title;
private string author;
private int pages;
private int wordCount;
```

So your current code should look something like this:

```
using System;
using System.Collections.Generic;
using System.Linq;
using System.Text;
using System.Threading.Tasks;

namespace CreatingClasses
{
    class Book
    {
        private string title;
```

```
        private string author;
        private int pages;
        private int wordCount;
    }
}
```

Now let's look at what we've done. For each instance variable, we start off with the **private** modifier, which we'll discuss in a second. The rest is just like creating any other variable—we state the type of variable, followed by the name we want to give it.

These variables are a little different from what we've seen up until now, in terms of how they're used. The variables we've been creating have been *local variables*, meaning that they can be accessed locally, within a single method, or sometimes just within a single loop. We've also already seen a second type of variable called *parameters*, which are like local variables, only they are declared as a part of a method header, which means the calling method gets to assign the values. *Instance variables*, like these new ones we just created, belong to an object. They can be accessed anywhere you want throughout the class, spanning all of the methods it contains, and all of the constructors it has. They're not just limited to a single method.

Access Modifiers: private and public

Let's go back to that **private** keyword. Everything in a class has what is called an *accessibility level*. This indicates where it can be accessed from. The **private** keyword indicates that this variable can only be accessed (used or changed) from inside of the class. Anything outside of the class doesn't even know it exists. It's private.

Another accessibility level, which we'll see shortly, is **public**. The **public** access modifier means that it can be seen or used from anywhere, including outside of the class. We have the option of making these variables public, but that would mean that anyone, anywhere, can directly modify them, which could have bad consequences. For instance, that would mean that someone could change **wordCount** to a negative value, which wouldn't make sense.

By making things private, we keep them safe from outside meddling. That's nearly always what we want for instance variables. If there's a need for things outside of the class to access this data or modify it, we can provide methods that are public, which can handle outside requests and ensure that those requests don't do something dangerous or dumb with our class's data.

Generally speaking, a class should be responsible for handling changes to its instance variables. This keeps everything that the class is responsible for safe and protected inside of the class. This idea is called *encapsulation*.

By the way, if you don't put an access modifier, these variables would be private by default; **private** is the default accessibility level for all members of a class. Because of this, it wasn't technically necessary to use **private** in this case, but I think it's a good idea to put it in there, to make it obvious.

Adding Constructors

Now that we've got instance variables ready, our second step is to indicate how you can create new instances of our class. This is done by adding in special methods called constructors.

When it comes time to make constructors, we usually want to look back at the instance variables we have and think about which of those variables must be set to get the new instance into a valid state, as well as what things people will commonly want to supply at creation time.

For instance, it probably doesn't make sense to create a book without a title. So we'll probably want to make a constructor that has a title as a parameter. And most of the time, we will also want the name of the author, so let's make a second constructor that has both a title and an author. Finally, we may also want a way to create a new book that supplies a value for all four of our instance variables. So that's three constructors that we've described.

Constructors are special methods, and while they look and work like normal methods in most aspects, there are a couple of differences. Below is the code for the first constructor, which requires the user to pass in only the book's title:

```
public Book(string title)
{
    this.title = title;
}
```

We add this code inside of our class, which officially makes it a member of the class, as shown here:

```
using System;
using System.Collections.Generic;
using System.Linq;
using System.Text;
using System.Threading.Tasks;

namespace CreatingClasses
{
    class Book
    {
        private string title;
        private string author;
        private int pages;
        private int wordCount;

        public Book(string title)
        {
            this.title = title;
        }
    }
}
```

Let's take a minute and look at what we've created, and how constructors are different from normal methods. For starters, notice that here, we use the **public** access modifier. We want people outside of the class to be able to create **Book** objects, so we've made this constructor public. (Constructors do not need to be public. There are rare times where it makes sense to make them private or other accessibility levels that we have yet to talk about, but public is probably the most common.)

There are two things that we can see here that make constructors different from a normal method. For one, the constructor will always have the exact same name as the class. **Book**, in this case. Second, you can see that our constructor doesn't have a return type (we don't say **public void Book()** or anything). Using the name of the class, and providing no return type tell the compiler that this is a constructor, not a normal method.

We then put the parameters for our constructor in parentheses, just like a method.

By the way, if you don't put any constructors in your class, the C# compiler will put one in for you. That constructor is parameterless, and if you wrote it yourself, it would look like this:

```
public Book()
{
}
```

There are a few other important diversions that our first constructor brings up, which we'll discuss in turn: variable scope, and the **this** keyword.

Variable Scope

One thing about variables that we haven't talked about yet is *scope* or *variable scope*. The scope of a variable refers to the places in the code that a variable is accessible.

Let's say you create a method, and inside of it, you create a variable like this: **int aNumber = 3;**. This variable can be used anywhere inside of that method. However, you can't use that variable inside of another method. It has *method scope*.

Let's look at another example:

```
public void DoSomething()
{
    for(int x = 0; x < 10; x++)
        Console.WriteLine(x);

    x = -5;  // This won't compile--the variable 'x' is "out of scope".
}
```

In this method, we have a **for** loop. Inside of the **for** loop, we create a variable called **x**. This variable is only valid inside of the **for** loop—once we're outside of the loop, the variable is no longer valid. It is "out of scope." We won't be able to use it anymore. It has what's called *block scope*, or *inner block scope*.

Interestingly, we can create a second **for** loop, reusing the same variable name (**x**) for a different variable, and the two won't conflict:

```
public void DoSomething()
{
    for(int x = 0; x < 10; x++)
        Console.WriteLine(x);

    for(int x = 50; x < 60; x++)
        Console.WriteLine(x);
}
```

The key to scope is, let the curly braces be your guide. A good rule of thumb is that anything created inside of a set of curly braces can be used anywhere inside of those curly braces.

A step larger than method scope is *class scope*, or *class-level scope*. If you look at the instance variables we created, they were put inside of the class's curly braces, and can be used anywhere inside of the class, including the constructors that we made, and the class's methods which we'll create in a minute.

To summarize:

- **Class Scope:** Accessible throughout the class.

- **Method Scope:** Accessible throughout the method.
- **Block Scope:** Accessible through only a particular block (like a **for** loop).

Name Hiding

One interesting thing that you can do in C# is name a parameter or local variable (which has method scope or block scope) the exact same thing as something that has class scope. While making our class, remember that we made an instance variable called **title**, which has class scope, and is accessible throughout the class. Then, in our constructor, we had a parameter that is also called **title**, which has method scope and is valid only throughout the method/constructor.

This leads to a situation called *name hiding*. We have a class-scoped instance variable called **title**, but we also have a method-scoped parameter called **title**. So when we refer to **title** inside of this constructor (or method), which of the two is it referring two? The one with the smaller scope—in this case, the parameter, which has method scope.

So we create a situation where one variable "hides" another variable, because they have the same name. That's all fine and good, but what if we want to access the one in the larger scope? The instance variable?

Well, there are a few ways to address this. One is to just call the two variables different names, so there's no hiding going on. For instance, we could have called the parameter **bookTitle** instead of **title**. But it is a bit of a hassle to always be trying to think of similar but different names for two variables that store the same information, but at different levels.

A second approach is to name all of your instance variables in a specific way, so that you always know what you're referring to, while still being consistent. For instance, start all of your instance variable names with **m_**. (The 'm' is for "member" here.) So your instance variables would be **m_author**, **m_title**, **m_pages**, and **m_wordCount**. If you do this, you'll successfully avoid name hiding, and because it is systematic, you'll always know what belongs where.

Let me be clear on this point: the **m_** approach is fairly common, and a lot of people use it. But it has a few certain disadvantages that tend to drive me away from it. Most importantly, it kills readability. One of the biggest challenges with programming is that it is much easier to write code than it is to read code. When you see all of these **m_**'s floating around, you have to mentally remove it to know what the variable does. And if you're ever reading the code out loud (discussing it with another programmer) you always have to say "em underscore title" when you're referring to **m_title**. But like I said, this is fairly common, and if that's what you like, go for it.

There's a third approach that I, personally, prefer. This is the way that I'll be doing things throughout this book. This third approach is to name the instance variable and the local variable or parameter the same thing, allow name hiding to happen, and use the **this** keyword whenever you want to refer to the instance variable if there happens to be a name collision. Which brings us to...

The 'this' Keyword

Any time that you are inside of a class, you can get access to all of the class's members, including its methods and instance variables, by using the **this** keyword. **this** is a reference to the current object that you are inside of. So coming back to our **Book** example, anywhere throughout the class, including the constructor that we just added, you can say **this.author** to get access to the **author** instance variable. This also works with methods, so you can call another method that belongs to the class by doing **this.MethodName();**.

In many cases, **this** isn't needed, because C# already assumes you are referring to the stuff inside of the class if there's not anything by that name in the method scope or block scope.

But connecting it all together, if you have name hiding going on, the **this** keyword allows you to get back to the instance variable which has class scope, even though there's another variable or parameter with method scope that would otherwise be hiding it.

So here's another look at the constructor we just added:

```
public Book(string title)
{
    this.title = title;
}
```

In this constructor, there's a parameter with the same name as the instance variable (**title**). Any time we just use **title**, we're using the method-scoped **title** parameter. Any time we use **this.title**, we're going around the name hiding and directly accessing the instance variable called **title**. When we say **this.title = title** we're assigning the value in the parameter to the instance variable.

Anyway, that's enough of a detour for now. Let's add the other two constructors we mentioned. These will look like this:

```
public Book(string title, string author)
{
    this.title = title;
    this.author = author;
}

public Book(string title, string author, int pages, int wordCount)
{
    this.title = title;
    this.author = author;
    this.pages = pages;
    this.wordCount = wordCount;
}
```

Adding Methods

So now we need to figure out what methods our class might want to provide. Perhaps people using our new **Book** type will want a way to request the book's title. We also might want to give a book a new title, or figure out a new word count, based on actual text. The list of methods we might want to add can go on and on.

We've created methods before, so this should be fairly familiar to you. We'll add methods for each of these with the following code:

```
public string GetTitle()
{
    return title;
}

public void SetTitle(string title)
{
    this.title = title;
}
```

```
public void AssignWordCountFromText(string text)
{
    wordCount = text.Split(' ').Length;
}
```

So now our completed **Book** class should look like this:

```
using System;
using System.Collections.Generic;
using System.Linq;
using System.Text;
using System.Threading.Tasks;

namespace CreatingClasses
{
    class Book
    {
        private string title;
        private string author;
        private int pages;
        private int wordCount;

        public Book(string title)
        {
            this.title = title;
        }

        public Book(string title, string author)
        {
            this.title = title;
            this.author = author;
        }

        public Book(string title, string author, int pages, int wordCount)
        {
            this.title = title;
            this.author = author;
            this.pages = pages;
            this.wordCount = wordCount;
        }

        public string GetTitle()
        {
            return title;
        }

        public void SetTitle(string title)
        {
            this.title = title;
        }

        public void AssignWordCountFromText(string text)
        {
            wordCount = text.Split(' ').Length;
        }
    }
}
```

There are a few things that ought to be discussed with this new code.

First, you'll note that all of these methods are marked **public**. That means that anyone outside of the class can call these methods. We could have marked these methods **private** if we had wanted to, which would make it so you could only call the method from inside of the class. In our specific case, we wanted everyone to be able to call these methods, so we made them **public**.

Second, we see here methods that start with **Get** and **Set**. It is extremely common to have methods that more or less just return the value of an instance variable, and matching ones that sets the values of the variables. Additionally, while setting the value of an instance variable, we often check to ensure that the new value is valid. So you may have lots of **GetSomething** methods and **SetSomething** methods.

"Getter" and "setter" methods, as they're called, are incredibly common. So common, in fact, that C# provides a feature in the language that makes this much more readable and simple to use (called *properties*) which we'll talk about in the next chapter.

Perhaps you also noticed that we didn't add **static** to these methods, like we've done in the past. It's time for us to take a look at what **static** actually means, and see why we didn't add it here.

The 'static' Keyword

So far, in this chapter, we've been working with instance variables and methods that belong to an instance of the class. This means you can create different instances of the class, which each have their own set of data. When you call the methods using them, the methods work on that individual instance's data. For example, you can create multiple **Book** objects that each have their own title, author, page count, and word count. Calling **GetTitle** with a specific instance will get the title of that specific book.

In previous chapters, we've been marking our methods **static**. So let's talk about what that means. Any class member that is marked with the **static** keyword belongs to the class as a whole, rather than any particular instance. In fact, in C#'s evil twin language, Visual Basic.NET, instead of using "static" they use the word "shared." That's actually a much clearer way of thinking about it. Anything that is static doesn't belong to a specific object, but rather it is shared among all of the objects of that particular class.

Static Variables

For a class-level variable, marking it with **static** means that it will have the same value for all instances. If instance #1 changes a static class variable to the value of 4, then instance #2 will also see that change. Static variables are shared between all instances of the class. This means that the variable is no longer an instance variable, but rather, a *static class variable*, or *class variable* for short. It belongs to the class as a whole, not to any particular instance of it.

Static Methods

If a method is marked with **static**, as we have done before, this means that you don't need an instance of the class to call it, since it belongs to the class as a whole. To illustrate, a "normal" method that isn't static might be accessed like this:

```
Book book = new Book("Ender's Game");
book.AssignWordCountFromText(
        "I've watched through his eyes, I've listened through his ears, " +
        "and I tell you he's the one.");
```

Here you have an instance of the **Book** class, and you call the method by using the dot operator on the actual instance or object.

A method marked with the **static** keyword is accessed by using the class name instead. You don't need to have an instance of the class to use the static methods. We saw this with the **Console** class:

```
Console.WriteLine("Hello World!");
```

WriteLine is a static method, so we don't need to create a **Console** object (an instance of the **Console** class) in order to use the **WriteLine** method.

All of the methods we've made up until now have been static. But we're now ready to drop the static part and apply methods directly to instances of the class. While there is still value to static methods, most of the methods we write from here on out will not be static.

Static Classes

While you probably won't find uses for this right away, since we're already talking about **static**, I should point out that you can also use the **static** keyword on a class. If you do this, this means that the class *cannot* have instance variables or methods. They must all be static, and the C# compiler will check this for you. The **Console** class is one example of this—it is impossible to create an instance of the **Console** class, but there's also no need to do so, because all of its methods are static. This is great for classes that are used simply to group together a collection of utility methods.

Static Constructors

It will probably be a while before you'll need static constructors, but while we're on the static topic, I should also tell you that you can apply the **static** keyword to constructors. Or rather, to a single constructor. A static constructor allows you to set up any static class variables that you may have, so that they're ready to go whenever anyone wants to use them.

As soon as a program first attempts to use your class, the CLR will stop for a second and check for any static constructors that it may have. If one exists, it will run it. Because of this, you can be sure that before your class is used in any fashion, this code has already executed. A simple example of a static constructor is below:

```
public class Something
{
    public static int sharedNumber;

    static Something()
    {
        sharedNumber = 3;
    }
}
```

Of course, this is probably overkill in this specific situation, because if all you want to do are simple assignments, you could do this instead, which performs the same work in less space:

```
public class Something
{
    public static int sharedNumber = 3;
}
```

Having said that, there will still be times where your class-level setup is much more involved, and you'll still want to use a static constructor.

The 'internal' Access Modifier

There is one other type of access modifier that we should probably discuss at this point. That is the **internal** access modifier. If something is marked as **internal**, this means that it can be used anywhere inside the current project or *assembly*. While there are some exceptions to this, you can generally think of an assembly as a single project in compiled form, resulting in a single EXE or DLL.

Things that are public are available to anyone who has access to the assembly, even if it is lives in a different assembly or project. If you intend for something to be used widely, it should be public.

But **public** is very... accessible. Anyone who has the compiled code can do things with it. In some cases, you're building your types for your own personal use, but your intention isn't to give it all out to anyone who has your program. Sometimes, public feels like *over*sharing.

And this isn't just about being greedy and selfish, either. Anything that is a part of your public interface—anything that you present to the outside world with the **public** keyword—is something that you are going to have to maintain and support. Any changes you make to them will directly affect the people using them. It is good advice to limit the things you want to make public to as few things as possible.

This is where the **internal** access modifier comes into play. The **internal** modifier means that it can be used by anything in the same project or assembly, but anything outside of the assembly does not have access to it.

The reason I bring this all up here and now is because we have already seen and used the internal accessibility level without realizing it. I indicated before that instance variables or methods that don't have an explicitly stated accessibility level default to **private**. Well classes default to **internal**.

If the top of your class says:

```
class Book // Or any other name...
{
```

This means the class is has the **internal** accessibility level. If you want to be able to use the class outside of the original project, you'll need to change that to:

```
public class Book
{
```

Alternatively, you can also explicitly state that a class is internal, or that a method or instance variable is internal by using the **internal** keyword:

```
internal class AnyClass
{
    internal int aNumber;

    internal void DoSomething()
    {
    }
}
```

I should mention that if your class is internal, but a method or instance variable is public, that people outside of the assembly still won't be able to get access to it. The type itself must be public first, in order to even check if the member is publicly available.

Finishing the Sample

OK, there's one more final thing we want to do here before we wrap up this chapter: make sure we see how to use our new **Book** class.

This class will work very much like what we saw in the previous chapter about using classes. In your **Main** method, you can create an instance of your **Book** class, and work with it like this:

```
Book book = new Book("Harry Potter", "J.K. Rowling");

// Changed my mind. Let's use the full name.
book.SetTitle("Harry Potter and the Half-Blood Prince.");

// Now I forgot. What was the title again?
Console.WriteLine(book.GetTitle());
```

This chapter has described the basics of building classes. You'll keep learning more as you keep going through the next few chapters, but the real learning will come over the next few weeks, months, and even years. Splitting apart your program into classes is not always easy or intuitive.

I always tell people that class design is one part science, one part art, and one part black magic.

The science comes from learning basic principles and rules that tend to lead to better overall design (see the entire discipline of software engineering).

The art comes from practice and experience. After creating 10000 classes, you'll have a natural, intuitive sense of what will work and what won't.

The black magic part comes from the fact that no matter how much you know and how experienced you are, you'll still get it wrong sometimes. Especially as you develop a program and try to add features you had never thought of. Fortunately, software can easily be refactored and rearranged from the wrong organization to the right organization.

There's no such thing as getting it right all the time, so don't beat yourself up when you do. That applies to both beginners and professionals. Just change it to make it work like you think it should and keep moving forward.

Try It Out!

Designing and Building Classes. Try creating the two classes below, and make a simple program to work with them, as described below.

Create a Color class:

- On a computer, colors are typically represented with a red, green, blue, and alpha (transparency) value, usually in the range of 0 to 255. Add these as instance variables. (What type did you chose to represent these and why?)
- A constructor that takes a red, green, blue, and alpha value.
- A constructor that takes just red, green, and blue, while alpha defaults to 255 (opaque).
- Methods to get (retrieve) the red, green, blue, and alpha values from a color object, as well as set new values for each.
- A method to get the grayscale value for the color, which is the average of the red, green and blue values.

Create a Ball class:

- The **Ball** class should have a size/radius as well as a color instance variable. You should use the **Color** type you just created. Let's also add an instance variable that keeps track of the number of times it has been thrown.
- Create any constructors you feel would be useful.
- Create a **Pop** method, which changes the ball's size to 0.
- Create a **Throw** method that adds 1 to the throw count, but only if the ball hasn't been popped (has a size of 0).
- A method that returns the number of times the ball has been thrown.

Write some code in your **Main** method to create a few balls, throw them around a few times, pop a few, and try to throw them again, and print out the number of times that the balls have been thrown. (Popped balls shouldn't have changed.)

Try It Out!

Classes Quiz. Answer the following questions to check your understanding. When you're done, check your answers against the ones below. If you missed something, go back and review the section that talks about it.

1. **True/False.** Classes are reference types.
2. (Classes/Objects) define what a particular type of thing can do and store, while a/an (class/object) is a specific instance that contains its own set of data.
3. Name three types of members that can be a part of a class.
4. **True/False.** If something is static, it is shared by all instances of a particular type.
5. What is a special type of method that sets up a new instance is called?
6. Where can something **private** be accessed from?
7. Where can something **public** be accessed from?
8. Where can something **internal** be accessed from?

Answers: (1) True. **(2)** Classes, object. **(3)** Instance variables, methods, constructors. **(4)** True. **(5)** Constructor. **(6)** Only within the class. **(7)** Anywhere. **(8)** Anywhere inside of the project it is contained in.

19

Properties

In a Nutshell

- Properties provide a quick and effective approach to creating getters and setters for instance variables.
- You can create a property with code like the following:

```
public int Score
{
    get
    {
        return score;
    }
    set
    {
        score = value;
        if (score < 0)
            score = 0;
    }
}
```

- Not all properties need both a setter and a getter.
- Getters and setters can have different accessibility levels.
- Auto-implemented properties can be created that allow you to quickly define a simple property with default behavior (**public int Score { get; set; }**).
- Auto-implemented properties can have a default value: **public int Score { get; set; } = 10;**

In the last chapter, I talked about how it is very common to have an instance variable, and then want to provide methods to get its value and set its value. This leads to lots of **GetSomething** and **SetSomething** methods. This is so common that C# provides a very powerful feature that makes it easy to access the value of an instance variable called *properties*. This chapter will discuss why properties are so helpful, and several ways to create them.

The Motivation for Properties

Imagine you are creating a simple class to represent a player in a video game, which looks like this so far:

```
using System;
using System.Collections.Generic;
using System.Linq;
using System.Text;
using System.Threading.Tasks;

namespace Properties
{
    class Player
    {
        private int score;
    }
}
```

Let's say that you want to be able to have something outside of the class be able to get and set the value of the player's score.

The "classic" way (which we just learned last chapter) is to create a method that returns the value of **score** and another one that sets the value of **score**, possibly with some extra logic to check to make sure that the score is valid (like non-negative). This would be done like this:

```
public int GetScore()
{
    return score;
}

public void SetScore(int score)
{
    this.score = score;
    if(this.score < 0)
        score = 0;
}
```

The need for getter and setter methods is extremely common. However, working with this is a little cumbersome. You always need to say **player.SetScore(10)** and it is kind of cluttered and less readable.

It is tempting to just make our **score** instance variable **public** instead of **private**. Then you could simply say **player.score = 10**. That would be very convenient, except for one problem: now the instance variable is public, and anyone can mess it up by putting bad data in there. That is one of the things that the **SetScore** method took care of for us.

C# has a feature that solves all of these issues, creating a clean way to get or set the value of an instance variable without publicly exposing it. This feature is called a *property*.

Creating Properties

A property is simply an easy way to create get and set methods, while still maintaining a high level of readability. Instead of the two methods that we showed in that last example, we could do this instead:

```
public int Score
{
    get
    {
        return score;
    }
    set
    {
        score = value;
        if (score < 0)
            score = 0;
    }
}
```

This creates a **Score** property. We can see that it is **public**, so anyone can use it (though we could also use **private** if we had wanted). We also specify the type of the property, which in this case is an **int**. Then we give it a name.

We then use the **get** and **set** keywords to show what should happen when someone tries to read the value of the property or set the value of the property. The code inside of the curly braces is nearly the same code that was in the **GetScore()** and **SetScore(int)** methods.

With a normal method (like **SetScore**) we were able to list parameters for the method to use. In a property, we can't list parameters. Instead, we're given exactly one, which we can access with the keyword **value**. This takes on the value that the caller is trying to set. The type of **value** is the same type as the property. We can call our property using something like this:

```
Player player = new Player();
int currentScore = player.Score;
player.Score = 50;
```

When we try to read from the **Score** property, the code inside of the property's **get** block is executed and returned. When we assign a value to the **Score** property, the code inside of the **set** block gets executed. In this case, the value **50** gets put into the **value** keyword inside of the **set** block.

Properties allow us to get and set the value of instance variables, including doing extra work to validate what the user is trying to set, without making our code harder to read.

It is worth mentioning that properties are just syntactic sugar. Behind the scenes, the C# compiler is turning these back into normal method calls. But since it is all happening behind the scenes, we don't really care about that. We just care that it makes our code cleaner.

Whenever we have a property that gets or sets the value of a specific instance variable, that instance variable is called a *backing field* of the property. In our example above, the **Score** property uses the **score** instance variable as its backing field.

Properties are not required to have a backing field. Let's look at another example where a property does not have a backing field:

```
using System;
using System.Collections.Generic;
using System.Linq;
using System.Text;
using System.Threading.Tasks;

namespace Properties
```

```
{
    class Time
    {
        private int seconds;

        public int Seconds
        {
            get
            {
                return seconds;       // Seconds is the backing field here...
            }
            set
            {
                seconds = value;
            }
        }

        public int Minutes
        {
            get
            {
                return seconds / 60; // But there's no 'minutes' backing field here.
            }
        }
    }
}
```

You can see here that we have a **seconds** instance variable, along with a **Seconds** property that uses it as a backing field. (By the way, using lower case names for instance variables, and upper case names for the matching property is very common.)

In addition, we also have a **Minutes** property that doesn't have a backing field at all. Instead, we just do some other work—in this case, return the number of seconds divided by 60.

Another thing this example shows us is that we don't necessarily need both **set** and **get** blocks for a property. You can have one or the other or both. If you only have a **get** block, then the property is *read-only*, because you can't set it. This is the case with the **Minutes** property we just saw. If you have only a **set** block, the property would be write-only. (Though write-only properties are rarely actually useful.)

Different Accessibility Levels

So far, our properties have all been public. This means that both the getter and the setter are accessible to everyone. However, it is possible to make the two have different accessibility levels from each other. For example, we might want the getter to be public, so that everyone can read it, while making the setter private, so that it can only be changed from within the class.

To make this happen, you can specify an access modifier before the **get** or **set** keywords, like this:

```
public int Score
{
    get  // This is public by default, because the property is marked 'public'.
    {
        return score;
    }
    private set // This, however, is now private.
    {
```

```
        score = value;
    }
}
```

Auto-Implemented Properties

You will probably find yourself making a lot of things that look something like this:

```
private int score;

public int Score
{
    get
    {
        return score;
    }
    set
    {
        score = value;
    }
}
```

For simple cases like this, C# also has the ability to create auto-implemented properties that do the exact same thing. Instead of any of the code above (including the private instance variable **score**), you could simply say:

```
public int Score { get; set; }
```

This creates a backing field behind the scenes, and simple **get** and **set** code. You won't have access to the backing field in this case. This is a nice shorthand way to create very simple properties.

Read-Only Auto-Implemented Properties

You can also create read-only auto-implemented properties. These can only be written to in a constructor (or with a default value, which we'll look at next). After that, their value cannot be changed:

```
public class Vector2
{
    public double X { get; }
    public double Y { get; }

    public Vector2(double x, double y)
    {
        X = x;
        Y = y;
    }
}
```

It may seem a little odd to restrict a property from being written to. But making all data of a type be read-only after creation (making the type "immutable") has certain advantages. For instance, you know that if you've got a shared copy of the object that nobody is going to change its values without you knowing because they simply can't be changed. Immutability is a broader topic than just read-only properties, and we'll discuss it again later through this book.

The key point to remember here is that you can make your auto-implemented property read-only by simply having only a getter with no setter. This causes it to be writeable while the object is being created, but not afterwards.

Default Values

For an auto-implemented property, you can also assign a default value to the property:

```
public class Vector2
{
    public double X { get; set; } = 0;
    public double Y { get; set; } = 0;
}
```

Of course, the default value of any number will always be 0, so in this case we haven't actually changed anything, but it conveys the idea and the syntax correctly.

You can also supply default values for the read-only properties we just described in the last section.

If your property has a default value, it will get assigned to the property before the constructor runs. That means that at the time the constructor runs, it will already have a specific value, and if you give the property a value in the constructor, the value assigned in the constructor will be the actual value stored in the property, simply because it happens last.

Object Initializer Syntax

There's one other interesting thing that we should discuss while we're on the subject of properties. Let's say you have a **Book** class like in the previous chapter that has instance variables for the book's author, title, number of pages, and word count. Let's also say that you create properties for each of those, called **Author**, **Title**, **Pages**, and **WordCount**.

When you create a new **Book** object, you can use the following syntax to assign values to the properties, right when you create the object, like this:

```
Bookbook = new Book() { Title = "Frankenstein", Author = "Mary Shelley" };
```

This prevents us from needing to do this:

```
Book book = new Book();
book.Title = "Frankenstein";
book.Author = "Mary Shelley";
```

Instead, we can combine it all into one line. This particular setup is called *object initializer syntax*. It can be a convenient way to set up objects without needing to define tons of different constructors.

> **Try It Out!**
>
> **Playing with Properties.** At the end of the last chapter, we created a couple of classes as practice. Go back to those classes and add properties for the various instance variables that they have. While doing so, make sure you play around with at least one auto-implemented property.

20

Structs

In a Nutshell

- A *struct* or *structure* is similar to a class in terms of the way it is organized, but a struct is a value type, not a reference type.
- Structs should be used to store compound data (composed of more than one part) that does not involve a lot of complicated methods and behaviors.
- All of the simple types are structs.
- The primitive types are all aliases for certain pre-defined structs and classes.

A couple of chapters ago, we introduced classes. These are complex reference types that you can define and build from the ground up. C# has a feature call *structs* or *structures* which look very similar to classes organizationally, but they are value types instead of reference types.

In this chapter, we'll take a look at how to create a struct, as well as discuss how to decide if you need a struct or a class. We'll also discuss something that may throw you for a loop: all of the built-in types that we've been working with since we first learned about types are actually all aliases for structures (or a class in the case of the **string** type).

Creating a Struct

Creating a struct is very similar to creating a class. The following code defines a simple struct, and an identical class that does the same thing:

```
struct TimeStruct
{
    private int seconds;

    public int Seconds
    {
        get { return seconds; }
        set { seconds = value; }
    }
```

```
    public int CalculateMinutes()
    {
        return seconds / 60;
    }
}

class TimeClass
{
    private int seconds;

    public int Seconds
    {
        get { return seconds; }
        set { seconds = value; }
    }

    public int CalculateMinutes()
    {
        return seconds / 60;
    }
}
```

You can see that the two are very similar—in fact the same code is used in both, with the single solitary difference being we use the **struct** keyword to create a struct, while we use the **class** keyword to create a class.

Structs vs. Classes

Since the two are so similar in appearance, you're probably wondering how the two are different.

The answer to this question is simple: structs are value types, while classes are reference types. If you didn't fully grasp that concept back when we discussed it in Chapter 16, it is probably worth going back and taking a second look.

While this is a single difference in theory, this one change makes a world of difference. For example, a struct uses value semantics instead of reference semantics. When you assign the value of a struct from one variable to another, the entire struct is copied. The same thing applies for passing one to a method as a parameter, and returning one from a method.

Let's say we're using the struct version of the **TimeStruct** we just saw, and did this:

```
public static void Main(string[] args)
{
    TimeStruct time = new TimeStruct();
    time.Seconds = 10;

    UpdateTime(time);
}

public static void UpdateTime(TimeStruct time)
{
    time.Seconds++;
}
```

In the **UpdateTime** method, we've received a copy of the **TimeStruct**. We can modify it if we want, but this hasn't changed the original version, back in the **Main** method. We've modified a copy, and the original still has a value of 10 for **seconds**.

Had we used **TimeClass** instead, handing it off to a method copies the reference, but that copied reference still points the same actual object. The change in the **UpdateTime** method would have affected the time variable back in the **Main** method.

Like I said back when we were looking at reference types, this can be a good thing or a bad thing, depending on what you're trying to do, but the important thing is that you are aware of it.

Interestingly, while we get a copy of a value type as we move it around, it doesn't necessarily mean we've duplicated everything it is keeping track of entirely. Let's say you had a struct that contained within it a reference type, like an array, as shown below:

```
struct Wrapper
{
    public int[] numbers;
}
```

And then we used it like this:

```
public static void Main(string[] args)
{
    Wrapper wrapper = new Wrapper();
    wrapper.numbers = new int[3] { 10, 20, 30 };
    UpdateArray(wrapper);
}

public void UpdateArray(Wrapper wrapper)
{
    wrapper.numbers[1] = 200;
}
```

We get a copy of the **Wrapper** type, but for our **numbers** instance variable, that's a copy of the reference. The two are still pointing to the same actual array on the heap.

Tricky little things like this are why if you don't understand value and reference types, you're going to get bit by them. If you're still fuzzy on the differences, it's worth a second reading of Chapter 16.

There are other differences that arise because of the value/reference type difference:

- Structs can't be assigned a value of **null**, since null indicates a reference to nothing.
- Because structs are value types, they'll be placed on the stack when they can. This could mean faster performance because they're easier to get to, but if you're passing them around or reassigning them a lot, the time it takes to copy them could slow things down.

Another big difference between structs and classes is that in a struct, you can't define your own parameterless constructor. For both classes and structs, if you don't define any constructors at all, one still exists: a default parameterless constructor. This constructor has no parameters, and is the simplest way to create new objects of a given type, assuming there's no special setup logic required. With classes, you can create your own parameterless constructor, which then allows you to replace the default one with your own custom logic. This cannot be done with structs. The default parameterless constructor creates new objects where everything is zeroed out. All numbers within the struct start at 0, all **bool**s start at false, all references start at **null**, etc. While you can create other constructors in your struct, you cannot create a parameterless one to replace this default one.

Deciding Between a Struct and a Class

Despite the similarities in appearance, structs and classes are designed for entirely different purposes. So when it comes time to create a new type, which one do you choose?

Here are a few things to think about as you decide between the two. For starters, do you have a particular need to have reference or value semantics? Since this is the primary difference between the two, if you've got a good reason to want one over the other, your decision is basically already made.

If your type is not much more than a compound collection of a small handful of primitives, a struct might be the way to go. For instance, if you want something to keep track of a person's blood pressure, which consists of two numbers (systolic and diastolic pressures) a struct might be a good choice. On the other hand, if you think you're going to have a lot of methods (or events or delegates, which we'll talk about in Chapters 30 and 31) then you probably just want a class.

Also, structs don't support inheritance which is something we'll talk about in Chapter 21, so if that is something you may need, then go with classes.

In practice, classes are far more common, and probably rightly so, but it is important to remember that if you choose one way or the other, and then decide to change it later, it will have a huge ripple effect throughout any code that uses it. Methods will depend on reference or value semantics, and to change from one to the other means a lot of other potential changes. It's a decision you want to make consciously, rather than just always defaulting to one or the other.

Prefer Immutable Value Types

In programming, we often talk about types that are *immutable*, which means that once you've set them up, you can no longer modify them. (As opposed to mutable types, which you can modify parts of on the fly.) Instead, you would create a new copy that is similar, but with the changes you want to make. All of the built-in types (including the **string** type, which is a reference type) are immutable.

For value types like the structs you create, there is a danger to allowing them to be mutable. Because they have value semantics, any time you pass a value from one variable to another, or to a different method as a parameter, you end up with a copy of the original. It is far too easy and common to think that the value you got, which is actually a copy, could be used to modify the original. If your type doesn't allow you to modify the individual instance variables that make up your type, then the only way to make a change is by creating a new one with the changes you need applied to it.

Making a value type immutable will save you a great deal of trouble in the long run.

The Built-In Types are Aliases

Back in Chapter 6, we took a look at all of the primitive or built-in types that C# has. This includes things like **int**, **float**, **bool**, and **string**. In Chapter 16, we looked at value types vs. reference types, and we discovered that these primitive types are value types, with the exception of **string**, which is actually a reference type.

It turns out that not only are they value types, but they are also struct types. This means that everything that applies to structs that we've been learning about also applies to these primitive types.

Even more, all of the primitive types are *aliases* for other structs (or class, in the case of the **string** type).

While we've been working with say, the **int** type, behind the scenes the C# compiler is simply changing this over to a struct that is defined in the same kind of way that we've seen here. In this case, it is the **Int32** struct (**System.Int32**).

So while we've been writing code that looks like this:

```
int x = 0;
```

We could have also used this:

```
Int32 x = new Int32();
Int32 y = 0;              // Or combined.
int z = new Int32();      // Or combined another way. It's all the same thing.
int w = new int();        // Yet another way...
```

The **int** type and the **Int32** struct are identical. There is no difference at all between the two. Which brings us to a slightly updated definition of a *primitive type*, or *built-in type*: a type that the compiler has special knowledge about, and allows for special, simplified syntax to use it.

All of the primitive types have aliases, shown in the table below:

Primitive Type	Alias For:
bool	System.Boolean
byte	System.Byte
sbyte	System.SByte
char	System.Char
decimal	System.Decimal
double	System.Double
float	System.Single
int	System.Int32
uint	System.UInt32
long	System.Int64
ulong	System.UInt64
object	System.Object
short	System.Int16
ushort	System.UInt16
string	System.String

I want to point out a couple of things about the naming here. Nearly all of these types have the same name, except with a capital letter. Keywords in C# are all lowercase by convention, but nearly everybody will capitalize type names, which explains that difference.

You'll also see that instead of **short**, **int**, and **long**, the structs use **Int** followed by the number of bits they use. This is far more explicit than the keyword versions. There's no confusing how big each type is.

And last, you'll notice that the **float** type is **Single**, instead of **Float**. The word "float" is not very accurate, since both **double** and **float** are both technically floating point types. The term "single" is perhaps more correct, simply because it is more precise. The people who made C# chose to use **float** because it is similar to the languages that it is based on (C/C++/Java) all of which have a **float** type.

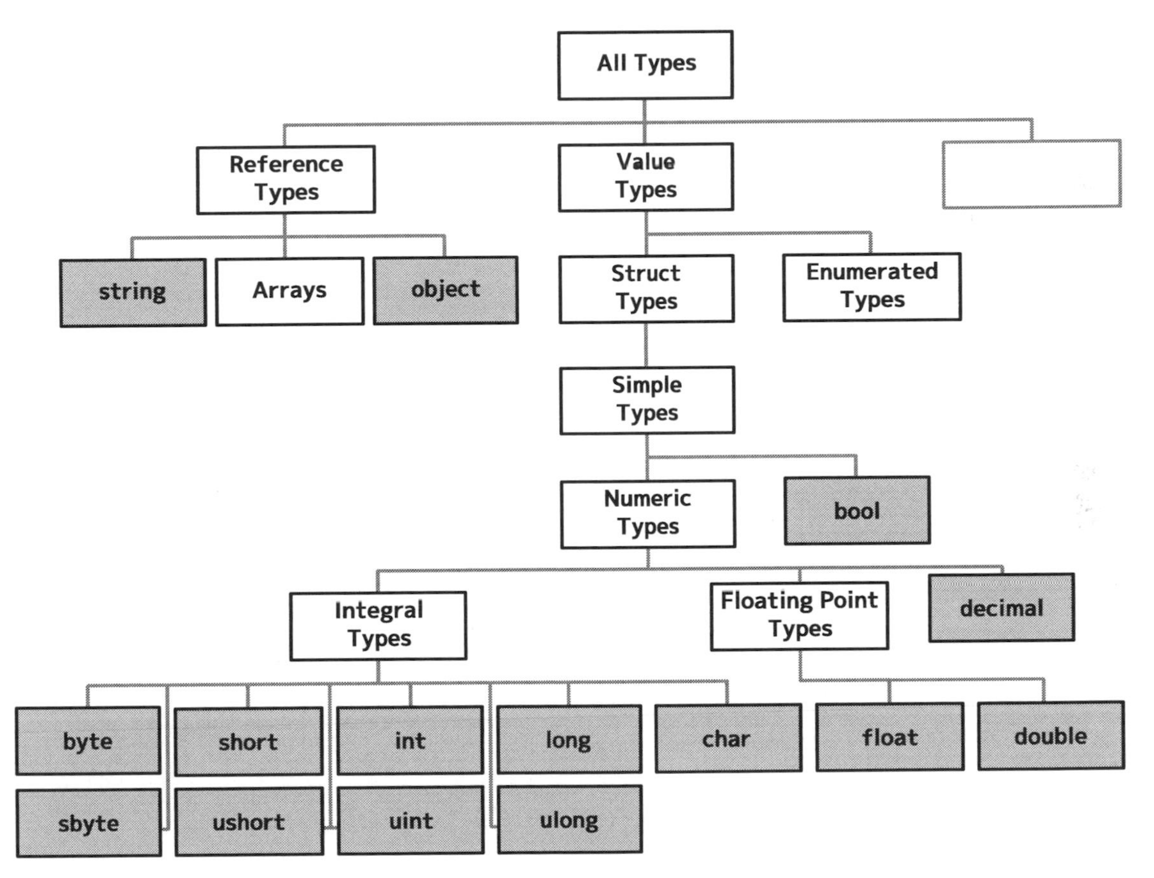

Try It Out!

Structs Quiz. Answer the following questions to check your understanding. When you're done, check your answers against the ones below. If you missed something, go back and review the section that talks about it.

1. Are structs value types or reference types?
2. **True/False.** It is easy to change classes to structs, or structs to classes.
3. **True/False.** Structs are always immutable.
4. **True/False.** Classes are never immutable.
5. **True/False.** All primitive/built-in types are structs.

Answers: (1) Value types. **(2)** False. **(3)** False. **(4)** False. **(5)** False. string and object are reference types.

21

Inheritance

In a Nutshell

- Inheritance is a way to reuse classes by expanding them into more specific types of classes. For instance, a **Square** class can be derived from (or inherit from) a **Polygon** class, giving it all of the features that the **Polygon** class has.
- Any class can be used as a base class, except for sealed classes.
- To indicate a class is derived from another, you use the colon followed by the base class name. For example class **Square : Polygon { /* ... */ }.**
- The **protected** access modifier means that anything within the class can use it, as well as anything in any derived class.

Imagine that you are trying to keep track of geometric polygons, like triangles, rectangles, squares, pentagons, and so on. You can probably imagine creating a **Polygon** class that stores the number of sides the polygon has, and maybe also a way to keep track of the actual positions of the vertices (corners) of the polygon.

Your **Polygon** class could be used to represent any of these polygons, but let's imagine that for a square, you want to be able to do some extra things with it. For example, you might want to create one using only a size (since the width and height are the same length in a square), or maybe you want to be able to create a method that returns the area of the square.

One approach would be to create a **Square** class with instance variables to keep track of the number of sides and the locations of the vertices, like the **Polygon** class has, plus the extra stuff you want. The problem with this is that now you have two entirely different classes, and if you want to make a change in the way people work with polygons, you'll also need to make a change in the way they work with squares. More than that, while you could create an array of polygons (**Polygon[] polygons**) you wouldn't be able to put squares in that list, even though in your head, they *are* polygons, and should be able to be treated as polygons.

After all, a square is just a special type of a polygon.

And in fact, that is getting right to the heart of what we're going to discuss in this chapter. A square is a polygon, but more specifically, it is a special type of polygon that has its own special things. Anything a polygon can do, a square can do too, because a square is a polygon.

This is what programmers call an *is-a relationship*, or an *is-a-special-type-of relationship.* It is so common that programming languages have a special way to facilitate this kind of thing.

This is called *inheritance*. Inheritance allows us to create a **Polygon** class, and a **Square** class that is based on **Polygon**, This makes it *inherit* (or reuse) everything from the **Polygon** class that it is based on. Any changes we make to the **Polygon** class will also be automatically changed in the **Square** class.

Classes support inheritance, but structs and other types do not. It us a key difference between the two.

Base Classes

A *base class* (or *superclass* or *parent class*) is any normal class that happens to be used by another for inheritance. In our discussion, the **Polygon** class would be a base class.

For instance, we could have a class that looks like this:

```
using System;
using System.Collections.Generic;
using System.Linq;
using System.Text;

namespace Inheritance
{
    // Nothing special about this class. It will be used as the base class.
    class Polygon
    {
        public int NumberOfSides { get; set; }

        public Polygon()
        {
            NumberOfSides = 0;
        }

        public Polygon(int numberOfSides)
        {
            NumberOfSides = numberOfSides;
        }
    }
}
```

Derived Classes

A *derived class* (or *subclass*) is one that is based on, and expands upon another class. In our example, the **Square** class is a derived class that is based on the **Polygon** class. We would say that the **Square** class is *derived* from the **Polygon** class.

You can create a derived class just like any other class with one small addition to indicate which class is its base class. To indicate that the **Square** class is derived from the **Polygon** class, we use the colon (':') and name the base class:

```
class Square : Polygon  // Inheritance!
{
    //...
}
```

Inside of the class, we simply indicate any new stuff that the **Square** class has that the **Polygon** class doesn't have. Our completed **Square** class might look like this:

```
using System;
using System.Collections.Generic;
using System.Linq;
using System.Text;

namespace Inheritance
{
    class Square : Polygon
    {
        public float Size { get; set; }

        public Square(float size)
        {
            Size = size;
            NumberOfSides = 4;
        }
    }
}
```

So now a square will have a **Size** property, but in addition, it inherits instance variables, properties, and methods that were in the **Polygon** class. So the **Square** class also has a **NumberOfSides** property that it inherits from the **Polygon** class.

One other thing that we should mention here is that a class can inherit from a class that inherits from another class. You're allowed to have as many layers of inheritance as you want.

Using Derived Classes

Derived classes are just like any other class, and can be used as a type just like any other class. From a basic perspective, there's nothing special about how you use them:

```
Polygon polygon = new Polygon(3); // a triangle
Square square = new Square(4.5f); // a square, which is a polygon with 4 sides of length 4.5.
```

But here's the interesting thing. Because the computer knows that a **Square** is just a special type of **Polygon**, you can actually use the **Square** type anywhere that you could use a **Polygon** type. After all, a **Square** is a **Polygon**!

```
Polygon polygon = new Square(4.5f);
```

As the program is running, when we're using the **polygon** variable, we only have access to the stuff that a **Polygon** has. So we can check to see the **NumberOfSides** it has. But as the program is running, we can only treat it as a **Polygon**. There's a chance that it is a **Square** (or another derived class), but the computer won't assume it is. So we won't be able to automatically use stuff created in the **Square** class in this particular case. Because our variable uses the **Polygon** type, we can only work with the things the **Polygon** type defines.

While we can put a **Square** in a variable with the **Polygon** type, you can't go the other way around. You can't assign a **Polygon** object to a variable with the type **Square**, like this:

```
Square square = new Polygon(3);  // Does not compile.
```

This is because while all squares are polygons, not all polygons are squares.

Checking an Object's Type and Casting

As I just described, it is possible to be actually using an instance of a derived type like the **Square**, but only know that it is a **Polygon**. Sometimes we want to be able to figure out what type an object is specifically, and work with it differently.

It is possible to check to see if an object is a specific type, with the **is** keyword and casting:

```
Polygon polygon = new Square(4.5f);

if (polygon is Square)
{
    Square square = (Square)polygon;
    // We can now do what we want with the square.
}
```

The 'as' Keyword

In addition to casting, there is another way we can convert one object type to another. This is done by using the **as** keyword, which looks like this:

```
Polygon polygon = new Square(4);
Square square = polygon as Square;
```

There are a couple of differences between using the **as** keyword and casting, and it turns out, using the **as** keyword in this type of a situation is usually a better choice.

To fully understand this, let's assume that in addition to the **Polygon** and **Square** classes we've been looking at, there is a **Triangle** class as well. Let's say we have a variable with the **Polygon** type, and assign it a new **Triangle**, but then we inadvertently try to turn it into a **Square**:

```
Polygon polygon = new Triangle();
Square square = (Square)polygon;
```

In this case, the program will crash, giving us an error that says it was an invalid cast. On the other hand, if we used the **as** keyword, instead of crashing, we would get **null**, which we can then respond to:

```
Polygon polygon = new Triangle();
Square square = polygon as Square;

if(square != null)
{
    // ...
}
```

Using Inheritance in Arrays

In light of what we've just been discussing, it is worth mentioning that you can create an array of the base class and put any derived class inside of it. This is just like what we already saw with variables that weren't arrays, so it shouldn't be too surprising:

```
Polygon[] lotsOfPolygons = new Polygon[5];
lotsOfPolygons[2] = new Square(2.1f);
lotsOfPolygons[3] = new Triangle();
```

In any case where a base class can be used, you can always substitute a derived class.

Constructors and Inheritance

I mentioned earlier that in a derived class, all properties, instance variables, and methods are inherited. Constructors, on the other hand, do not.

You can't use the constructors in the base class to create a derived class. For instance, our **Polygon** class has a constructor that takes an **int** that is used for the polygon's number of sides. You can't create a **Square** using that constructor. And for good reason. Remember, constructors are special methods that are designed to outline how to build new objects of a particular type. With inheritance, a constructor in the base class doesn't have any clue how to set up the new parts of the derived class. To create an instance of a class, you must use one of the constructors that are defined in that specific class.

When a derived class defines a constructor, it needs to call one of the constructors from the base class. By default, a constructor in the derived class will use the base class's parameterless constructor. (Remember, if you don't define any constructor yourself, a default parameterless constructor will be created for you.) If the base class doesn't have a parameterless constructor, or if you would like a constructor in the derived class to use a different one, you will need to indicate which constructor to use. This is done by using a colon and the **base** keyword along with the parameters for the constructor we want to use, like this:

```
public Square(float size)
    : base(4) // Uses the Polygon(int numberOfSides) constructor in the base class.
{
    Size = size;
}
```

The 'protected' Access Modifier

In the past, we discussed the **private** and **public** access modifiers. To review, remember that **public** meant anyone could get access to the member (variable, method, property, etc.), while **private** means that you only have access to it from inside of the class that it belongs to.

With inheritance we add another option: **protected**. If a member of the class uses the **protected** accessibility level, then anything inside of the class can use it, as well as any derived class. It's a little broader than **private**, but still more restrictive than **public**.

If the **Polygon** class made the **NumberOfSides** property **protected** instead of **public**, then you could access it from anywhere in the **Polygon** class, as well as the **Square** class, and any other derived class, but not from outside of those.

The Base Class of Everything: object

Without specifying any class to inherit from, any class you make still is derived from a special base class that is the base class of everything. This is the **object** class. All classes are derived from this class by default, unless you state a different class. But even if you state a different class as the base class, that class will be derived from the **object** class. If you go up the inheritance hierarchy, everything always gets back to the **object** class eventually.

You are allowed to use this type in your code, using the **object** keyword (which is an alias for **System.Object**). Remember what I said earlier about how you can store a derived class in a base type like the code below?

```
Polygon polygon = new Square(4.5f);
```

You can also put it in the **object** type:

```
object anyOldObject = new Square(4.5f);
```

Since the **object** type is the base class for all classes, any and every type of object can be stored in it.

Sealed Classes

There may come a time where you want to prevent a class from being inherited. This might happen if you want to ensure that no one derives anything from a specific class.

You can use the **sealed** keyword to make it so nothing can inherit from a specific class by adding it to your class definition:

```
sealed class Square : Polygon
{
    // ...
}
```

Sealing a class can result in a performance boost.

Partial Classes

Sometimes classes get really big. Generally speaking, if this happens, you'll want to try to break the class down into smaller classes that do less if possible. If not, you can split a class into multiple files or sections using the **partial** keyword:

```
public partial class Class
{
    public void DoSomething()
    {
    }
}

public partial class Class
{
    public void DoSomethingElse()
    {
    }
}
```

Even if the two parts are in different files, they'll be combined into one when you compile your program. You can also use the **partial** keyword on structs and interfaces (Chapter 23).

C# Does Not Support Multiple Inheritance

Some object-oriented programming languages (C++, for example) allowed you to choose more than one base class to derive from. There are some benefits to this, because it allows one type of object to have an *is-a* relationship with more than one other type. For example, if you have a **Knife** class and a **Saw** class, you could create a **PocketKnife** class that is both a knife and a (really crappy) saw along with a dozen other things.

Unfortunately, having multiple inheritance tends to make things complicated, and can lead to situations where it is ambiguous how the class should behave. It is because of this that C# has decided not to allow multiple inheritance. You must pick only one base class.

You can sometimes mimic multiple inheritance by using interfaces, which we'll talk about in Chapter 23.

Try It Out!

Inheritance Quiz. Answer the following questions to check your understanding. When you're done, check your answers against the ones below. If you missed something, go back and review the section that talks about it.

1. **True/False.** A derived class adds functionality on to a base class.
2. **True/False.** Everything that is in a base class can be accessed by the derived class.
3. **True/False.** Structs support inheritance.
4. **True/False.** All classes are derived from the **object** class.
5. **True/False.** A class can be a derived class and a base class at the same time.
6. **True/False.** You can prevent a class can't be derived from (can't be used as a base class).

Answers: **(1)** True. **(2)** False. Not private members. **(3)** False. **(4)** True. **(5)** True. **(6)** True.

Polymorphism, Virtual Methods, and Abstract Classes

In a Nutshell

- Polymorphism means that you can create derived classes that implement the same method in different ways from each other. This allows different classes to do the same task in their own custom way.
- A method in a base class can be marked **virtual**, allowing derived classes to use the **override** keyword and provide an alternative implementation.
- Classes can be marked as **abstract**, making it so you can't actually create an instance of the class. You have to create an instance of a derived class instead.
- In an **abstract** class, you can mark methods as **abstract** as well, and then leave out the method implementation. Derived classes will need to provide an implementation for any abstract methods.
- The **new** keyword, when attached to a method, means that a new method is being created, unrelated to any methods in the base class with the same name, resulting in two methods with the same name, with the one in the base class being hidden.

In the last chapter, we talked about how it is possible to create a class that is based on another, adding more stuff to it. It turns out inheritance lets us do a lot more than just add new things to an existing class. In this chapter, we'll look at how a derived class can provide its own implementation for a method that was defined in the base class, allowing it to handle the same task in its own way.

Polymorphism

"We do things differently around here." Maybe you've heard people say things like that before. What they're implying is they do the same kind of thing as somebody else, but it is done in a better way.

Take a web search for instance. Everybody knows about Google, but there are plenty of other web-based search engines, like Bing and Yahoo! Search. All of them do the same basic task, but each of them do it in their own way. You give them a word, they go off and dig around in their notes that's they've been furiously taking as they frantically searched for the end of the Internet, and give you back a list of anything that could be useful.

This is an excellent example of something programmers call *polymorphism*, a fancy Greek word meaning "many forms." Polymorphism is a feature that allows derived classes to *override* or change the way the base class does things, and do them in their own way. Alternatively, instead of changing the way the base class does things, the base class may not even *have* a way to do things, forcing derived classes to find their own way to do things. But when all is said and done, these derived classes will have a different way of doing things from the base class and from each other. The fact that these classes can all do the same task (by calling the same method) differently is what gives us "many forms" and polymorphism.

In the last chapter, with our basic inheritance model, derived classes could only add things to a type. With polymorphism, we can do more than just add stuff. We can also alter how the derived class will handle the things the base class supplies.

Going with our search engine example, we might have a **SearchEngine** class like this:

```
class SearchEngine
{
    public virtual string[] Search(string findThis)
    {
        return new string[0]; // I'm terrible at searching... I give up.
    }
}
```

Our **SearchEngine** class happens to define a single method. Here in the base class, we've provided a default (and terrible) implementation of the **Search** method.

You'll also notice the **virtual** keyword. This is what gives derived classes permission to change the way this method works. If you want to allow a derived class to change the way a method is used, you simply need to stick this on your method. (This also can be used for properties, as well as indexers and events, which we haven't discussed yet.)

In a derived class, we now have the option to provide an alternative implementation for the method:

```
class GoogleSearch : SearchEngine
{
    public override string[] Search(string findThis)
    {
        // Google Search is, of course, way better than 3 hard-coded results like this…
        return new string[] {
            "Here are some great results.",
            "Aren't they neat?",
            "I found 1.2 billion things, but you will only look at the first 10." };
    }
}
```

To change the implementation for the derived class, we simply create a new copy of the method, along with our new custom-made implementation, and add in the **override** keyword.

If the base class didn't use the **virtual** keyword for a method you want to override, you're not going to be able to override it. The **virtual** keyword is the base class's way of saying "I give you permission to do something different with this method," while the **override** keyword is the derived class's way of saying "You gave me permission, so I'm going to use it and take things in a different direction."

You don't *need* to override virtual methods, you're just *allowed* to. And if a class has more than one virtual method, you can override some and not others.

Naturally, you can make any number of derived classes that do things their own way:

```
class RBsSearchEngine : SearchEngine
{
    public override string[] Search(string findThis)
    {
        return new string[] {
                "I found something.",      // Thanks EDI
                "I found this for you."    // Thanks SIRI
            };
    }
}
```

Here's where it gets interesting. Because our derived classes **RBsSearchEngine** and **GoogleSearch** are inherited from the **SearchEngine** class, we can use them anywhere a **SearchEngine** object is required:

```
SearchEngine searchEngine1 = new SearchEngine(); // The plain old original one.
SearchEngine searchEngine2 = new GoogleSearch();
SearchEngine searchEngine3 = new RBsSearchEngine();
```

But now we can use that **Search** method, and depending on the *actual* type, the various implementations will be called:

```
string[] defaultResults = searchEngine1.Search("hello");
string[] googleResults = searchEngine2.Search("hello");
string[] rbsResults = searchEngine3.Search("hello");
```

We'll get different results for each of these methods, because each of these three types define the method differently. The default **SearchEngine** class provides a default implementation of the method but the **GoogleSearch** and **RBsSearchEngine** "alter the deal" and override the original implementation, swapping it out for their own. The CLR is smart enough to know which method to use at the right time.

This type of setup, where calling the same method signature results in different methods actually being executed, is the crux of polymorphism. Polymorphism means "many forms," and in this situation, we can see that calling the same method results in different behaviors or forms.

Revisiting the 'base' Keyword

Last chapter, when we brought up creating constructors in a derived class, we brought up the **base** keyword, which lets you access the constructors from the base class. We can also use the **base** keyword elsewhere to access non-private things from the base class, including its methods and properties.

While most of the time, you can directly access things in the base class without actually needing to use the **base** keyword, when you override a method, you'll be able to use the **base** keyword to still get access to the original implementation of the method.

This is convenient for the cases in which you want to override a method by doing everything the old method did, and then doing a little more. This could work something like this:

```
public override string[] Search(string findThis)
{
    // Calls the version in the base class.
    string[] originalResults = base.Search(findThis);

    if(originalResults.Length == 0)
        Console.WriteLine("There are no results.");

    return originalResults;
}
```

Abstract Base Classes

Let's go back to our **SearchEngine** example. The **SearchEngine** class indicates that any and every **SearchEngine** type object has the ability to search. But looking back at our code, our base **SearchEngine** class didn't do anything intelligent. We provided a dummy implementation that didn't do anything besides waste time writing useless code.

In cases like this, instead of providing a dummy implementation, we have an option to provide no implementation whatsoever. With no implementation or body for a method, the method in question becomes *abstract*.

To accomplish this in code, we need to add the keyword **abstract** to our class, as well as to any methods that we wish to keep unimplemented in the base class:

```
public abstract class SearchEngine
{
    public abstract string[] Search(string findThis);
}
```

When a class is abstract, it is allowed to include abstract methods. Abstract methods cannot go inside of regular, non-abstract classes. When a class is marked **abstract**, you can't create any instances of that class directly. Instead, you need to create an instance of a derived class that isn't abstract.

So with the code above, you won't be able to actually create an instance of **SearchEngine**. Instead, you'd need to create an instance of one of the derived classes, like **GoogleSearch** or **RBsSearchEngine**.

In the code above, you can also see that the method is marked with **abstract** as well. In this case, the method doesn't need a method body, and isn't even allowed to have one. Instead of providing an implementation in curly braces below it, you simply put a semicolon at the end of the line.

Abstract classes can also have as many virtual or normal methods as you want.

In a derived class, to provide an implementation for an abstract method, you use the **override** keyword in the exact same way as we did with virtual methods.

The 'new' Keyword with Methods

There's one other topic that a lot of people seem to get tangled up with all of this **override** and **virtual** stuff, and that is putting the **new** keyword on a method.

You are allowed to use the **new** keyword on a method like this:

```
public new string[] Search(string findThis)
{
    //...
}
```

When you use the **new** keyword here, what you are saying is, "I'm making a brand new method in the derived class that has absolutely nothing to do with any methods by the same name in the base class." In this example, this **Search** method is completely unrelated to the **Search** method in the **SearchEngine** class.

This is usually a bad idea because it means that you now have two unrelated methods with the same name and same signature. This name conflict *can* be resolved, but it is best to just avoid it altogether.

People often get this **new** keyword confused with overriding virtual and abstract methods for two reasons. One, it looks similar. The **new** keyword here gets used in the exact same way as **override** would. Second, if you forget to put anything in at all (**new** or **override**) when you use the same method signature, you'll get a compiler warning that tells you to add the **new** or **override** keyword. (Or in some cases, it only mentions the **new** keyword.) People see the warning and think, "Hmm... I guess I should add the **new** keyword to get the warning to go away."

The bottom line is, you rarely want to use the **new** keyword. It's just not what you're looking for. If you really are trying to come up with a new unrelated method, consider using a different name instead.

Try It Out!

Polymorphism Quiz. Answer the following questions to check your understanding. When you're done, check your answers against the ones below. If you missed something, go back and review the section that talks about it.

1. **True/False.** Polymorphism allows derived classes to provide different implementations of the same method.
2. **True/False.** The **override** keyword is used to indicate that a method in a derived class is providing its own implementation of a method.
3. **True/False.** The **new** keyword is used to indicate that a method in a derived class is providing its own implementation of a method.
4. **True/False.** Abstract methods can be used in a normal (non-abstract) class.
5. **True/False.** Normal (non-abstract) methods can be used in an abstract class.
6. **True/False.** Derived classes can override methods that were **virtual** in the base class.
7. **True/False.** Derived classes can override methods that were **abstract** in the base class.
8. **True/False.** In a derived class, you can override a method that was neither **virtual** nor **abstract** in the base class.

Answers: (1) True. **(2)**True. **(3)** False. **(4)** False. **(5)** True. **(6)** True. **(7)** True. **(8)** False.

23

Interfaces

In a Nutshell

- Interfaces define the things that something makes available to the outside world.
- Interfaces in C# define a set of methods, properties, events, and indexers that any class that uses the interface are required to provide.
- Interfaces are built in way that looks like classes or structs: **public interface InterfaceName { /* Define members in here. */ }**
- Members of an interface are implicitly **public** and **abstract**, and are required to be so.
- A class indicates that it implements an interface in the same way it shows it has a base class: **public class AwesomeClass : InterfaceName { /* Implementation Here */ }**
- You can implement multiple interfaces at the same time. You can also implement interfaces and declare a base class at the same time.

In this chapter, we'll cover a feature of C# called an interface, which outlines certain things that something provides. Interfaces are a fairly straightforward concept compared to some of the things we've been looking at in the last few chapters. We'll outline what an interface does, how to create one, and how to set up a class or struct to use one.

What is an Interface?

At a conceptual level, an interface is basically just the boundary it shares with the outside world. Your TV, for instance, probably has several buttons on it (and if it is really old, maybe even a dial or two). Each of these things allows the outside world to interact with it. That's its interface.

Interfaces are everywhere in the real world. They point out to users of the object how it can be used or controlled and also provides feedback on its current status. A couple of examples are in order here.

Cars (all vehicles in general) have a specific interface that they present to their users (the driver). While there are some nuances and details that we'll skip over, the key components are a steering wheel, a gas pedal, and a brake pedal. A car will also provide certain bits of feedback information to the driver on the dashboard, including the vehicle's ground speed and engine speed, as well as its fuel level.

Likewise, the keys on a piano are an interface that a pianist can access to make music. While pianos come in different forms (grand pianos, upright pianos, player pianos, electronic keyboards, etc.) they present a similar interface to the user of the system.

Let's take a moment and pick out a few key principles that apply to interfaces in the real world, like the car and keyboard interfaces. We'll soon see that C# interfaces provide these same key principles.

1. Interfaces define how outside users make requests to the system. By rotating the steering wheel on a car clockwise, we're making a request to turn the vehicle to the right. By pressing Middle C key on the piano's keyboard, we're asking the piano to make a sound at a particular pitch.
2. Interfaces define the ways in which feedback is presented to the user.

This interface defines how other objects can interact with the car, without prescribing the details of how it works behind the scenes. For example, in very old cars, this was all done in a purely mechanical way. Newer cars have things like power steering, and perform some of their feedback to the user digitally. And consider an electric car, which has a "gas" (acceleration) pedal but drives the vehicle using an electric motor instead of a gas motor.

You can even go to an amusement park and find bumper cars and speedway/raceway type rides that use an identical interface to control vehicles. One of the nice things about this shared interface is that as soon as you understand how to use one object through the interface, you generally understand how to use all objects that share the same interface.

Interfaces work both ways. When something presents a particular interface to the user, the user assumes that it's capable of doing the things the interface promises. When you've got a brake pedal in a car, you assume it will let you stop the vehicle quickly. When the brake pedal doesn't fulfill the promised interface, bad things happen. It's why it is frustrating when an elevator has a "Close Door" button, that doesn't actually close the door. The elevator presents an interface that promises the ability to close the door, but fails to actually do so.

The broader lesson here is that interfaces define a specific set of functionality. Users of the interface can know what to expect of the object from the interface, without having to know the details of how the object actually makes it work, and in fact, as long as the interface remains intact, the details on the inside could be swapped out and the user wouldn't have to change anything.

C# has a construct called an *interface* that fills the same role as these interfaces in the real world. : listing a set of members (specifically methods, properties, events (Chapter 31) and indexers (Chapter 33)) that any type claiming to have the interface must provide.

Creating an Interface

Creating an interface is actually very similar to creating a class or a struct. As an example, let's say our program has a need to write out files (see Chapter 28) but you want to write to lots of different file formats. We can create a different class to handle the nuances of each file type. But they are all going to be used in the same way as far as the calling code is concerned. They all have the same interface.

To model this, we can define an interface in C# for file writers. It is worth pointing out that Visual Studio has a template for creating new interfaces. This can be accessed by going to the Solution Explorer and right-clicking on the project or folder that you want to add the new interface to, and choosing **Add > New Item**, bringing up the Add New Item dialog box. In the list of template categories on the left side, choose **Visual C#**, which will show the list of all C# templates in the middle of the dialog. Up near the top, you'll see a template called **Interface**. This will get you set up with a new interface in a manner similar to how the **Class** template got us set up with a new class.

Interfaces are structured in a way that looks very much like a class or struct, but instead of the **class** or **struct** keyword, you will use the **interface** keyword:

```
using System;
using System.Collections.Generic;
using System.Linq;
using System.Text;
using System.Threading.Tasks;

namespace AwesomeProgram
{
    public interface IFileWriter
    {
        string Extension { get; }

        void Write(string filename);
    }
}
```

You can also see that all of the members of our interface have no actual implementation. They look like the abstract methods that we talked about in the last chapter. But also notice that we didn't use the **virtual** keyword here. Nor do we say that it is **public** (or **private** or **protected** for that matter). All members of an interface are public and virtual by default, and cannot be altered into anything else. You cannot add a method body, nor can you use a different accessibility level.

I also want to point out that I started the name of my interface with the capital letter **I**. It is incredibly popular in the C# world to start names of interfaces with the letter I. People do this because it makes it easy to pick out what things are interfaces, and what things are classes. Not everyone does this, and you don't need to either if you don't want, but it is fairly widespread so expect to see it.

Using Interfaces

Now we can go ahead and create classes that use this interface. For example, we might create a **TextFileWriter** class, which knows how to write to a text file. Because we want it to do the things that any **IFileWriter** can do, we'll want to signify that this class implements the **IFileWriter** interface that we just defined. This can be done like this:

```
using System;
using System.Collections.Generic;
using System.Linq;
using System.Text;

namespace AwesomeProgram
{
    public class TextFileWriter : IFileWriter
    {
        public string Extension
```

```
        {
            get { return ".txt"; }
        }

        public void Write(string filename)
        {
            // Do your file writing here...
        }
    }
}
```

To make a class use or *implement* an interface, we use the same notation that we used for deriving from a base class. We use the colon (':') followed by the name of the interface that we're using.

If a class implements an interface like this, we are required to include an implementation for all of the members that the interface specifies. The class can also include a lot more than that, but at a minimum, it will need to implement the things in the interface.

In most ways, an interface will work in the same way that a base class does. For instance, we can have an array of our **IFileWriter** interface, and put different types of objects that implement the **IFileWriter** interface in it:

```
IFileWriter[] fileWriters = new IFileWriter[3];
fileWriters[0] = new TextFileWriter();
fileWriters[1] = new RtfFileWriter();
fileWriters[2] = new DocxFileWriter();

foreach(IFileWriter fileWriter in fileWriters)
{
    fileWriter.Write("path/to/file" + fileWriter.Extension);
}
```

Multiple Interfaces and Inheritance

As I mentioned at the end of the Chapter 23, C# doesn't allow you to inherit from more than one base class. However, you are allowed to implement more than one interface. Having your type implement multiple interfaces just means your type will need to include all of the methods for all of the interfaces that you are trying to implement.

In addition to implementing multiple interfaces, you can also still derive from a base class (and only one). While some languages allow you to have multiple base classes, C# and many other languages don't. This is because allowing you to choose two or more base classes tends to make the language, the overall structure of your code, and the intuitiveness of your code much more complicated.

While there are definitely a few cases where multiple inheritance would be ideal, these cases are fairly rare, and can usually be taken care of with either implementing multiple interfaces, or by using a technique called *composition*, where one class or object simply has another object inside of it (as an instance variable).

Indicating that a class should implement more than one interface is pretty simple. You just list the interfaces you want to implement after the class name, separating them by commas, in any order you want. If there is a base class, it must come first in the list.

```
public class AnyOldClass : RandomBaseClass, IInterface1, IInterface2
{
```

```
    // ...
}
```

Try It Out!

Interfaces Quiz. Answer the following questions to check your understanding. When you're done, check your answers against the ones below. If you missed something, go back and review the section that talks about it.

1. What keyword is used to define an interface?
2. What accessibility level are members of an interface?
3. **True/False.** A class that implements an interface does not have to provide an implementation for all of the members of the interface.
4. **True/False.** A class that implements an interface is allowed to have other members that aren't defined in the interface.
5. **True/False.** A class can have more than one base class.
6. **True/False.** A class can implement more than one interface.

Answers: (1) interface. **(2)** Public. **(3)** False. **(4)** True. **(5)** False. **(6)** True.

24

Using Generics

In a Nutshell

- Generics in C# are a similar concept to generics in Java and templates in C++. They are a way to create type safe classes and structs without needing to commit to a specific data type.
- When you create an instance of a type that uses generics, you'll need to indicate which type (or types) you are using it for in any particular instance using the angle brackets ('<' and '>'): **List<string> listOfStrings = new List<string>();**
- This chapter also covers some of the details of using the **List** class, the **IEnumerable** interface, and the **Dictionary** class, all of which use generics.

In this chapter, we'll take a look at a powerful feature in C# called *generics*. We'll start by taking a look at why generics even exist in the first place, looking at the actual problem they solve. We'll then look at a few classes that come with the .NET Framework that use generics. These classes are the **List** and **Dictionary** classes, as well as the **IEnumerable<T>** interface. These types have many uses, and as we make software, we'll definitely be putting them and other generic classes to good use.

We'll look at how to actually make your own generic classes in the next chapter.

The Motivation for Generics

Before we can really discuss generics, we need to discuss why they exist in the first place. So I'm going to start by having you think about the underlying problem that generics will address.

Let's say you wanted to make a class to store a list of numbers. While an array could do this, perhaps we could make a class that wraps an array and creates a new larger one when we run out of space. Then we never need to worry about how long the list is. (There is already a **List** class that does this, but pretend it doesn't for the sake of this discussion. We'll revisit **List** before the end of this chapter.)

Imagine for a second, what you'd need to do to make a class to accomplish this task. For example:

```
public class ListOfNumbers
{
    private int[] numbers;

    public ListOfNumbers()
    {
        numbers = new int[0];
    }

    public void AddNumber(int newNumber)
    {
        // Add some code in here to make a new array, slightly larger
        // than it already was, and add your number in.
    }

    public int GetNumber(int index)
    {
        return numbers[index];
    }
}
```

This isn't complete, but you get the idea. You make the class and use the **int** type all over the place.

But now let's say you want a list of **string**s. Isn't it a shame that you can't use your **ListOfNumbers** class, and put **string**s in it?

So what now? Any ideas?

I suppose we could make a very similar class called **ListOfStrings**, right? It would end up looking basically the same as our **ListOfNumbers** class, only with the **string** type instead of the **int** type. In fact, we could go crazy making all sorts of **ListOf...** classes, one for every type imaginable. But that's kind of annoying, because we could have almost a limitless number of those classes. Lists of **int**s, lists of **string**s, lists of lists....

Maybe there's another way. What if we simply made a list of **object**s? Remember how **object** is the base class of everything?

We could create just a simple **List** class that uses the **object** type:

```
public class List
{
    private object[] objects;

    public List()
    {
        objects = new object[0];
    }

    public void AddObject(object newObject)
    {
        // Add some code in here to make a new array, slightly larger
        // than it already was, and add your object in.
    }

    public object GetObject(int index)
    {
        return objects[index];
    }
}
```

Now we can put any type of object we want in it. We only need to create one class (just the **List** class) for any type of object we want.

This might seem to work at first glance, but it's got a few problems as well. For instance, any time you want to do anything with it, you'll need to cast stuff to the type you're working with. If we have a list has only **string**s, when we want to call the **GetObject** method, we'll need to cast it like this:

```
string text3 = (string)list.GetObject(3);
```

Casting takes time to execute, so it slows things down a bit. Worse than that, this is less readable.

There's a bigger problem here though. Let's assume we're sticking **string**s in our list. We can cast all we want, but since it's a list of **object**s, we could theoretically put *anything* in the list, not just **string**s.

There's nothing preventing us from accidentally putting something else in there. After all, the compiler can't tell that there's anything wrong with saying **list.AddObject(new Random())** even though we intended it to be a list for only **string**s. We could try to be extremely careful and make sure that we don't make that mistake, but there are no guarantees. And even if we are careful ourselves, other people using the code might make that mistake.

We can never know for sure what type of object we're pulling out. We'll always have to check and make sure that it is the type we think it is, because it might *possibly* be something else. Programmers have a name for this by the way. They say it isn't *type safe*. Type safety is where you always know what type of object you are working with.

Type safety wasn't a problem in our first approach because we weren't using **object**. We were using a specific type, like **string**. We knew exactly what we were working with, so we could ensure that we were using the right thing all the time. There was no need for casting, and no possibility of mistakes.

So we've got two bad choices here. Either we make lots of type safe classes of lists, one for each kind of thing we want to put in it, or we make a single class of plain old **object**s that isn't type safe, but doesn't require making lots of different versions.

Here is where generics come in to save the day, fixing this tricky dilemma for us.

What are Generics?

Generics are a clever way for you to define special *generic* types (classes or structs) which save a spot to place a specific type in later. When it comes time to use the generic type, you can choose what type you want to use in that specific instance. So one time, you'll use the generic class and say "this time it is a list of **int**s," and another time you'll say, "Now I want a list of **Hamburger** objects."

In short, generics provide a way to create type safe classes, without having to actually commit to any particular type when you create the class.

In the next chapter, I'll show you how to actually create your own generic types, but right now, we'll start by taking a look at a few of generic types that already exist as a part of the .NET Framework, and that I'm sure you'll find very valuable. We'll start with the **List** class, which is the "official" version of what we've been describing up until now in this chapter. Then we'll look at a generic interface (**IEnumerable**) which is an interface that allows you to look at all items in a collection, one at a time, and is used by nearly all collection classes. Finally, we'll look at the **Dictionary** class, which is a bit more advanced and shows off a few more of the features that generics has to offer.

The List Class

Lists are so commonly used that, as you can probably guess, a **List** class has been made for you. Let's take a look at that class, and explore how lists work a bit further.

This class is called **List**, so you might think you can create a list like this:

```
List listOfStrings = new List(); // This doesn't quite work...
```

But that doesn't quite work. The reason why this doesn't work is because this class uses generics, which the above code ignores. To create an instance of the **List** class, you need to specify the type of stuff you're putting into the list. To do this, you put the type you want to use inside of angle brackets ('<' and '>'). If you want a list of **string**s, you would do this:

```
List<string> listOfStrings = new List<string>();
```

You now have a list containing strings!

For the most part, that's all you need to know about using generics. It's fairly simple to use. However, it's worth taking some time to learn more about this **List** class, because it is incredibly versatile. So let's take some time and do that now.

There's an **Add** method that you can use to add items to the end of the list:

```
listOfStrings.Add("Hello World!");
```

If you look closely, you'll notice that since the type we are using is **List<string>**, this method *requires* that we put only **string**s in. Try anything else, and the compiler will catch you. So it is type safe. But we've only had to define a single generic list class. It's the best of both worlds.

If adding items to the end isn't what you need, you can use the **Insert** method, which allows you to put in the index to add the item at, pushing everything else back:

```
listOfStrings.Insert(0, "text3");
```

Remember that indices typically start at 0, so the above code inserts at the front of the list.

You can also get an item out of the list using the **ElementAt** method.

```
string firstItem = listOfStrings.ElementAt(0);
```

There's another way to get and set items in the list. You can also use the square brackets (**[** and **]**) with the **List** class, just like an array. (This is done through an indexer, which we'll see in Chapter 33.)

So you can say:

```
string secondItem = listOfStrings[1];
```

And you can also say:

```
listOfStrings[0] = "This message replaces whatever was already at index 0";
```

Also, the **RemoveAt** method allows you to remove items from the list:

```
listOfStrings.RemoveAt(2);
```

While we're talking about deleting stuff, you can remove *everything* from the list with the **Clear** method:

```
listOfStrings.Clear();
```

While arrays have the **Length** property to determine how many items are in the array, the **List** class has a **Count** property instead (there's no **Length** property).

```
int itemsInList = listOfStrings.Count;
```

If we're using a **List**, but we want to get a copy of the contents as an array, there's an easy way to do this. There's a **ToArray** method that will make this conversion for us, turning our generic **List** instance into an array of the appropriate type:

```
List<int> someNumbersInAList = new List<int>();
someNumbersInAList.Add(14);
someNumbersInAList.Add(24);
someNumbersInAList.Add(37);

int[] numbersInArray = someNumbersInAList.ToArray();
```

We are also able to loop over all of the items in a **List**, just like with arrays:

```
List<int> someNumbersInAList = new List<int>();
someNumbersInAList.Add(14);
someNumbersInAList.Add(24);
someNumbersInAList.Add(37);

foreach(int number in someNumbersInAList)
{
    // ...
}
```

Since the **List** class is generic, you can create an instance of the **List** class using any other type you want. And when you do, all of the methods like **Add** and **ElementAt** will work *only* for the type that you are using, keeping it type safe like we wanted.

Using Collection Initializer Syntax

When we first introduced arrays, we also discussed *collection initializer syntax* (or simply a *collection initializer*) that allowed us to set up an array using simplified syntax:

```
int[] numbers = new int[] { 1, 2, 3, 4, 5 };
```

We can do this same thing with the **List<T>** class as well. The earlier code that added multiple items to a list could be simplified using collection initializer syntax:

```
List<int> someNumbersInAList = new List<int>() { 14, 24, 37 };
```

This trick works on anything that implements the **IEnumerable<T>** interface (see the next section) and has an **Add** method. The C# compiler simply turns that into calls to **Add**.

The IEnumerable<T> Interface

One of the generic types that you'll encounter the most is the **IEnumerable<T>** interface. If you recall from the chapter on enumerations, the definition of the world "enumerate" means to count off, one-by-one, each item in a group or collection. This is essentially what the **IEnumerable<T>** interface is for. It allows a collection to give others the ability to look at each item contained in it, one at a time.

Nearly every class in the .NET Framework that contains multiple items implements this interface. In that sense, it serves as the base or lowest level of defining a collection. Both the **List** class and **Dictionary** class that we'll see next implement it. Even arrays implement **IEnumerable<T>**.

IEnumerable<T> provides the foundation for all sorts of interesting and important features in C#. In Chapter 13 when we were talking about arrays and the **foreach** loop, we saw that if something implements **IEnumerable<T>**, it can be used in a **foreach** loop. And **IEnumerable** is used as the foundation for LINQ queries, which we'll talk about in Chapter 36.

All sorts of things implement **IEnumerable<T>**. If all you care about is the ability to do something with each item in a collection, you can get away with treating it as simply an **IEnumerable<T>**:

```
IEnumerable<int> numbers = new int[3] { 1, 2, 3 };
```

You'll see some methods returning **IEnumerable<T>** if they don't want you to know the actual concrete type being used (it could be an array or a **List** or something else) and if they want you to have the ability to loop through the items (not add or remove items).

The Dictionary Class

To wrap up our introduction on using generics, let's look at another generic class that is slightly more complex than the **List** class: the **Dictionary** class. This class isn't quite as versatile as the **List** class, but you'll definitely find uses for it, so it is a good choice to look at.

Let's start by describing what the **Dictionary** class actually is. To practically nobody's surprise, the name of the class is actually a pretty good description of what this is. This class works like a dictionary does. In a dictionary, you have a large list of words and a definition that belongs to each of them. You use one piece of text (the word) to look up another piece of text (the definition).

The **Dictionary** class does a very similar thing. We use one piece of information (the *key*) to store and look up another piece of information (the *value*). So it is a mapping of key/value pairs. Because the class is generic, we're not limited to **string**s. We can use any type that we want.

For example, let's look at how you could use a **Dictionary** to create a phone book. We'll use **string**s for people's names, and **int**s for their phone numbers. (Perhaps **int**s aren't the best choice, but I want a simple example that uses two different types. A better choice might be to create a **Person** or **Contact** class that stores names and other personal information, and a **PhoneNumber** class that deals with all of the details of phone numbers, like country and area codes. But let's keep it simple.)

The **Dictionary** class is generic, just like the **List** class, but it has *two* different types that are generic: the key, which we use to do the lookup, and the value that belongs to it. Instead of just putting one type in the angle brackets like we did before, the **Dictionary** class requires that we put two in, one for each of the types that are generic:

```
Dictionary<string, int> phoneBook = new Dictionary<string, int>();
```

The **Dictionary** class allows us to use the indexing operator ('[' and ']') to get or set values in it:

```
phoneBook["Gates, Bill"] = 5550100;
phoneBook["Zuckerberg, Mark"] = 5551438;

int billsNumber = phoneBook["Gates, Bill"];
```

We haven't covered everything **Dictionary** can do. Feel free to explore and see what else is there.

Making Generic Types

In a Nutshell

- You can create your own class that uses generics by placing the generic type in angle brackets after the class definition: **public class PracticeList<T> { /* ... */ }**.
- Using single capital letters for generic types is common, so **T** is a common in generic classes.
- You can have multiple generic types: **public class PracticeDictionary<K, V> { /* ... */ }**.
- With a generic type defined, you can use that type anywhere throughout your class including the type of instance variables, the return types of methods, or the types of parameters for methods. For example: **public T GetItem(int index) { /* ... */ }**
- Placing constraints on a generic type variable restrict the kinds of types that can be used, but you know more about the generic type and can do more with it. For example, **public class PracticeList<T> where T : IExampleInterface { /* ... */ }** forces **T** to be an **IExampleInterface**, but you can now use methods that are defined in **IExampleInterface**.

In the last chapter, we took a look at how generics work. We'll now turn our attention to how to actually make them in your own type. Structs and classes can both use generics.

In this chapter, we'll outline how to make our own generic **List** class. While this class already exists in the .NET Framework, this is still a good example that will help us understand how generics work. Obviously, the "official" **List** class is going to be better (more features and better tested) than the one we make here. When making a real program, you should use the pre-existing one, instead of making your own.

To keep our own example list looking as different as possible from the standard **List** class, I'm going to call our list **PracticeList** instead of just **List**.

> **Try It Out!**
> **Building Your Own Generic List Class.** To help you get a hold of the idea of generics, follow through the contents of this chapter and build your own **PracticeList** class that uses generics.

Creating Your Own Generic Types

Creating a generic type is very similar to creating a plain old class or struct. Like with any new class, you can start by adding a new .cs file to your project for your class, just like we did in Chapter 18. I've called mine **PracticeList** (in PracticeList.cs).

On the line where you declare your class, you will also put the angle brackets ('<' and '>') with the name that you are going to give to your generic type (**T** in this case), as shown below:

```
using System;
using System.Collections.Generic;
using System.Linq;
using System.Text;
using System.Threading.Tasks;

namespace Generics // Your namespace might be different.
{
    public class PracticeList<T>
    {
    }
}
```

You can see that this is the same as a normal class, with the exception of having that **<T>** in there. This is what allows the class to use generics.

Now we have a **T** type, which is currently undetermined. When actually creating instances of our **PracticeList** class, we'll be able to supply the type we want to use. By the way, **T** is what we call a *generic type parameter*. We can now use this type parameter anywhere we want to throughout our class to represent the type that will eventually replace it.

We could put multiple types in the angle brackets, separated by commas, to have multiple generic types in our class. For example, we could use **<K, V>** for a key/value pair, just like the **Dictionary** class has.

There's nothing special about the **T** name. It is just very common if there is only one generic type, so I'm following suit here. The **K** and **V** are also very common. In general, people tend to use single letters for their generic types. It helps make it obvious that something is referring to a generic type, rather than just another type called **T**, since most classes or structs have longer names than a single letter. But you could technically call your generic type parameter just about anything.

Using Your Generic Type in Your Class

Now that we've created a class with a generic type parameter, we can use this type throughout our class. For instance, since this is a list, we probably want to store the items of our list somewhere. So we can add the following as an instance variable:

```
private T[] items;
```

We create a **private** array called **items**, but you'll see that the type of that array is **T**—our generic type, rather than any specific type. So whenever someone uses our **PracticeList** class, they'll choose what type to use, and the type they choose will go in here. If they create a **PracticeList** of **int**s (**PracticeList<int> list = new PracticeList<int>();**) then this will be an array of **int**s (**private int[] items**).

Let's keep building our class.

We probably also want a constructor that will set up our array to have 0 items in it to start:

```
public PracticeList()
{
    items = new T[0];
}
```

This should be simple enough. We create a new array with our generic type, with 0 items in it. True, having an array with 0 items is pretty worthless, but we'll grow the array as items are added, so it won't stay empty forever.

We should also create a method that returns a particular item from the list. In this method, we'll return the generic type:

```
public T GetItem(int index)
{
    return items[index];
}
```

This should be pretty straightforward. The return type is simply the generic type parameter. If someone used it and made a **PracticeList** of **int**s, this would be returning **int**s.

Let's add one more method that adds items to the list. This method is the most complicated method in this class, but the generics aren't the cause of that. It's just a bit trickier than the other things the class needs to be able to do.

```
public void Add(T newItem)
{
    T[] newItems = new T[items.Length + 1];

    for (int index = 0; index < items.Length; index++)
        newItems[index] = items[index];

    newItems[newItems.Length - 1] = newItem;

    items = newItems;
}
```

Here, our method's parameter is the generic **T** type that we defined for the class, which works like we've seen in the other pieces of the code.

Let me also take a second to briefly describe what the above method is actually doing. Arrays can't grow in size, which makes adding new items to our **PracticeList** class a little tricky. Instead, what we're doing here is creating a new array that is one item bigger than the original array. The **for** loop then goes down the entire length of the original array and copies the values it contains over to the new, slightly larger array. We finally put the new value in the last spot of the new array, and update our **items** instance variable to use the new, slightly longer array.

Our completed class will look something like this:

```
using System;
using System.Collections.Generic;
using System.Linq;
using System.Text;
using System.Threading.Tasks;

namespace Generics
{
    public class PracticeList<T>
    {
        private T[] items;

        public PracticeList()
        {
            items = new T[0];
        }

        public T GetItem(int index)
        {
            return items[index];
        }

        public void Add(T newItem)
        {
            T[] newItems = new T[items.Length + 1];

            for (int index = 0; index < items.Length; index++)
                newItems[index] = items[index];

            newItems[newItems.Length - 1] = newItem;

            items = newItems;
        }
    }
}
```

Constraints for Type Parameters

For any generic type parameter, you can place constraints or limitations on what types you can actually use for it. For instance, you can state that a generic type must be derived from a particular base class, or that it must implement a particular interface.

While these constraints limit the actual types that can be used in the place of the generic type, it also makes it so that the C# compiler and the CLR know more about the type you are actually using. If the C# compiler can be sure that the generic type will be derived from a particular base class, it knows you can call the methods that belong to that base class.

In short, while constraints limit the types you can use, it gives you a lot more power in what you can actually do with the type parameter.

To specify a constraint, you use the **where** keyword and the colon (':') operator:

```
public class PracticeList<T> where T : IComparable
{
    //...
}
```

Now, since we know that **T** must use the **IComparable** interface, we can use the methods of the **IComparable** interface anywhere we want to throughout the class. Whatever type we ultimately end up supplying for the generic type parameter **T** must implement the **IComparable** interface (though it is OK if it does a lot more than that too). These constraints specify the *minimum* that is needed.

Multiple constraints for a single type can be added by separating them with commas, and if you have multiple generic types, you can specify type constraints for each of them using a new **where** keyword.

```
public class PracticeDictionary<K, V> where K : SomeRandomInterface, SomeBaseClass
                                      where V : SomeOtherInterface
{
    //...
}
```

You can also specify a parameterless constructor constraint, which requires that the type used has a public parameterless constructor. This is done by adding **new()** as a constraint:

```
public class PracticeList<T> where T : new()
{
    //...
}
```

With this, we know that our type parameter **T** has a parameterless constructor, and we can use that to our advantage:

```
T newObject = new T();
```

Interestingly, you cannot add "parameterful" constructor constraints. (You can't say **where T : new(int)**, or anything like that.)

You can also specify that a type must be a value or reference type with the struct or class constraint:

```
public class PracticeList<T> where T : class     // Must be a reference type.
{
    //...
}
```

Or:

```
public class PracticeList<T> where T : struct    // Must be a value type.
{
    //...
}
```

You can also indicate that one type parameter must be derived from another type parameter:

```
public class GenericClass<T, U> where T : U
{
    //...
}
```

Generic Methods

In addition to generic classes, you can have individual methods that use generics as well. These can be a part of any class or struct, generic or regular. Like with generic classes, this is useful when you want to have a set of methods that are identical, except for the type that it works on. To do this, you simply list the generic type parameters after the method name but before the parentheses like this:

```
public T LoadObject<T>(string fileName) { ... }
```

Like with classes, you can use as many generic type parameters as you want, apply generic constraints, and also use the generic type parameters in the method's parameter list and return type:

```
public T Convert<T, U>(T item) where T : class
{
    // ...
}
```

To use a generic method, you would do the following:

```
Person person = LoadObject<Person>("person1.txt");
```

Try It Out!

Generics Quiz. Answer the following questions to check your understanding. When you're done, check your answers against the ones below. If you missed something, go back and review the section that talks about it.

1. Describe the problem generics address.
2. How would you create a list of **string**s, using the generic **List** class?
3. How do you indicate that a class has a generic type parameter?
4. **True/False.** Generic classes can only have one generic type parameter.
5. **True/False.** Generic type constraints limit what can be used for the generic type.
6. **True/False.** Constraints let you use the methods of the thing you are constraining to.

Answers: (1) Type safety without creating a lot of types that differ only by the types they use. **(2)** List<string> strings = new List<string>(); **(3)** In angle brackets by the class name. **(4)** False. **(5)** True. **(6)** True.

Part 4

Advanced Topics

We've covered much of the C# language. By this point, you will likely be able to make most of the programs that you'd want to make, and understand most of the code that other people write. But there's still more to learn.

Feel free to keep reading; it is all valuable information. But if you want to go off now and write your own cool programs, and come back and read the chapters in Part 4 as you find a need for them, or as you have time for them, that's fine too.

In Part 4, we'll look topic-by-topic at more advanced features of the C# language, and tackle some common tasks that you will likely need to do before too long. This includes:

- More about namespaces and **using** directives (Chapter 26).
- More about methods (Chapter 27).
- Reading from and writing to files (Chapter 28).
- Handling errors using exceptions (Chapter 29).
- Delegates (Chapter 30).
- Events (Chapter 31).
- Overloading operators (Chapter 32) and creating indexers (Chapter 33).
- Creating extension methods for existing types (Chapter 34).
- Lambda expressions (Chapter 35) and query expressions (Chapter 36).
- Multi-threading your application (Chapters 37 and 38).
- A quick pass at a few other features in C# that are worth knowing a bit about (Chapter 39).

26

Namespaces and Using Directives

In a Nutshell

- Namespaces are simply collections of types that are assigned a name collectively. They are typically collections of related code.
- A type's fully qualified name is the combination of the namespace name and the type name. You can always use a type's fully qualified name.
- **using** directives tell the compiler that it should look in a particular namespace if it finds a name that is not fully qualified.
- Name collisions are when the compiler is aware of two different types with the same name. They can be resolved by either using fully qualified names or by using an alias.
- A **using static** directive can be added for any static class, allowing you to use the method without needing the class name: **using static System.Console;** and later: **WriteLine(x);**

Back in Chapter 3, when we made our Hello World program, we first saw **using** directives and the **namespace** keyword. At the time, I mentioned that there's more to understanding these two closely related concepts, but that it was a discussion for another day. That day has finally come!

In this chapter, we'll take a look at what a namespace is, and look in-depth at the use of type names, spread throughout your program. We'll then look at what **using** directives actually do. All of this will give you a much better idea of what is going on with these two important parts of your code.

Namespaces

In a program, it is possible to have two types with the exact same name. Consider this: How many **Point** classes do you think exist out there in the world?

Lots.

OpenGL libraries (a very common 3D graphics library) will often have one, every UI framework (like Windows Forms and WPF) has one. The list goes on and on.

You might be thinking, why can't they all just use the same class? Reuse code, and all of that good software engineering mumbo-jumbo.

That's a good idea. In theory. But each of these different libraries needs different things from their **Point** class. Some want it in 2D, others want it in 3D. Some want to use **float**s while others want **double**s. And they each want their **Point** class to be able to do different things. They're fundamentally different types, which just happen to have the same name.

And here's where namespaces come in to play.

A *namespace* is simply a grouping of names.

Usually, people will put related code in the same namespace, so you can also think of a namespace as a module or package for related code. Namespaces are a little like last names. They separate one type with a certain name from other types with the same name.

Fully Qualified Names

Looking back at that **Point** class, the Windows Forms **Point** class is in the **System.Drawing** namespace. The WPF version is in the **System.Windows** namespace. The namespace allows you to distinguish which of the two you have.

When combined, the namespace name and the class name is called a *fully qualified name*. **System.Drawing.Point** is a fully qualified name, as is **System.Windows.Point**. There's no mistaking which of the two you mean.

Up until now, we haven't been using fully qualified names. But we could have been. For example, every place that we've used the **Math** class, we could have used the fully qualified name **System.Math**. We could have said **System.Math.PI** and **System.Math.Sin(1.2)**.

Using Directives

You can always use a fully qualified name, but in most cases, that's just too much typing, and it tends to make your code less readable.

In any particular file, we have the ability to point out a namespace and say, "I'm using the stuff in that namespace, so if I don't use a fully qualified name, that's where you'll find it." This is done by putting in a **using** directive, which we've been doing since the beginning.

At the top of a file, we can list all of the names of the namespaces that we will be frequently using. With a **using** directive for **System** (**using System;**) we can use the name of anything in the **System** namespace without needing to use the fully qualified name. (This particular **using** directive, along with a few others, is added to our file for us when we create a new file in Visual Studio.)

This is the reason that we've been able to get away with just saying **Math.PI** all along, instead of **System.Math.PI**. Our program already had a **using System;** statement at the top of the file.

The Error 'The type or namespace X could not be found'

If you try to use the unqualified name for a type, leaving out the namespace, and you don't have an appropriate **using** directive, you'll run into the following error:

```
The type or namespace 'X' could not be found (are you missing a using directive or an assembly
reference?)
```

This lists two possible causes for the problem: missing a **using** directive or missing an assembly reference. The first part of that is usually what's happening. (Missing assembly references are covered in Chapter 42.) This can easily be solved by simply adding the appropriate **using** directive in. (Or use the fully qualified name instead.)

There are two ways you can add a missing **using** directive. First, you could scroll up to the top of the file and manually type in a **using** directive for the right namespace. That works, but it has a few small problems. If you don't know what namespace you actually need, you have to hunt around the Internet to figure it out, and even when you know the right namespace, you still have to leave what you were doing and then find your way back to it when you're done.

There's a shortcut that is much easier.

To illustrate, I've gone into my program and deleted the **using System;** statement in my program, so that C# doesn't know what to do with **Math** anymore.

Once you type something like this:

```
double pi = Math.PI;
```

You'll see Visual Studio will underline **Math** in red, because it doesn't know what to do with it:

```
class Program
{
    static void Main(string[] args)
    {
        double pi = Math.PI;
    }
}
```

When this happens, use the mouse to hover over the underlined word, and two extra pieces of UI will pop up: a lightbulb icon and a message describing the error with a link to show possible fixes.

```
class Program
{
    static void Main(string[] args)
    {
        double pi = Math.PI;
    }
}
```

The name 'Math' does not exist in the current context

Show potential fixes (Ctrl+.)

By either clicking on the lightbulb or clicking the link to show potential fixes, you'll get a list of ways this (or any other) problem could be resolved.

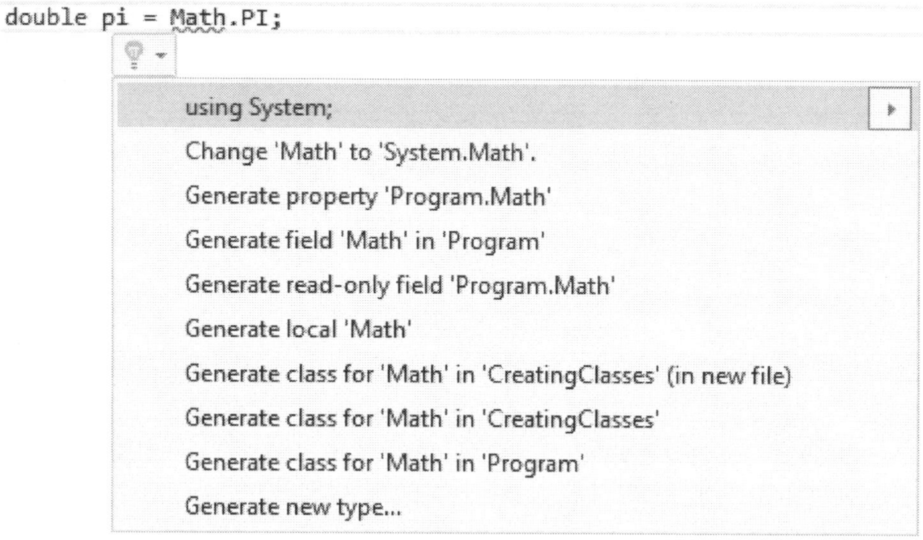

```
double pi = Math.PI;
```

using System;
Change 'Math' to 'System.Math'.
Generate property 'Program.Math'
Generate field 'Math' in 'Program'
Generate read-only field 'Program.Math'
Generate local 'Math'
Generate class for 'Math' in 'CreatingClasses' (in new file)
Generate class for 'Math' in 'CreatingClasses'
Generate class for 'Math' in 'Program'
Generate new type...

Clicking on **using System;** will automatically add a **using** directive for you. (Clicking on the second option there will automatically change your code to use the fully qualified name.)

This makes it so you don't have to memorize or hunt down the namespace that things are in because Visual Studio figures it out for you. You don't even need to leave what you were working on to do it.

I should also point out that there's a keyboard shortcut to get this to pop up: **Ctrl + .** (the period key).

If you are using a type name that is in multiple namespaces, the little drop down box may have multiple choices in it. If so, you'll need to be sure to choose the right one. Choosing the wrong one is often a quick way to a wasted hour and a bad headache.

Name Collisions

On a few rare occasions, you'll add **using** directives for two namespaces that both contain a type with the same name. This is called a *name collision* because the unqualified name is ambiguous. In this case, even though you have **using** directives for an unqualified name, it still can't figure it out.

One solution to this is to just go back to fully qualified names. That tends to be my preferred solution. It keeps things clear and unambiguous. But there's another way to deal with name collisions: aliases.

Using Aliases to Solve Name Collisions

You can also use an alias to solve the problem of a name collision. An alias is a way to create a new name for an existing type name. You can think of it like a nickname.

To illustrate, imagine you have the following code, which won't compile because **C** is a name collision:

```
using N1;
using N2;

namespace N1
{
    public class C { }
}

namespace N2
{
    public class C { }
```

```
}

namespace NamespaceExample
{
    class Program
    {
        static void Main(string[] args)
        {
            C c = new C(); // Compiler can't tell if you're using N1.C or N2.C.
        }
    }
}
```

You can define an alias by using the **using** keyword (yes, there's a lot of uses for the **using** keyword):

```
using CFromN1 = N1.C;
```

This line would go inside of the namespace you want to use it in (**NamespaceExample**, in this case). You can now use **CFromN1** as a class name instead of just **C**. The full code might look like this:

```
using N1;
using N2;

namespace N1
{
    public class C { }
}

namespace N2
{
    public class C { }
}

namespace NamespaceExample
{
    using CFromN1 = N1.C; // The name here is arbitrary. You could even use 'C'.

    class Program
    {
        static void Main(string[] args)
        {
            CFromN1 c = new CFromN1();
        }
    }
}
```

The drawback to this approach, of course, is that you now have an additional name floating around that doesn't actually exist anywhere. It's an alias.

If you can avoid a name collision altogether, do that. Choose names for your own classes that don't conflict with common type names. But sometimes, you have no choice and you'll need to choose between fully qualified names and aliases, both of which have advantages and disadvantages.

Static Using Directives

While we're on the subject of **using** directives, there's another new feature of C# 6.0 that should be discussed: **using static** directives.

As we mentioned earlier in this chapter, when you put a **using** directive at the top of your file, you allow your code to use just the class name to access all of the types contained within a namespace.

When working with static classes, which we talked about in Chapter 18, you can also use a special type of using directive called a **using static** directive. Remember a static class is one that is marked with **static**. This forces all of its members to be static. Static classes are primarily used for utility classes, or ones where you want to group together a bunch of related methods under a shared name. Two very common examples of this are the **Console** and **Math** classes that we've spent a lot of time talking about.

For these static classes, you can add a using directive with the static keyword for these classes, and you'll have access to its methods without even needing to name the class it belongs to:

```
using static System.Math;
using static System.Console;

namespace StaticUsingDirectives
{
    public static class Program
    {
        public static void Main(string[] args)
        {
            double x = PI;
            WriteLine(Sin(x));
            ReadKey();
        }
    }
}
```

The above code illustrates how once the using directives for a static class have been added, you can simply reference its members directly. Instead of **Math.PI** and **Math.Sin**, we can simply use **PI** and **Sin**. Instead of **Console.WriteLine** and **Console.ReadKey**, we can simply use **WriteLine** and **ReadKey**.

Like most language features, this can be a bit of a double-edged sword. It's nice to have your code that much more concise. But at the same time, you strip out meaningful information about which type each static method, property, or variable came from. It's possible to get carried away with this. Use it when the extra simplicity makes things clearer, and avoid it when it makes it more ambiguous.

Try It Out!

Namespaces Quiz. Answer the following questions to check your understanding. When you're done, check your answers against the ones below. If you missed something, go back and review the section that talks about it.

1. **True/False. using** directives make previously inaccessible code accessible.
2. **True/False. using** directives make it so you do not need to use fully qualified names.
3. **True/False.** Two types are allowed to have the same name.
4. **True/False.** A name collision is when two types have the same name.
5. Name two ways to resolve a name collision.
6. What **using** directive would need to be added so that you can use **System.Math**'s **Abs** (absolute value) method as **Abs(x)** instead of **Math.Abs(x)**?

Answers: (1) False. **(2)** True. **(3)** True. **(4)** True. **(5)** Use fully qualified names or aliases. **(6)** using static System.Math;

Methods Revisited

In a Nutshell

- Parameters can be made optional by providing a default value for the parameter: **public void DoSomething(int x = 6) { ... }**
- When calling a method, you can name your parameters: **DoSomethingElse(x: 4, z: 2, y:1);** When doing this, you can put the parameters in any order.
- By using the **params** keyword, methods can have a variable number of parameters.
- The **ref** and **out** keyword can be used on a method parameter to pass the actual variable, rather than just the contents of the variable, allowing the called method to modify the contents of variables in the calling method.

We first looked at methods in Chapter 15. We'll now take a second look at methods and discuss a few more advanced features, because we're ready for it now.

Optional Parameters

Let's say you are making a method that simulates a die roll by picking a random number between 1 and the number of sides on the die. Something like this:

```
private Random random = new Random();

public int RollDie(int sides)
{
    return random.Next(sides) + 1;
}
```

It's entirely possible that 99% of the time, you're going to be passing in 6 for the **sides** parameter. It will be kind of annoying to always need to say **RollDie(6)**. C# provides a way to specify a default value for a parameter, making it optional as long as you're OK with the default. Doing this is pretty easy:

```
public int RollDie(int sides = 6)
{
    return random.Next(sides) + 1;
}
```

With this code, you can now call **RollDie()** and the value of 6 will be used, or you can still fall back to putting a value in if you would like:

```
RollDie();   // Uses the default value of 6.
RollDie(20); // Uses 20 instead of the default.
```

While you're allowed to have as many optional parameters as you want, and you can mix and match them with non-optional parameters, all optional parameters must come at the end, after all of the "normal" parameters.

Named Parameters

Occasionally, you'll go to use a method, but you don't remember what order the parameters are in. This is especially true when the method requires several parameters of the same type. As an example, look at the method below, which takes an input value and clamps it to a particular range, returning the result:

```
public int Clamp(int value, int min, int max)
{
    if(value < min) { return min; } // Bump the value up to the min if too low.
    if(value > max) { return max; } // Move the value down to the max if too high.
    return value;                   // Otherwise, we're good with the original value.
}
```

When you go to use this method, you might see something that looks like this:

```
Clamp(20, 50, 100);
```

When you see this, you always need to do a little digging to figure out what's going on. Is it going to clamp the value of 100 to the range 20 to 50? Or clamp the value of 20 to the range 50 to 100?

To avoid this ambiguity, C# allows you to supply names for parameters. This is done by putting the parameter name, followed by a colon, followed by the value for that parameter:

```
Clamp(min: 50, max: 100, value: 20);
```

With named parameters, it is very clear what value belongs to which parameter. Furthermore, as long as the compiler can figure out where everything goes, this allows you to put the parameters in any order, making it so that you don't need to worry about the official ordering, as long as they all have a value.

When you call a method, you can use regular unnamed parameters, or named parameters, but all regular parameters must come first, and each variable must be assigned to only once.

You can also combine optional parameters with named parameters.

One side effect of this feature is that parameter names are now a part of your program's public interface. Changing a parameter's name, can break code that calls the method using named parameters.

Variable Number of Parameters

Let's say you want to average some numbers together. It is easy to write a method to average two numbers:

```
public static double Average(int a, int b)
{
    return (a + b) / 2.0;
}
```

What if you want to average three numbers? You could write a similar method for that:

```
public static double Average(int a, int b, int c)
{
    return (a + b + c) / 3.0;
}
```

What if you wanted 5? Or 10? You can imagine that you could keep adding more and more methods, but eventually it gets out of hand.

C# provides a way to supply a variable number of parameters to a method, by using the **params** keyword with an array variable:

```
public static double Average(params int[] numbers)
{
    double total = 0;

    foreach (int number in numbers)
        total += number;

    return total / numbers.Length;
}
```

To the outside world, the **params** keyword makes it look like you can supply any number of parameters:

```
Average(2, 3);
Average(2, 5, 8);
Average(41, 49, 29, 2, -7, 18);
```

However, behind the scenes, the C# compiler will turn these into arrays to use in the method call.

It is also worth pointing out that you can directly pass in an array to this type of method:

```
int[] numbers = new int[5] { 5, 4, 3, 2, 1 };
Average(numbers);
```

You can combine a **params** argument with other parameters, but the **params** argument must be the last one, and there can be only one of them.

The 'out' and 'ref' Keywords

Back in Chapter 16, we talked about value and reference semantics. We looked at how whenever you pass something into a method as a parameter, the contents of the variable are copied. With value types, this meant we got a complete copy of the original data. For reference types, we got a copy of the reference, which was still pointing to the same object in the heap.

Methods have the option of handing off the actual variable, rather than copying its contents, by using the **ref** or **out** keyword. Doing this means that the called method is working with the *exact same variable* that existed in the calling method.

This sort of creates the illusion that value types have been turned into reference types, and sort of takes reference types to the next level, where it feels like you've got a reference to a reference.

To do this, you add either the **ref** keyword or the **out** keyword to a particular parameter.

```
public void MessWithVariables(out int x, ref int y)
{
    x = 3;
    y = 17;
}
```

We can then call this code like this:

```
int banana = 2;
int turkey = 5;

MessWithVariables(out banana, ref turkey);
```

At the end of this code, the **banana** variable will contain a value of 3, and the **turkey** variable will contain a value of 17.

If you look at this code, you'll see that you need to put **out** or **ref** when you call the method as well. There's a good reason for this. Handing off the actual variables from one method to another is a risky game to play. By requiring these keywords in both the called method and the calling method, we can ensure that the variables were handed off intentionally.

There is a small difference between using **ref** and **out**, and that difference has to do with whether the calling method or the called method is responsible for initializing the variable. With the **ref** keyword, the calling method needs to initialize the variable before the method call. This allows the called method to assume that it is already initialized. These are sometimes called *reference parameters*.

With the **out** keyword, the compiler ensures that the called method initializes the variable before returning from the method. These are sometimes called *output parameters*.

Using reference or output parameters does a couple of things. Because we're passing the actual variable, instead of just the contents of the variable, this means that value types won't need to be completely copied when passed into a method. This can speed things up because it could save us from copying a lot of value types.

You can also use this to get back multiple values from a method. You're only allowed to return one value from a method, but multiple parameters can be returned as an output parameter instead. (Though, there are other ways to do this that are usually preferred, including building a class that can store all of the data of interest, or using the **Tuple** class.)

Sending the actual variable to a method, instead of just the contents of the variable is a dangerous thing to do. We're giving another method control over our variables, and if used carelessly, it could cause a lot of trouble. Use it sparingly and wisely.

Try It Out!

Methods Revisited Quiz. Answer the following questions to check your understanding. When you're done, check your answers against the ones below. If you missed something, go back and review the section that talks about it.

For questions 1-3, consider the following code: **void DoSomething(int x, int y = 3, int z = 4) { ... }**

1. Which parameters are optional?
2. What values do **x**, **y**, and **z** have if called with **DoSomething(1, 2);**
3. What values do **x**, **y**, and **z** have if called with the following: **DoSomething(x : 2, z : 9);**
4. **True/False.** Optional parameters must be after all required parameters.
5. **True/False.** A parameter that has the **params** keyword must be the last parameter.
6. Given the method **void DoSomething(int x, params int[] numbers) { ... }** which of the following are allowed?
    a. **DoSomething();**
    b. **DoSomething(1);**
    c. **DoSomething(1, 2, 3, 4, 5);**
    d. **DoSomething(1, new int[] { 2, 3, 4, 5 });**
7. **True/False.** Parameters that are marked with **out** result in handing off the actual variable passed in, instead of just copying the contents of the variable.
8. **True/False.** Parameters that are marked with **ref** result in handing off the actual variable passed in, instead of just copying the contents of the variable.
9. **True/False.** Parameters that are marked with **out** must be initialized inside the method.
10. **True/False.** Parameters that are marked with **ref** must be initialized inside the method.

Answers: (1) y and z. **(2)** x=1,y=2,z=4. **(3)** x=2,y=3,z=9. **(4)** True. **(5)** True. **(6)** b, c, d. **(7)** True. **(8)** True. **(9)**True. **(10)** False.

Reading and Writing Files

In a Nutshell

- The **File** class is a key part to any file I/O.
- You can write out data to a file all in one go, using **File.WriteAllText** or **File.WriteAllLines**.
- You can read the entire contents of a file all at once using **File.ReadAllText** or **File.ReadAllLines**.
- You can also read from and write to files a little at a time, in either text format or binary format, by wrapping the **FileStream** that is returned by **File.OpenRead** or **File.OpenWrite** in a **TextReader**/**TextWriter** or **BinaryReader**/**BinaryWriter**.
- When writing a little at a time, you need to clean up when you're done, to ensure that unmanaged resources are cleaned up.

Working with files is a very common task in programming. Reading from and writing to files is often called *file input* and *file output* respectively, or simply *file I/O*.

There is a collection of types that will make it very easy to work with files. In this chapter, we'll look at how to write to a file or read from a file all in one step. We'll then look at how to work with files doing a little reading or writing at a time, in both text and binary formats.

All At Once

Writing the File

There are two simple ways to write a bunch of stuff to a file all at once. These ways are defined in the **File** class, which is going to be the starting point for all of our file I/O.

To write a bunch of text to a file, we can use the **File** class's static **WriteAllText** method:

```
string informationToWrite = "Hello persistent file storage world!";
File.WriteAllText("C:/Place/Full/Path/Here.txt", informationToWrite);
```

Note that the **File** class is in the **System.IO** namespace which isn't included by default, so you'll have to add a **using** directive (**using System.IO;**) in as we described back in Chapter 26.

Alternatively, we can take an array of **string**s and write them out to a file, with each item in the array placed on its own line in the file, using the **WriteAllLines** method:

```
string[] arrayOfInformation = new string[2];
arrayOfInformation[0] = "This is line 1";
arrayOfInformation[1] = "This is line 2";
File.WriteAllLines("C:/Place/Full/Path/Here2.txt", arrayOfInformation);
```

You can see that both of these require the path to the file to write to, along with the text to write out. (Relative paths work too, though.)

Reading the File

Reading a file back in is just as easy:

```
string fileContents = File.ReadAllText("C:/Place/Full/Path/Here.txt");
string[] fileContentsByLine = File.ReadAllLines("C:/Place/Full/Path/Here2.txt");
```

This does the same thing, but in reverse. In the first part, the entire contents of the file are "slurped" into the **fileContents** variable. On the second line, the whole thing is pulled in as an array of **string**s, where each line in the file is a different item in the **string** array.

Assembling and Parsing File Contents

Using **File.ReadAllText** or **File.ReadAllLines** is simple and easy to work with, but usually, your work doesn't end there. A **string**, or an array of **string**s is often not the final format of our information.

As an example, let's say you have a file that contains high scores for a game. This might look like this (this is a CSV file, by the way):

```
Arwen,2778
Gimli,140
Bilbo,129
Aragorn,88
Sam,36
```

Let's say we have a matching **HighScore** class that we created with a **Name** property and a **Score** property:

```
public class HighScore
{
    public string Name { get; set; }
    public int Score { get; set; }
}
```

To go from our program's internal representation (say, an array of this **HighScore** type) to a file, we might do something like this:

```
public void SaveHighScores(HighScore[] highScores)
{
    string allHighScoresText = "";
    foreach(HighScore score in highScores)
        allHighScoresText += $"{score.Name},{score.Score}\n"; // string interpolation again.
```

```
    File.WriteAllText("highscores.csv", allHighScoresText);
}
```

To read the file back in and reassemble your high scores list, you may start off by reading in the entire text of the file, then *parse* (the process of breaking up a block of text into smaller, more meaningful components) the text and turn it back into our list of high scores.

The following code reads in the high scores file we just created and turns it back into a high scores list:

```
public HighScore[] LoadHighScores(string fileName)
{
    string[] highScoresText = File.ReadAllLines(fileName);

    HighScore[] highScores = new HighScore[highScoresText.Length];

    for (int index = 0; index < highScoresText.Length; index++)
    {
        string[] tokens = highScoresText[index].Split(',');

        string name = tokens[0];
        int score = Convert.ToInt32(tokens[1]);

        highScores[index] =new HighScore() { Name = name, Score = score };
    }

    return highScores;
}
```

You can see here that the **Split** method is going to be our good friend when we're reading in stuff from a file. This breaks one string into smaller strings, splitting it where it runs into a particular character (the commas in this case).

The **Convert** class has lots of different methods to convert to different types, so pick the ones you need to convert the incoming strings into the types that you need.

Text-Based Files

There's another way to handle text-based file I/O that is quite a bit more complicated, but allows us to do the writing or reading a little at a time instead of all at once.

Writing the File

We'll start with the **File** class again, but this time, instead of writing out the entire file all at once, we'll just open the file for writing. In doing so, we'll end up with a **FileStream** object, which we'll wrap in a **TextWriter** object to simplify the process, and use that to write out stuff as needed.

```
FileStream fileStream = File.OpenWrite("C:/Place/Full/Path/Here3.txt");
StreamWriter writer = new StreamWriter(fileStream);

writer.Write(3);
writer.Write("Hello");

writer.Close();
```

The **OpenWrite** method returns a **FileStream** object. We can work with the **FileStream** object directly, but it is really low level. Like writing individual bytes. Since we don't want to spend our lives pushing bytes around, we wrap the **FileStream** object in a **StreamWriter** object, which works at a much higher level where we can ask it to write **string**s, **int**s, or other things with the **Write** method.

When we're done writing, we call the **Close** method to release any connections we have to the file.

Reading the File

Reading a file this way is actually more troublesome than writing it was. The problem is that when we write stuff out to a file, there's no real way to know how it was structured to read stuff back in. So one approach is to fall back to the **File.ReadAllLines** method we saw before. You can mix and match, as you see fit. But it is still worth looking at how to read files in a way that mirrors how we wrote out these files.

Just keep in mind that this will be a bit more complicated, because we'll need to work at a bit lower level than we wrote our file out in. In most cases, going back to the earlier version will be the simplest.

```
FileStream fileStream = File.OpenRead("C:/Place/Full/Path/Here3.txt");
StreamReader reader = new StreamReader(fileStream);

// Read a single character at a time.
char nextCharacter = (char)reader.Read();

// Read multiple characters at a time.
char[] bufferToPutStuffIn = new char[2];
reader.Read(bufferToPutStuffIn, 0, 2);
string whatWasReadIn = new string(bufferToPutStuffIn);

// Read a full line at a time.
string restOfLine = reader.ReadLine();

reader.Close();
```

Like with writing, we start by opening the file, this time with the **OpenRead** method. We then wrap the **FileStream** object in a **TextReader**, and then we're ready to start reading.

To read in a single character, you can use the **Read()** method. It returns an **int**, so you'll need to cast it to a **char**.

You can also read in a whole bunch of text using a different overload of the **Read** method. You can see from the example above that to use this method you need to create an array of **char**s, and pass that in to the **Read** method. You also indicate what index to start writing in the array (we use 0 here) and how many characters to read in (2, in this case). This code also shows how you can easily convert the character array to a **string**.

The **TextReader** class has a few other methods that you may find valuable, so if you really want to go this route, take a look at what other methods it has.

When we're done, we call the **Close** method to make sure that we are no longer connected to the file.

Binary Files

Instead of writing a text-based file, the other choice is to write a binary file. In binary format, you won't be able to open the file in a simple text editor and make sense of it.

Binary files have a couple of advantages. First, binary files usually take up less space than text-based files. Second, the data is "encrypted" to some extent. (I'm using that term very loosely here, since this isn't true encryption.) Since it isn't text, people can't just open it and read it. It sort of protects your data.

Writing the File

You'll see that this is very similar to the text-based version in the previous section. The code to write to a binary file is the same, with the exception of using a **BinaryWriter** instead of a **StreamWriter**:

```
FileStream fileStream = File.OpenWrite("C:/Place/Full/Path/Here4.txt");
BinaryWriter writer = new BinaryWriter(fileStream);

writer.Write(3);
writer.Write("Hello");

writer.Close();
```

Like before, we open a **FileStream** that's connected to the file we want with **OpenWrite**. This time though, we wrap it in a **BinaryWriter** instead of a **TextWriter**.

We can then call as many **Write** methods as we want. When we're done, we call **Close** to release our connection to the file.

Reading the File

Reading a binary file is actually quite a bit simpler to work with than the text-based version was.

```
FileStream fileStream = File.OpenRead("C:/Place/Full/Path/Here4.txt");
BinaryReader reader = new BinaryReader(fileStream);

int number = reader.ReadInt32();
string text = reader.ReadString();

reader.Close();
```

We open the file and wrap the **FileStream** object with a **BinaryReader**. Now though, we can simply call the **BinaryReader**'s various versions of **Read***()**— **ReadInt32** for **int**s, **ReadString** for strings, **ReadInt16** for **short**s, and so on.

When we're done, we call **Close** to close our connection to the file.

29

Error Handling and Exceptions

In a Nutshell

- Exceptions are C#'s built-in error handling mechanism.
- Exceptions package up information about a problem in an object, and are "thrown" from the method that discovered the problem back up the call stack until it finds a place that can catch/handle the error, or until it reaches the top of the call stack and kills your program.
- If you know you are running code that could potentially cause a problem, you wrap it in a **try-catch** block, putting the type of exception in the parentheses of the **catch** block: **try { /* exception thrown here? */ } catch(Exception e) { /* handle problem here */}**
- You can catch **Exception** or any type derived from **Exception** (e.g., **FormatException**). Catching a more specific type will catch only that type of error, leaving other types to other catch blocks.
- You can string together multiple catch blocks, going from most specific to least specific types of exceptions: **try { } catch(FormatException e) { /* handle format errors here */ } catch(Exception e) { /* handle all other errors here */ }**
- You can create your own types of exceptions by deriving from the **Exception** class.
- If your code discovers a problem, you can start an exception with the **throw** keyword: **throw new Exception("An error has occurred.");**
- Exception filters allow you to catch an exception only if certain conditions are met, letting others go on to another catch clause: **catch(Exception e) if(e.Message == "Message")**

We've spent the entirety of this book so far pretending that everything in our program went according to plan. But the reality is, that's just not what happens. Sometimes a method is running, trying to do its job, when it discovers that something has gone terribly wrong.

As an example, think back to Chapter 15, where we looked at a method that was designed to get a number from the user that was between 1 and 10. That initial version of the code looked like this:

```
static int GetNumberFromUser()
{
    int usersNumber = 0;

    while(usersNumber < 1 || usersNumber > 10)
    {
        Console.Write("Enter a number between 1 and 10: ");
        string usersResponse = Console.ReadLine();
        usersNumber = Convert.ToInt32(usersResponse);
    }

    return usersNumber;
}
```

This code works perfectly, as long as things go according to plan. But what happens when somebody types in "asdf" instead of an actual number?

When we get to the point where we call **Convert.ToInt32**, things are going to fall apart. We're asking it to do something that is an exception to the normal circumstances: turn text that has nothing to do with numbers into a number. Not surprisingly, **Convert.ToInt32** is going to fail at this task. If we don't find a way to successfully handle this error, it will ultimately bring down our whole application.

C# provides a powerful way to trap or "catch" these errors and recover from them gracefully. C# uses an approach called *exception handling*. It is similar to many other languages, like C++, Java, and VB.NET.

How Exception Handling Works

When an error occurs, the code that discovered the problem will package up all of the information it knows about the problem into a special object called an *exception*. These exceptions are either an instance of the **Exception** class or a class derived from the **Exception** class.

Since the method that discovered the error does not know how to recover from it (otherwise it would just handle it itself, without creating an exception), it takes this **Exception** object and *throws* it up to the method that called it, hoping it knows how to resolve the issue. When this happens, it is important to know that once an exception is thrown, the rest of the method will not get executed.

Hopefully, the calling method will know what to do in the event that this particular problem occurs. If it does, it will *catch* the exception, and do whatever needs to be done to recover from the problem. If this method isn't able to provide a solution, the exception will continue on to the method that called it, making its way higher and higher up the call stack.

If the exception makes it all the way back to the top of the call stack, the **Main** method, and it is unable to handle the exception, your program will have a fatal error. If you're running in release mode, this means your program will crash and die. This is called an *unhandled exception*. Clearly, you want to avoid unhandled exceptions whenever possible, but from a practical standpoint, some inevitably get missed.

On the other hand, if you're running in debug mode, Visual Studio will intercept these unhandled exceptions at the last minute. This is actually really convenient. Worst case scenario, this works a little like an autopsy. You'll be able to dig around and see what the actual conditions were when the problem occurred, so that you can know what changes you need to make to fix the problem. On the other hand, in a best case scenario, you can actually make those changes immediately, while your program is still running, and continue on. This is like being able to change a flat tire on your car while racing down the freeway. The details of how to do all of this can be found in Chapter 44.

Through the rest of this chapter, we'll take at what C# code you need to catch exceptions, handle different types of exceptions in different ways, and how to throw your own exceptions if you discover a problem yourself. Lastly, we'll discuss the **finally** keyword, and how it ties in to exception handling.

Catching Exceptions

Before we can catch exceptions, we have to be aware that they could occur. In other words, we need to know that what we're doing may cause problems that we could handle and recover from.

To catch exceptions, we will take this potentially dangerous code and place it in a **try** block. After the **try** block, we place a **catch** block that identifies the exceptions we can handle, and how to resolve them.

Going back to the example that we started with at the beginning of this chapter (asking the user for a number between 1 and 10) we know that there are potentially two problems that could arise. First, as we already discussed, the user could type in text instead. Secondly, they could type in a number that is so large that it can't actually be converted into something that fits into an **int**.

To catch these errors, we can add in code that looks like this:

```
static int GetNumberFromUser()
{
    int usersNumber = 0;

    while(usersNumber < 1 || usersNumber > 10)
    {
        try
        {
            Console.Write("Enter a number between 1 and 10: ");
            string usersResponse = Console.ReadLine();
            usersNumber = Convert.ToInt32(usersResponse);
        }
        catch(Exception e)
        {
            Console.WriteLine("That is not a valid number. Try again.");
        }
    }

    return usersNumber;
}
```

Inside of the **try** block, we put the potentially dangerous code. After the **try** block, we add in a **catch** block to handle the situation if things go wrong. You can see here that this will catch any and all exceptions that occur, placing it in the variable named **e**. Any code that we've discussed throughout this book can go inside of the **try** or **catch** block.

In this particular case, we handle it by simply telling the user that what they entered didn't make any sense. Because the **usersNumber** variable wasn't ever updated, it will still be 0, and the **while** loop will cause the flow of execution to loop back around and ask for another number.

It's worth pointing out that in the code above, the actual information about the exception that occurred is captured in a variable named **e** that has the type **Exception**. You're allowed to use that variable inside of the **catch** block as needed. It's just a normal variable.

Not Giving the Exception Variable a Name

I mentioned earlier that with a **catch** block, you can use the exception you're catching, (for example, **Exception e**) as a variable inside of the **catch** block. On the other hand, if you aren't going to use it, you don't need to even give the variable a name:

```
try
{
    //...
}
catch(Exception) // this Exception variable has no name
{
    //...
}
```

Without a name, it is impossible to use the exception variable, but it also frees up the name you would have used for something else, and you won't get that compiler warning that says, "The variable **e** is declared but never used."

Handling Different Exceptions in Different Ways

I mentioned earlier that exceptions are packaged up into an instance of the **Exception** class (or a derived class). This is how different *types* of errors can be handled differently. Looking back at the previous exception handling code, because we're using the base **Exception** type, this will catch them all, regardless of what the actual error was. (I like to call this "Pokémon exception handling.") In most cases, you'll actually want to handle different exception types in different ways. They represent different problems, and will likely have different error handling code.

Instead, you can handle different types of errors in different catch blocks like this:

```
try
{
    int number = Convert.ToInt32(userInput);
}
catch (FormatException e)
{
    Console.WriteLine("You must enter a number.");
}
catch (OverflowException e)
{
    Console.WriteLine("Enter a smaller number.");
}
catch (Exception e)
{
    Console.WriteLine("An unknown error occured.");
}
```

When you do this, only *one* of the **catch** blocks will actually be executed. Once the exception is handled, it will skip the rest of the **catch** blocks. So if a **FormatException** is thrown, it enters the first **catch** block and run the code there to fix the problem, but the **catch(Exception e)** block will *not* get executed, even though the exception was technically of the type **Exception**. It was already handled.

Doing this makes it so you can handle different types of exceptions in different ways. You just need to be sure to put the more specific type—the derived type—before the more general, base type. If your first block was the **catch(Exception e)** block, it would catch everything, and nothing would ever get into the **FormatException** or **OverflowException** blocks. So ordering is important.

It's also worth pointing out that you don't need to catch all exceptions that might come up. You can build your **catch** blocks in a way that some are handled while others are allowed to propagate further up the call stack, possibly to a **catch** block in another method, or possibly to the point where it kills the program. You're not required to catch all exceptions if you don't want to.

Throwing Exceptions

It's time to look at the other side of this equation: creating and throwing exceptions yourself! Don't get carried away with throwing exceptions. If the method that discovers a problem knows the solution to the problem, don't bother throwing anything. Just handle the problem and continue on. To generalize that statement, the closer you handle an error to the place it was detected the better.

To throw an exception, simply use the **throw** keyword:

```
public void CauseTrouble()  // Always up to no good...
{
    throw new Exception("Just doing my job!");
}
```

It kind of feels like the **return** statement, but it throws exceptions instead of returning a value. You create the **Exception** object, just like any other object, with the **new** keyword, and away it goes.

You're not limited to using the **Exception** class. (It's preferable not to.) There are other options already defined. For instance, here's a small list of some common exceptions and what they're for:

- **NotImplementedException**: If a method has been defined but not implemented yet, you can use this. When Visual Studio automatically generates something, it puts this in the body.
- **IndexOutOfRangeException:** You tried to access an array or something at an index that is beyond how big the array is.
- **InvalidCastException:** You tried to cast something to another type, but the type you tried to cast to wasn't the right kind of object.
- **FormatException:** The text you had is not in the right format for converting to something else (like letters in a string that is supposed to be turned into a number).
- **NotSupportedException:** You tried to do an operation that wasn't supported. For instance, make a method call at a time that didn't allow it.
- **NullReferenceException:** A reference type contained **null** instead of an actual object.
- **StackOverflowException:** You see this all the time when you run out of space on the stack from calling too many methods. This is usually a result of recursion that went bad.
- **DivideByZeroException:** You tried to divide by zero and got caught.
- **ArgumentNullException:** One of the arguments or parameters that you gave to a method was **null**, but the method requires something besides **null**.
- **ArgumentOutOfRangeException:** An argument contained a value that the method couldn't intelligently deal with (e.g., it required a number between 1 and 10, but you gave it 13).

And of course, if you can't find an existing exception type that fits, you can create your own. All you need to do is create a class that is derived from the **Exception** class or from another exception class.

This will look something like this:

```
// For all of you on the other side of the pond, if it makes you feel any better, you
// can still call them "beef burgers".
public class AteTooManyHamburgersException : Exception
{
    public int HamburgersEaten { get; set; }

    public AteTooManyHamburgersException(int hamburgersEaten)
    {
        HamburgersEaten = hamburgersEaten;
    }
}
```

With this class, you can now say:

```
throw new AteTooManyHamburgersException(125);
```

And:

```
try
{
    EatSomeHamburgers(32);
}
catch(AteTooManyHamburgersException e)
{
    Console.WriteLine($"{e.HamburgersEaten} is too many hamburgers.");
}
```

As a general rule, you should be throwing different exception types whenever the calling code would want to handle the error in different ways. If your method could fail in two different ways, you should be throwing two different exception types. This allows people to use different **catch** blocks to handle them differently. If there's no good pre-made exception type for you to use, make your own.

I bring this up because it is really easy to get lazy and start always throwing just the plain old **Exception** type. Don't get lazy; use the right exception types, even if that means creating your own.

The 'finally' Keyword

Remember that when you throw an exception, the flow of execution stops going through the method immediately? Well what if you had done something like opened a file? If everything had gone according to plan, you would have been able to close the file in an intelligent way, but since a problem came up, and an exception is being thrown, that code wouldn't ever be executed, leaving our file open. The **finally** keyword allows us to address this problem.

At the end of a **try-catch** block, you can have a **finally** block (making it a **try-catch-finally** block). The code inside of the **finally** section will get executed no matter what. If it ran through the **try** block without errors, it will get executed. If it ran into a **return** statement, it will get executed. If there was an exception, it will get executed (just before jumping up to the calling method with the exception).

Here's an example of how this works:

```
try
{
    // Do some stuff that might throw an exception here
}
catch (Exception)
```

```
{
    // Handle the exception here
}
finally
{
    // This code will always get execute, regardless of what
    // happens in the try-catch block.
}
```

This is important, so let me repeat myself: the stuff in the finally block gets executed no matter how you leave the **try-catch** block. Error, return statement, reaching the end. It doesn't matter.

Related to this, you're not allowed to put **return** statements inside of **finally** blocks. We may already be returning when we hit the finally block, or we may be in the process of handling an exception. Allowing return statements here just doesn't make any sense so it's banned.

This brings up an interesting thing, though. **finally** blocks get attached to exception handling in many people's minds. (Perhaps it's because it's always discussed in chapters about exception handling, like this one.) People start to think of it as a way to ensure that a chunk of code always gets executed, regardless of whether there was an error or not.

That much is true, but interestingly, you can use a **try-finally** block all by itself, with no connections to exceptions or catching them:

```
private static int numberOfTimesMethodHasBeenCalled = 0;

public static int RandomMethod(int input)
{
    try
    {
        if(input == 0) return 17;
        if(input == 1) return -2;
        if(input == 2) return -11;

        return 5;
    }
    finally
    {
        numberOfTimesMethodHasBeenCalled++;
    }
}
```

In this case, no matter how we return from the method, the code in the **finally** block will always get executed. (There's a lot of better ways to write this particular code, but it illustrates the point.)

Exception Filters

Sometimes, you only want to catch an exception in certain specific cases. For instance, some libraries utilize a set of error codes, a specific value in a property on a custom-made Exception type, or just use different text in the **Message** property of the exception to determine what went wrong.

In this case, there's a high likelihood that you don't want to handle every single error state in the exact same way. You may not even want to handle some at all, or leave for something higher up in the call stack to worry about. The reality is, you don't always want to handle all exceptions of the same type in the exact same way or at the exact same time.

One option in this case is to catch the exception, check for certain conditions and handle the things you want, and then rethrow anything left over:

```
try
{
    throw new Exception("Message");
}
catch(Exception e)
{
    if (e.Message == "Message") { Console.WriteLine("I caught an exception."); }
    else { throw; }
}
```

The above is called "rethrowing the exception." It allows you to throw the exact same exception that was caught, but with one limitation. The exception's stack trace has now been moved away from the original place it was thrown (the third line in the sample above) and the system now thinks it was thrown at the site of the rethrow (the second to last line in the sample). That can be a little frustrating. It makes it more difficult to track down where a problem really originated.

In C# 6.0, there's a better way: exception filters.

Exception filters allow you to add a bit of code to a catch block that can be used to filter whether an exception is actually caught or not. The syntax looks like the if-statements we introduced in Chapter 10 and have been using ever since:

```
try
{
    throw new Exception("Message");
}
catch(Exception e) if(e.Message == "Message")
{
    Console.WriteLine("I caught an exception.");
}
```

This approach is preferable to filtering by placing an if-statement inside of the catch block. The exception filter syntax is usually clearer, and it also avoids the problem we had in the earlier version where the stack trace was changed.

Some Rules about Throwing Exceptions

When you are throwing exceptions, there are a few guiding principles that you should try to follow. Throwing exceptions can really do some crazy things to the flow of execution in your program, and so we want to take care to prevent bad things from happening when you do so.

At a bare minimum, you should not leave resources open that were opened before the exception was thrown. They should be closed first. This can usually be handled in a **finally** clause. But at a minimum, make sure that your program is still in a safe state.

Second, you want to revert back to the way things were before the problem occurred, so that once the error has been handled, people know the system is still in a valid state. A **finally** block can do this too.

Lastly, if you can avoid throwing an exception in the first place, that is what you should do.

Delegates

In a Nutshell

- A delegate is essentially a way to treat a method like an object, allowing you to store it in a variable, pass it as a parameter to a method, or return it from another method.
- To create a delegate, you use the **delegate** keyword, like this: **public delegate float MathDelegate(float a, float b);**. This delegate can now be used to keep track of any method with the same return type and the same parameter list.
- You can then use a delegate in a way that is very similar to a variable: **MathDelegate mathOperation = Add;**
- You can also call the method that is being stored in the delegate variable: **float result = mathOperation(5, 7);**

Delegates: Treating Methods like Objects

We've done just about everything you can think of with variables. We've put numbers in them, words in them, and true/false values in them. We've created enumerations to define all possible values a variable can take on, and we've created structs and classes to store complicated variables with multiple parts, including methods to operate on that data.

Now we're going to take it a step further. What if we had a type of variable that we could put a method in? What if we could assign a method to a variable and pass it around, just like we've done with all of our other types of data? C# has this feature, and it is called a *delegate*. Treating methods like data may seem kind of strange at first, but there are many good uses for it.

Creating a Delegate

Having laid out the background, it is time to take a look at how you'd actually create a delegate. Not all methods are interchangeable with others. With regular data, when we have this problem, we simply use different data types. That's why we have all of the built-in types and can create our own types when those aren't enough. Methods have the same kind of limitations, and delegates will have to account for that. When we create a delegate, we're going to say, "This type of delegate can store any method that has these parameters, and this return type."

Setting up a delegate is relatively straightforward. Delegates are typically created directly inside of a namespace, just like we've done before with classes, structs, and enums. As such, it is also pretty typical to put them in their own file.

This will probably make more sense with an example, so let's do that by creating a **MathDelegate**, which can be used to keep track of any method that can take two **int**s as input, and return an **int**. This could be used for things like an **Add** method, which takes two numbers, adds them together, and returns the result. Or a **Multiply** method, which takes the two numbers and multiplies them.

So to start, we'd create a brand new file for our delegate with a matching name. (If we're going to call our delegate **MathDelegate**, we'd call the file MathDelegate.cs, like usual.) In that file, to create a delegate, we'd simply put something like the following in there:

```
public delegate int MathDelegate(int a, int b);
```

So our whole MathDelegate.cs file should look something like this:

```
using System;
using System.Collections.Generic;
using System.Linq;
using System.Text;
using System.Threading.Tasks;

namespace Delegates
{
    public delegate int MathDelegate(int a, int b);
}
```

Looking at this, you can probably see that this has a lot of similarities to declaring a method, with the exception of adding in the **delegate** keyword. As we'll soon see, we'll be able to create variables that use this delegate type (**MathDelegate**) to store any method that has this same parameter list and return type. (That is, requires two **int** parameters and returns an **int**.

Using Delegates

The first thing we need to actually use delegates is some methods that match the delegate's requirements: two **int** parameters and returns an **int**. I've gone back to my main **Program** class, where the **Main** method is located, and created these three methods, all of which match the requirements of the delegate:

```
public static int Add(int a, int b)
{
    return a + b;
}

public static int Subtract(int a, int b)
{
    return a - b;
}

public static int Power(int baseNumber, int exponent)
{
```

```
    return (int)Math.Pow(baseNumber, exponent);
}
```

To create a delegate-typed object, we simply create a variable with a delegate type, like the **MathDelegate** type we just created. Like with a normal variable, we can assign it a value, but in this case, that value will be the name of a method that matches the delegate:

```
MathDelegate mathOperation = Add;

int a = 5;
int b = 7;

int result = mathOperation(a, b);
```

On the last line, you see we can also use the variable, calling whatever method it happens to be storing at that particular point in time. Doing this looks like a normal method call, except that it "delegates to" or hands off execution to whatever method happened to be stored in the delegate variable at the time.

If you've never seen anything like this before, it may look kind of strange. But I hope you can start to imagine how delegates could be helpful, allowing you to dynamically change which method is called at runtime. As we move through the next few chapters, we'll see how they can be used as a part of events (Chapter 31), lambda expressions (Chapter 35), and query expressions (Chapter 36). They will be the foundation for a lot of really powerful features in C#.

The Delegate and MulticastDelegate Classes

Now that we've got a basic understanding of C#'s syntax when dealing with delegates, it's time to look under the hood and see what the C# compiler does with a delegate that you define. This will help explain some of the nuances of delegates, and should help you see delegates from another perspective.

A delegate is just syntactic sugar around a class. The .NET Framework defines a **Delegate** class. When you create a delegate, the C# compiler will turn it into a class that is derived from **Delegate**. Specifically, it will be derived from **MulticastDelegate**, which is derived from **Delegate**.

You are not allowed to create a class that directly derives from either of these manually (**class MyClass : MulticastDelegate** is not legal code). Only the compiler is allowed to derive from these special classes.

The **Delegate** and **MulticastDelegate** classes define the core functionality of delegates. On top of that, the C# language itself heaps a thick slab of magic and a pile of syntactic sugar on top to provide programmers with cleaner syntax for working with delegates.

The key thing to learn from this is that it's just a class, like the many other classes that we've now dealt with. This has several implications that are worthy of mention:

- You can declare a new delegate type anywhere you can declare a new class.
- A variable that stores a delegate can contain **null**.
- Before invoking the delegate, you should check for **null** first if there is any possibility that the variable actually contains a **null** reference.

When you create a delegate type, the class that the C# compiler will generate is derived from **MulticastDelegate**. As an example, the **MathDelegate** delegate we made in the last section is converted into something that looks like this:

```
public class MathDelegate : System.MulticastDelegate
{
    // Constructor
    public MathDelegate(Object object, IntPtr method);

    // This key method matches the definition of the delegate.
    public virtual int Invoke(int a, int b);

    // Two methods to allow the code to be called asynchronously.
    public virtual IAsyncResult BeginInvoke(int a int b);
    public virtual void EndInvoke(IAsyncResult result);
}
```

The constructor in that allows the system to create a new instance of this delegate class. The second parameter of the constructor, with the type **IntPtr**, references the method that you're trying to put into the delegate. If the method is an instance method (not static) the object parameter will contain the instance it is called with. If you create a delegate from a static method, this will be **null**.

Earlier, we had code that did this:

```
MathDelegate mathOperation = Add;
```

The compiler will roughly translate this into something more like this:

```
MathDelegate mathOperation = new MathDelegate(null, Add);
```

That's just an approximation, and there's a lot of hand waving in there. I'm completely skipping over the part where the compiler turns the method name of **Add** into an actual **IntPtr**, but that information just isn't available at the C# language level. It's only known by the compiler.

Earlier, we called the delegate method by simply using parentheses and putting our parameters inside:

```
int result = mathOperation(3, 4);
```

In this case, the C# compiler will turn this into a call to the **Invoke** method it created:

```
int result = mathOperation.Invoke(3, 4);
```

Written this way, it's a little easier to see why you might want to check for **null** before invoking the method, just in case the **mathOperation** variable is **null**.

While most C# programmers like the first version of calling a delegate (without the **Invoke**) the second version is both legal (the C# compiler will let you directly call **Invoke** if you want) and is also preferred by some C# programmers. So feel free to use it if it seems preferable or more readable to you.

We'll see more of this syntactic sugar in the next section when we look at chaining.

Delegate Chaining

With a name like "multicast," you'd imagine there'd be a way to have a delegate work with multiple methods. And you're right!

Things get a little complicated when our delegate returns a value, so let's come up with another example instead of the **MathDelegate** we were looking at earlier. This time, one that doesn't return a value. It just does something with some input.

Let's design a really basic logging framework. We can start with a simple **LogEvent** class that has just a **Text** property. Obviously, a real logging framework would probably have quite a bit more than just this, but let's keep it simple. We're just trying to do some delegate stuff here.

```
public class LogEvent
{
    public string Text { get; }
    public LogEvent(string text)
    {
        Text = text;
    }
}
```

We'll also define another simple delegate type that will handle or process these **LogEvent**s using whatever strategy it chooses to:

```
public delegate void LogEventHandler(LogEvent logEvent);
```

We could easily imagine all sorts of things we could do with an event handler method. Writing things to the console is an obvious first choice. We could also write them to a file or to a database. If we added a **Severity** property to our **LogEvent** class, we might even set something up so that if a **Fatal** level log event came in, we'd send an email somewhere.

Let's just stick with the low hanging fruit here and write a method that can write a **LogEvent** to the console and another one that can write it to a file:

```
private static void LogToConsole(LogEvent logEvent)
{
    Console.WriteLine(logEvent.Text);
}

private static void LogToFile(LogEvent logEvent)
{
    File.AppendAllText("log.txt", logEvent.Text + "\n");
}
```

These don't have to be static, of course, but since they don't rely on any instance variables or methods, I've made them static.

The first method simply writes to the console, which we've done a million times now.

The second one writes to a file using the static **AppendAllText** method on the **File** class. It writes to a file called **log.txt**, and since it's a relative path, you'll find it right next to the application's .exe file. I also stuck a new line character ('\n') at the end of that so that each log event will show up on its own line. In the case of the console version, **WriteLine** always appends a new line character at the end by default anyway, so we don't need to do that there.

OK, on to the good stuff. Multicast delegates.

Before, when we wanted the delegate to use just a single method, we assigned it the method like this:

```
LogEventHandler logHandlers = LogToConsole;
```

So how do we get a second method in there as well?

The easiest way is with the **+=** operator:

```
LogEventHandler logHandlers = LogToConsole;
logHandlers += LogToFile;
```

Now when we invoke the delegate, both methods in the delegate will be called:

```
logHandlers(new LogEvent("Message"));
```

Since both the **LogToConsole** method and **LogToFile** method have been added to the delegate, both things will happen, and we'll get the message pumped to the console and to the file.

Taking a particular method out of a delegate is simply a matter of using the **-=** operator:

```
logHandlers -= LogToFile;
```

It's useful to keep in mind that what we're doing here with the **+=** and **-=** operators is syntactic sugar, built around some static methods that belong to the **Delegate** class. These methods are the **Combine** and **Remove** methods respectively. The following two lines of code are functionally equivalent:

```
logHandlers += LogToFile;

logHandlers = (LogEventHandler)Delegate.Combine(logHandlers, new LogEventHandler(LogToFile));
```

Likewise, the following two lines of code are also functionally equivalent for removing a specific method from a delegate's chain:

```
logHandlers -= LogToFile;

logHandlers = (LogEventHandler)Delegate.Remove(logHandlers, new LogEventHandler(LogToFile));
```

Generally speaking, the simplified version using += or -= is clearer and simpler, so it is preferred.

Side Effects of Delegate Chaining

Adding multiple methods to a delegate literally creates a chain of objects, linked together by references. (If you're familiar with the concept of a linked list, that's essentially how it's implemented.) This has a couple of side effects that are worth noting.

The first thing is the return value of calling a multicast delegate. Earlier, we sidestepped that question by just using a delegate with a **void** return type, but it should be stated that the return value of a multicast delegate is the result of the final method. All of the other results are completely ignored. This makes multicast delegates somewhat less useful for non-void return values, but in practice, multicast delegates are generally only used for signatures with a void return type. So there is little functionality lost here. (You can technically go digging into the invocation list of a delegate and gather return values, but there is no simple syntactical sugar to do that, and it's rarely done.)

The second catch is in exception handling. In the event that a method that is being invoked by the delegate throws an exception, none of the handlers further down the chain will get a chance to be called. For this reason, you should do everything you can to avoid throwing exceptions in methods that get attached to a delegate.

The Action and Func Delegates

You can make your own delegates in C# whenever you want to, but it is likely that you won't actually *need* to create your own delegates very often. Included with the .NET Framework is a collection of delegates that are already defined for you. These delegates use generics, so they are incredibly flexible. These are the **Action** and **Func** delegates.

The **Action** delegates all have a **void** return type, while the **Func** delegates have a generic return type. There's a whole set of these that are each a little different. For instance, there's the plain old simple **Action** delegate, which looks like this:

```
public delegate void Action();
```

Then there's a version that has one generic parameter, two generic parameters, three generic parameters, and so on, up to 16. (That's crazy talk, right there!)

```
public delegate void Action<T>(T obj);
public delegate void Action<T1, T2>(T1 arg1, T2 arg2);
public delegate void Action<T1, T2, T3>(T1 arg1, T2 arg2, T3 arg3);
public delegate void Action<T1, T2, T3, T4>(T1 arg1, T2 arg2, T3 arg3, T4 arg4);
```

So as an example, if your method has a **void** return type, and requires two **int**s as parameters, you could use the **Action** delegate with two parameters, putting **int** in for both type parameters:

```
Action<int, int> myDelegate = MethodThatRequiresTwoInts;
myDelegate(3, 4);
```

If you need to return something, you can use one of the **Func** delegates instead. These have a similar pattern, but return a value (of a generic type):

```
public delegate TResult Func<TResult>();
public delegate TResult Func<T, Result>();
public delegate TResult Func<T1, T2, Result>();
```

Again, this goes up to 16 input type parameters, plus the generic result parameter. So instead of our little **MathDelegate** that we made, we could have used **Func<int, int, int>** instead:

```
Func<int, int, int> mathOperation = Add;
int a = 5;
int b = 7;
int result = mathOperation(a, b);
```

Events

In a Nutshell

- Events are a simple and easy way for one section of code to notify other sections that something has happened.
- Events rely on delegates to do their work.
- To create an event inside of a class, you would use code similar to this: **public event EventHandler PointChanged;**. In this case, **EventHandler** is the name of the delegate event handlers must use, and **PointChanged** is the name of the event.
- You raise events by first checking to make sure that there are event handlers attached then raising the event: **if(PointChanged != null) { PointChanged(this, EventArgs.Empty); }**
- Attach methods to an event with the **+=** operator: **PointChanged += HandlerMethod;**
- Detach methods from an event with the **-=** operator: **PointChanged -= HandlerMethod;**

One of the coolest things about C# is a feature called events. Events allow classes to notify others when something specific happens. This is extremely common in GUI-based applications, where there are things like buttons and checkboxes. These things have the ability to indicate when something of interest happens, like the button was pressed, or the user checked the checkbox, and other classes will be notified of the change and can handle the event in whatever way they need to.

User interfaces are not the only use for events. Any time that you have one class that has interesting things happening, other classes will want to know about it. Events are very useful in these situations. It keeps the two classes separated and allows the "client" or listener classes to attach themselves to specific events generated by other classes when they're interested, and detach themselves when they don't care anymore.

This is quite a bit like how we use things like Twitter, or an RSS reader to subscribe to blogs. When there is something out there that has interesting updates that we are interested in, we subscribe to it. While we're subscribed, we see any new updates they have. When we discover that what we were paying attention to was sacked by llamas, and there won't be any more notifications that we care about, we can unsubscribe and move on. The blog or Twitter account will continue making notifications for the sake of anyone else that's still listening.

The advantage of this kind of a model is that the object that is raising events doesn't need to know or care about the details of who is attached or listening to any particular event, nor is it responsible for getting listeners registered or unregistered. The listener can make its own decisions about when to start and stop listening.

In the previous chapter, we talked about delegates. Delegates are going to be a key part of events, so it is important to understand the basics of how they work before jumping into events.

Defining an Event

Our first step will be to create an event. Let's say we have a class that represents a point in two dimensions. So it has an x-coordinate and a y-coordinate. This class might look like this:

```
using System;
using System.Collections.Generic;
using System.Linq;
using System.Text;
using System.Threading.Tasks;

namespace Events
{
    public class Point
    {
        private double x;
        private double y;

        public double X
        {
            get { return x; }
            set { x = value; }
        }

        public double Y
        {
            get { return y; }
            set { y = value; }
        }
    }
}
```

To define an event inside a class, we'll need to add a single line of code as a member of the class:

```
public event EventHandler PointChanged;
```

So now our code should look like this:

```
using System;
using System.Collections.Generic;
using System.Linq;
using System.Text;

namespace Events
{
```

```
    public class Point
    {
        private double x;
        private double y;

        public double X
        {
            get { return x; }
            set { x = value; }
        }

        public double Y
        {
            get { return y; }
            set { y = value; }
        }

        public event EventHandler PointChanged;
    }
}
```

Like other members (properties, methods, and instance variables) we can use **public**, **internal**, **protected**, or **private** for events, but typically, **public** or **internal** is what you're looking for.

The **event** keyword is what makes this particular member an event that others can attach to.

We also specify the delegate type of methods that can be attached to the event. Remembering how delegates work, this means that all listener methods will be required to have a specific set of parameters, as well as a specific return type (though with events, it is almost invariably **void**). In this case, we use the **EventHandler** delegate, which is a pre-defined delegate specifically made for really simple events. This delegate has a return type of **void** and has two parameters, an **object**, which is the "sender" of the event, and an **EventArgs** object. The **EventArgs** object is really simple, but is designed to store some information about the event itself.

Any delegate type can be used here. You're not just limited to **EventHandler**.

Raising an Event

Now that we have an event, we need to add in code to *raise* the event in the right circumstances.

Don't add this into your code quite yet, but take a look at the following piece of code, which does the work of raising an event:

```
if(PointChanged != null)
    PointChanged(this, EventArgs.Empty);
```

This little bit of code is pretty simple. We check to see if the event is **null**. If the event is **null**, there are no event handlers attached to the event. (In a second, we'll see how to attach event handlers.) Raising an event with no event handlers results in a **NullReferenceException**, so we need to check this before we raise the event.

Once we know the event has event handlers attached to it, we can raise the event by calling the event with the parameters required by the delegate—in this case, a reference to the sender (**this**) and an **EventArgs** object (though we're just using the static **EventArgs.Empty** object in this case).

While this code to raise an event can technically go anywhere, I usually put it in its own method. It is very common to name these methods the same as the event, but with the word "On" at the beginning: **OnPointChanged**, in our particular case.

So we can add the following method to our code:

```
public void OnPointChanged()
{
    if(PointChanged != null)
        PointChanged(this, EventArgs.Empty);
}
```

When we detect that the conditions of the event have been met, we call this method to raise the event.

In this particular case, since we want to raise the event any time the point changes, we'll want to call this method when the value of **X** or **Y** gets set.

To accomplish this, we'll add a method call to the setters of both of these properties:

```
public double X
{
    get { return x; }
    set
    {
        x = value;
        OnPointChanged();
    }
}

public double Y
{
    get { return y; }
    set
    {
        y = value;
        OnPointChanged();
    }
}
```

So now, our completed code (including the event, the method to raise the event, and code to actually trigger the event) will look like this:

```
using System;
using System.Collections.Generic;
using System.Linq;
using System.Text;
using System.Threading.Tasks;

namespace Events
{
    public class Point
    {
        private double x;
        private double y;

        public double X
        {
            get { return x; }
            set
```

```
            {
                x = value;
                OnPointChanged();
            }
        }

        public double Y
        {
            get { return y; }
            set
            {
                y = value;
                OnPointChanged();
            }
        }

        public event EventHandler PointChanged;

        public void OnPointChanged()
        {
            if(PointChanged != null)
                PointChanged(this, EventArgs.Empty);
        }
    }
}
```

Attaching and Detaching Event Handlers

Now that we've got our **Point** class all set up with an event for whenever it changes, we need to know how to attach a method as an event handler for the event we've created.

The way we attach something to an event is giving the event a method to call when the event occurs. The method we attach is sometimes called an event handler. Because events are based on delegates we'll need a method that has the same return type and parameter types as the delegate we're using—**EventHandler** in this case. The **EventHandler** delegate we're using requires a **void** return type, and two parameters, an **object** that represents the sender (the thing sending the event) and an **EventArgs** object, which has specific information about the event being raised.

Anywhere that we want, we can create a method to handle the event that looks something like this:

```
public void HandlePointChanged(object sender, EventArgs eventArgs)
{
    // Do something intelligent when the point changes. Perhaps redraw the GUI,
    // or update another data structure, or anything else you can think of.
}
```

It is a simple task to actually attach an event handler to the event. Again, the following code can go anywhere you need it to go, where the method is in scope (accessible):

```
Point point = new Point();

point.PointChanged += HandlePointChanged;

// Now if we change the point, the PointChanged event will get raised,
// and HandlePointChanged will get called.
point.X = 3;
```

The key line there is the one that attaches the handler to the event: **point.PointChanged += HandlePointChanged;**. The += operator can be used to add the **HandlePointChanged** method to our event. If you try to attach a method that doesn't match the needs of the event's delegate type, you'll get an error when you go to compile your program.

You can attach as many event handlers to an event as you want. In theory, all event handlers attached to an event will be executed. However, if one of the event handlers throws and exception that isn't handled, the remaining handlers won't get called. For this reason, event handlers shouldn't throw any exceptions.

It is also possible to detach event handlers from an event, which you should do when you no longer care to be notified about the event. This is an important step. Without detaching old event handlers, it's easy to end up with tons of event handlers attached to a method, or even the *same* event handler attached to an event multiple times. In the best case scenario, things slow down a lot. In the worst case scenario, things may not even function correctly.

Note that you rarely want the same event handler method attached to an event more than once, but this is allowed. The method will be called twice, because it is attached twice.

To detach an event, you simply use the **-=** operator in a manner similar to attaching events:

```
point.PointChanged -= HandlePointChanged;
```

After this executes, **HandlePointChanged** will not be called when the **PointChanged** event occurs.

Common Delegate Types Used with Events

When using events, the event can use any delegate type imaginable. Having said that, there are a few delegate types that seem to be most common, and those are worth a bit of attention.

The first common delegate is one of the **Action** delegates, which we talked about in the last chapter. For instance, if you want an event that simply calls a parameterless void method when the event occurs (a simple notification that something happened, without relaying any sort of data or information to the listener) you could use plain old **Action**:

```
public event Action PointChanged;
```

You could subscribe the method below to that event:

```
private void HandlePointChanged()
{
    // Do something in response to the point changing.
}
```

If you need to get some sort of information to the subscriber, you could use one of the other variants of **Action** that has generic type parameters. For example, we might want to pass in the point that changed to the subscribers. So we could change the event declaration to use the generic **Action** that has an extra type parameter to include the point:

```
public event Action<Point> PointChanged;
```

Then we subscribe to the event with a method like this:

```
private void HandlePointChanged(Point pointThatChanged)
{
```

```
    // Do something in response to the point changing.
}
```

Using different versions of **Action**, we can pass any number of parameters to the event handlers.

Another common option is a delegate that works along the lines of this:

```
public delegate void EventHandler(object sender, EventArgs e);
```

In this example, we get a reference to the sender, who initiated the event, as well as an **EventArgs** object, which stores any interesting information about the event inside it. In other cases, you may be interested in using a more specialized type, derived from **EventArgs**, with even more information.

In this case, instead of using your own special delegate, you can use the **EventHandler<TEventArgs>** event handler. (To use this, you must use **EventArgs** or another type that is derived from it.)

So let's say you have an event that is keeping track of a change in a number, and in the notification process, we want to see the original number and what it was changed to. We can make a class derived from the **EventArgs** class like this:

```
public class NumberChangedEventArgs : EventArgs
{
    public int Original { get; }
    public int New { get; }
    public NumberChangedEventArgs(int originalValue, int newValue)
    {
        Original = originalValue;
        New = newValue;
    }
}
```

Then we'd define our event like this:

```
public event EventHandler<NumberChangedEventArgs> NumberChanged;
```

This event would be raised like we saw earlier, usually in a method called **OnNumberChanged**:

```
public void OnNumberChanged(int oldValue, int newValue)
{
    if(NumberChanged != null)
        NumberChanged(this, new NumberChangedEventArgs(oldValue, newValue));
}
```

Methods that want to handle this event would then look like this:

```
public void HandleNumberChanged(object sender, NumberChangedEventArgs args)
{
    Console.WriteLine($"The original value of {args.Original} is now {args.New}.");
}
```

You're not restricted to using any of these delegates (events can use any delegate you want) but it is both convenient and common, and worth considering.

The Relationship between Delegates and Events

It's a common misconception among C# programmers that events are just delegates that can be attached to multiple methods, or that an event is just an array of delegates. Both of these ideas are wrong, though it's an easy mistake to make.

In the vast majority of cases, delegates are not used to call more than one method at a time, while events quite often are. So many C# programmers don't see delegates that reference more than one method at a time, and make the incorrect assumption that they can't handle it at all.

A more accurate way to think of it is more along the lines of an event being a property-like wrapper around a delegate. Specifically, what we've seen up until now is similar to an auto-implemented property. Let's look back at our earlier example with our **NumberChanged** event:

```
public event EventHandler<NumberChangedEventArgs> NumberChanged;
```

The compiler expands this into code that looks more like this:

```
private EventHandler<NumberChangedEventArgs> numberChanged; // The wrapped delegate.

public event EventHandler<NumberChangedEventArgs> NumberChanged
{
    add { numberChanged += value; }    // Defines behavior when a method subscribes.
    remove { numberChanged -= value; } // Defines behavior when unsubscribing.
}
```

Wrapping the delegate like this prevents people from outside of the class from invoking the delegate directly (in other words, you can't raise an event from outside of the class that it exists in) as well as from mucking with the methods that are contained in the delegate chain.

Interestingly, the code above is actually completely legal code. If you don't like the default implementation of an event, you can write code that looks like the previous sample and do something different or additional with it. In practice, it's rare to have a real need to deviate from the standard pattern, but it is allowed.

One complication with explicitly implementing the **add** and **remove** parts for an event like this is that you can no longer invoke the event directly. Instead, you have to invoke the delegate that it wraps (**numberChanged** instead of **NumberChanged**, in this specific case). That's because you've created custom behavior for what it means to add or remove a method to an event and the normal rules aren't guaranteed to apply. For instance, it's possible that the event doesn't even wrap a delegate at all anymore, or that it subscribes or unsubscribes from multiple delegates, depending on certain conditions. Since the compiler can't guarantee what it means to raise the event anymore, the programmer must determine what it means.

Try It Out!

Delegates and Events Quiz. Answer the following questions to check your understanding. When you're done, check your answers against the ones below. If you missed something, go back and review the section that talks about it.

1. **True/False.** Delegates allow you to assign methods to variables.
2. **True/False.** You can call and run the method currently assigned to a delegate variable.
3. **True/False.** Events allow one object to notify another object when something occurs.
4. **True/False.** Any method can be attached to a specific event.
5. **True/False.** Once attached to an event, a method cannot be detached from an event.

Answers: (1) True. **(2)** True. **(3)** True. **(4)** False. **(5)** False.

32

Operator Overloading

In a Nutshell

- Operator overloading allows us to define how some operators work for types that we create.
- Operator overloading works for these operators: +, -, *, /, %, ++, --, ==, !=, >=, <=, >, and <.
- Operator overloading does not work for these operators: '&&' and '||', the assignment operator '=', the dot operator '. ', or the **new** operator (keyword).
- An example of overloading the '+' operator looks like this: **public static Vector operator +(Vector v1, Vector v2) { return new Vector(v1.X + v2.X, v1.Y + v2.Y); }**
- All operators must be **public** and **static**.
- The relational operators must be done in pairs. (== and !=, < and >, <= and >=.)
- Only overload operators when there is a single, unambiguous way to use the operation.

Throughout the course of this book, we've seen a lot of different operators. They all have built-in functionality that does certain specific things. Mostly, these are defined by math, going back thousands of years. In a few cases, we've seen some not-so-normal uses for these operators, like using the '+' operator for concatenating (sticking together) two **string**s (as in **string text = "Hello" + "World";**). In math, there's no way to add words together, but in C# we can do it.

When you create your own types, you can also define how some of these operators should work for them. For example, if you create a class called **Cheese**, you also define what the '+' operator should do, allowing you to add two **Cheese** objects together (Though if you do that, Pepper Jack and Colby better result in Colby-Jack!) This is called *operator overloading*, and it is a powerful feature of C#.

In a minute, we'll look at how to actually overload these operators, but for now, let's start by discussing what operators you're even allowed to overload. Many but not all operators can be overloaded. The creators of C# tried to allow you to overload as many as possible, but there are some that would just be too dangerous to overload.

The following operators can be overloaded: **+**, **-**, *****, **/**, **%**, **++**, **--**, **==**, **!=**, **>=**, **<=**, **>**, and **<**. Additionally, the operators **+=**, **-=**, **/=**, ***=**, and **%=** are not technically overloadable, but when we overload the **+** operator, the **+=** operator is automatically overloaded for us, and so on.

Oh, and it is also worth pointing out that the relational operators must be overloaded in pairs—if you overload the **==** operator, you must also overload the **!=** operator, and if you overload the **>** operator, you must also overload the **<** operator as well, and so on.

On the other hand, the following operators cannot be overloaded: the logical operators **&&** and **||**, the assignment operator (**=**), the dot operator (**.**), and the **new** operator. Being able to overload these operators would just be too dangerous and confusing to anyone trying to use them. Take the assignment operator, for example. We use it for things like **int x = 3;**. If you could make it do something else besides putting the value of 3 into the variable **x**, it could cause some serious issues.

Note that the array indexing operator (**[** and **]**) can be overloaded, but it is done using indexers, which we'll look at in the next chapter.

You also cannot create entirely new operators using operator overloading.

Overloading Operators

You can overload operators for any of your own types, but for the sake of simplicity, I'm going to pick a class that should make some sense for overloading operators. I'm going to do operator overloading for a **Vector** class, which stores an x and y coordinate. Vectors show up all over in math and physics, but if you're fuzzy on the concept, you can think of a vector as a point (in this case, a 2D point).

So let's start with the basic class as a starting point:

```
using System;
using System.Collections.Generic;
using System.Linq;
using System.Text;
using System.Threading.Tasks;

namespace OperatorOverloading
{
    public class Vector
    {
        public double X { get; set; }
        public double Y { get; set; }

        public Vector(double x, double y)
        {
            X = x;
            Y = y;
        }
    }
}
```

Let's say you want to add two vectors together. In math and physics, when we add two vectors, the result is another vector, but with the x and y components added together. For example, if you add the vector (2, 3) to the vector (4, 1), you get (6, 4), because we add the 2 + 4 is 6, and 3 + 1 is 4.

To overload the '+' operator, we simply add in the following code as a member of the **Vector** class:

```
public static Vector operator +(Vector v1, Vector v2)
{
    return new Vector(v1.X + v2.X, v1.Y + v2.Y);
}
```

So now your **Vector** class might look something like this at this point:

```
using System;
using System.Collections.Generic;
using System.Linq;
using System.Text;
using System.Threading.Tasks;

namespace OperatorOverloading
{
    public class Vector
    {
        public double X { get; set; }
        public double Y { get; set; }

        public Vector(double x, double y)
        {
            X = x;
            Y = y;
        }

        public static Vector operator +(Vector v1, Vector v2)
        {
            return new Vector (v1.X + v2.X, v1.Y + v2.Y);
        }
    }
}
```

All operator overloads must be **public** and **static**. This should make sense, since we want to have access to the operator throughout the program, and since it belongs to the type as a whole, rather than a specific instance of the type. We then specify a return type, **Vector** in this case. We use the **operator** keyword, along with the operator that we're overloading. We then have two parameters, which are the two sides of the **+** operator.

For a unary operator like **-** (the negation operator or negative sign) we only have one parameter:

```
public static Vector operator -(Vector v)
{
    return new Vector(-v.X, -v.Y);
}
```

Notice, too, that we can have multiple overloads of the same operator:

```
public static Vector operator +(Vector v, double scalar)
{
    return new Vector(v.X + scalar, v.Y + scalar);
}
```

Now you can add a vector and a scalar (just a plain old number). Though from a math standpoint, this doesn't make any sense.

The relational operators can be overloaded in the exact same way, only they must return a **bool**:

```
public static bool operator ==(Vector v1, Vector v2)
{
    return ((v1.X == v2.X) && (v1.Y == v2.Y));
}

public static bool operator !=(Vector v1, Vector v2)
{
    return !(v1 == v2); // Just return the opposite of the == operator.
}
```

Remember that I said that the relational operators must be overloaded in pairs, so if you overload **==**, you must also overload **!=**, as shown here.

If you look closely at these operator overloads, you can probably tell that they are basically just a method. (The only real difference is the **operator** keyword.) The C# compiler is actually going to turn this into a method for us as it compiles it.

Overloading operators is done primarily to make our code look cleaner. It is what's called *syntactic sugar*. Anything that you can do with an operator, you could have done with a method. Overloading an operator is sometimes more readable though.

Just because you *can* overload operators, doesn't mean you *should*. Imagine that you overload an operator to have it do something totally unexpected. If it isn't clear what your overloaded operator does, you'll cause yourself and others lots of problems. It's for this reason that Java and other languages, erring on the side of being overly cautious, have chosen to not even allow operator overloading. The rule to follow is to only overload operators that have a single, clear, intuitive, and widely accepted use.

Our completed class might look something like this:

```
using System;
using System.Collections.Generic;
using System.Linq;
using System.Text;
using System.Threading.Tasks;

namespace OperatorOverloading
{
    public class Vector
    {
        public double X { get; set; }
        public double Y { get; set; }

        public Vector(double x, double y)
        {
            X = x;
            Y = y;
        }

        public static Vector operator +(Vector v1, Vector v2)
        {
            return new Vector(v1.X + v2.X, v1.Y + v2.Y);
        }

        public static Vector operator +(Vector v, double scalar)
        {
            return new Vector(v.X + scalar, v.Y + scalar);
```

```
        }

        public static Vector operator -(Vector v)
        {
            return new Vector(-v.X, -v.Y);
        }

        public static bool operator ==(Vector v1, Vector v2)
        {
            return ((v1.X == v2.X) && (v1.Y == v2.Y));
        }

        public static bool operator !=(Vector v1, Vector v2)
        {
            return !(v1 == v2);
        }
    }
}
```

Now that we've defined our operator overloads, we can use them just like normal operators:

```
Vector a = new Vector(5, 2);
Vector b = new Vector(-3, 4);
Vector result = a + b;          // We use the operator overload here.
// At this point, result is <2, 6>.
```

Try It Out!

3D Vectors. Make a **Vector** class like the one we've created here, but instead of just **x** and **y**, also add in **z**. You'll need to add another property, and the constructor will be a little different. Add operators that do the following:

- Add two 3D vectors together. (1, 2, 3) + (3, 3, 3) should be (4, 5, 6).
- Subtract one 3D vector from another. (1, 2, 3) - (3, 3, 3) should be (-2, -1, 0).
- Negate a 3D vector. For example, using the negative sign on (2, 0, -4) should be (-2, 0, 4).
- Multiply a vector by a number (scalar) so (1, 2, 3) * 4 should be (4, 8, 12).
- Divide a vector by a number (scalar) so (2, 4, 6) / 2 should be (1, 2, 3).

Additionally, write some code to run some tests on your newly created 3D vector class and check to see if everything is working.

Try It Out!

Operator Overloading Quiz. Answer the following questions to check your understanding. When you're done, check your answers against the ones below. If you missed something, go back and review the section that talks about it.

1. **True/False.** Operator overloading means providing a definition for what the built-in operators do for your own types.
2. **True/False.** You can define your own, brand new operator using operator overloading.
3. **True/False.** All operators in C# can be overloaded.
4. **True/False.** Operator overloads must be **public**.
5. **True/False.** Operator overloads must be **static**.

Answers: **(1)** True. **(2)** False. **(3)** False. **(4)** True. **(5)** True.

33

Indexers

> **In a Nutshell**
> - An indexer is a way to overload the indexing operator (**[** and **]**).
> - Defining an indexer works like a cross between operator overloading and a property: **public double this[int index] { get { /* get code here */ } set { /* set code here */ } }**
> - Indexers can use any type as a parameter.
> - You can have multiple indices by simply listing multiple things where you define the indexer: **double this[string someTextIndex, int numericIndex]**.

In the previous chapter, we talked about how to overload operators. In this chapter, we'll talk about indexers, which is essentially a way to overload the indexing operator (**[** and **]**). Like with operator overloading, just because you *can* do indexing doesn't mean it is always the right solution.

How to Make an Indexer

Defining an indexer is almost like a cross between overloading an operator and a property. It is pretty easy to do, so we'll just start with some code that does this. This code could be added as a member of the **Vector** class that we made in the last chapter, though this same setup works in any class as well.

```
public double this[int index]
{
    get
    {
        if(index == 0) { return X; }
        else if(index == 1) { returnY; }
        else { throw new IndexOutOfRangeException(); }
    }
    set
    {
```

```
        if (index == 0) { X = value; }
        else if (index == 1) { Y = value; }
        else { throw new IndexOutOfRangeException(); }
    }
}
```

We first specify the access level of the indexer—**public** in our case—along with the type that it returns. (Note that we don't have to use **public** and **static**, unlike overloading other operators.) We then use the **this** keyword, and the square brackets (**[** and **]**), which indicate indexing. Inside of the brackets we list the type and name of the indexing variable that we'll use inside of this indexer.

Then, like a property, we have a **get** and a **set** block. Note that we do not need to have both of these if we don't want. We can get away with just one. Inside of the **get** and **set** blocks, we can use our **index** variable, and like properties, we can use the **value** keyword in the setter to refer to the value that is being assigned.

In this little example, I'm making it so that people can refer to the **x** and **y** components of the vector using the 0 and 1 index respectively. Admittedly, it's not too useful. With this in place, we would now be able to do this:

```
Vector v = new Vector(5, 2);
double xComponent = v[0]; // Use indexing to set the x variable.
double yComponent = v[1]; // Use indexing to set the y variable.
```

This is much clearer than if we had been forced to do it with methods. (**xComponent = v.GetIndex(0);**)

Using Other Types as an Index

We're not stuck with just using **int**s as an index. We can use any type we want. For example, strings:

```
public double this[string component]
{
    get
    {
        if (component == "x") { return X; }
        if (component == "y") { return Y; }
        throw new IndexOutOfRangeException();
    }

    set
    {
        if (component == "x") { X = value; }
        if (component == "y") { Y = value; }
        throw new IndexOutOfRangeException();
    }
}
```

This code is very similar to what we just saw, except that we're using a **string** for indexing. If they ask for "x", we return the x-component. If they ask for "y", we return the y-component.

So now we'd be able to do this:

```
Vector v = new Vector(5, 2);
double xComponent = v["x"]; // Indexing operator with strings.
double yComponent = v["y"];
```

There's still more! We can do indexing with *multiple* indices. Adding in multiple indices is as simple as listing all of them inside of the square brackets when you define your indexer:

```
public double this[string component, int index]
{
    get
    {
        // Do some work here to return a value from the two indices 'component' and 'index'.
        return 0;
    }
    set
    {
        // Do something intelligent here to assign the value passed in. Like with properties,
        // the value passed in can be accessed with the value keyword.
        components.Find(component).AtIndex(index) = value;
    }
}
```

Indexers can be very powerful, and allow us to make indexing or data access look more natural when working with our own custom made types. Like with operator overloading, we should take advantage of it when it makes sense, but be cautious about overusing it.

Try It Out!

Creating a Dictionary. Create a class that is a dictionary, storing words (as a **string**) and their definition (also as a **string**). Use an indexer to allow users of the dictionary to add, modify, and retrieve definitions for words.

You should be able to add a word like this: **dictionary["apple"] = "A particularly delicious pomaceous fruit of the genus Malus.";**

You should be able to change a definition by reusing the same "key" word: **dictionary["apple"] = "A fruit of the genus Malus that often times rots and is no longer delicious.";**

You should also be able to retrieve a definition using the indexer: **string definitionOfApple = dictionary["apple"];**

Note that the .NET Framework already defines a **Dictionary** class, which uses generics and in the real world could be used to do what we're trying to do here, plus a whole lot more. But we're trying to get the hang of indexers here, so don't use that class while doing this challenge.

Index Initializer Syntax

We've talked throughout this book about a lot of ways to initialize things. In Chapter 18 we introduced the simple constructor, where you pass in parameters to get things set up (**new Vector2(10, -5)**), there's object initializer syntax (Chapter 19) that additionally lets you assign values to the object's properties on the same line at the same time (**new Vector2() { X = 10, Y = -5 }**), there's collection initializer syntax that is used for setting up an array (Chapter 13) or a collection (Chapter 24) (**new Vector2[] { new Vector2(0, 0), new Vector2(10, 5), new Vector2(-2, -8) }**) and now we'll introduce one final option.

If a class defines an indexer, you can use *index initializer syntax* (or an *index initializer*). Going off of the *Try It Out!* problem above, with your own custom dictionary class, you could fill it up like this:

```
Dictionary dictionary = new Dictionary()
{
    ["apple"] = "A particularly delicious pomaceous fruit of the genus Malus.",
    ["broccoli"] = "The 7th most flavorless vegetable on the planet."
    // ...
};
```

This gets translated into direct use of the indexer, so the above code could be written like the code below without using index initializer syntax:

```
Dictionary dictionary = new Dictionary();
dictionary["apple"] = "A particularly delicious pomaceous fruit of the genus Malus.";
dictionary["broccoli"] = "The 7th most flavorless vegetable on the planet."
// ...
```

Because index initializer syntax can be used any time the type defines an indexer, and because lots of classes define indexers, there are quite a few places this can be used.

Like many other things we've discussed, sometimes index initializer syntax is more readable, while other times it's not. You should let the readability of the code dictate whether it's the right choice or not.

34 Extension Methods

In a Nutshell

- Extension methods let you define a method that feels like it belongs to a class that you don't have control over.
- Extension methods are defined as static classes as static methods. The first parameter of the method is the type of the class that you want to add the extension method to, and it must be marked with the **this** keyword: **public static class StringExtensions { public static string ToRandomCase(this string text) { /* Implementation here... */ } }**
- Once an extension method has been created, you can call it as though it is a part of the class: string **text = "Hello World!"; randomCaseText = text.ToRandomCase();**
- Extension methods are syntactic sugar to make your code look cleaner. The C# compiler rewrites into a direct call of the static method.

Let's say that you are using a class that someone else made. Perhaps one of the classes that comes with the .NET Framework, like the **string** type.

What if you're using that type, and you wish it had a method that it doesn't have? Particularly, if you don't have the ability to modify it? If it's one of your own classes, you can simply add it in. But if it's not one that you can modify, what then?

The **string** type has a **ToUpperCase()** method and a **ToLowerCase()** method, but what if we want to create a method to convert it to a random case, so each letter is randomly chosen to be upper case or lower case? (SomEtHIng LiKE tHiS.) Writing the method is relatively easy, but wouldn't it be nice if we could make it so that we can say something like **myString.ToRandomCase()** just like we can do with **myString.ToUpperCase();**? Without having access to the class, we wouldn't normally have the ability to add our **ToRandomCase()** method as a new method in the class.

Normally.

But there's a way in C#. It is called an *extension method*. Basically, we'll create a static method in a static class, along with the **this** keyword, and we can make a method that appears as though it were a member of the original class, even though it's technically not.

Creating an Extension Method

Creating an extension method is as simple as I just described; we'll make a **static** class, and put a **static** method in it that does what we want for the extension method.

We start by adding a new class file to your project (see Chapter 18). While it is not required, I typically call my class something like **StringExtensions** if I'm creating extension methods for the **string** class, or **PointExtensions**, if I'm creating extension methods for a **Point** class.

In addition, we want to stick the **static** keyword on our class, to make the whole class a static class. So to start, our class will look something like this:

```
using System;
using System.Collections.Generic;
using System.Linq;
using System.Text;

namespace ExtensionMethods
{
    public static class StringExtensions
    {
    }
}
```

To define our actual extension method, we simply create a **static** method here. The first parameter must be the type of object that we're creating the extension method for (**string**, in our case), marked with the **this** keyword:

```
public static string ToRandomCase(this string text)
{
    // The method implementation will go here in a second...
}
```

We can indicate a return type, and in addition to the first parameter that is marked with the **this** keyword, we could also have any other parameters we want.

Like with operator overloading and indexers, this is basically just syntactic sugar. The C# compiler will rework any place that we call our extension method. So when we're done, we'll be able to say:

```
string title = "Hello World!"
string randomCaseTitle = title.ToRandomCase();
```

But the compiler will rework the code above to look like this:

```
string title = "HelloWorld";
string randomCaseTitle = StringExtensions.ToRandomCase(title);
```

The extension method looks nicer and it feels like a real method of the **string** class, which is a nice thing.

But this can be a double-edged sword. The method *feels* like it is a part of the original class, but officially, it's not. For instance, you may move from one project to another, only to discover that what you thought was a part of the original class turned out to be an extension method written by someone else, and you can no longer use it.

Also, if your extension method is in a different namespace than the original class, you may have problems where the *actual* type is recognized, but the extension method can't be found. To get all of the pieces to come together, you may need to add in multiple **using** directives (Chapter 26).

Just be aware of the limitations an extension method has.

We can now finish up our example by completing the body of the **ToRandomCase** method:

```
string result = "";

for (int index = 0; index < text.Length; index++)
{
    if (random.Next(2) == 0) // We still need to create the random object.
        result += text.Substring(index, 1).ToUpper();
    else
        result += text.Substring(index, 1).ToLower();
}

return result;
```

This goes through the original string one character at a time, chooses a random number (0 or 1), and if it is 0, it makes it upper case, or if it is 1, it makes it lower case. So we end up with a random collection of upper and lower case letters, giving us the desired result.

So our complete code for the extension method class is this:

```
using System;
using System.Collections.Generic;
using System.Linq;
using System.Text;

namespace ExtensionMethods
{
    public static class StringExtensions
    {
        private static Random random = new Random();

        public static string ToRandomCase(this string text)
        {
            string result = "";

            for (int index = 0; index < text.Length; index++)
            {
                if (random.Next(2) == 0)
                    result += text.Substring(index, 1).ToUpper();
                else
                    result += text.Substring(index, 1).ToLower();
            }

            return result;
        }
    }
}
```

As we mentioned earlier, if your program is aware of the extension method, you can now do this:

```
string message = "I'm sorry, Dave. I'm afraid I can't do that.";
Console.WriteLine(message.ToRandomCase());
```

We can now use the extension method as though it is a part of the original type.

Try It Out!

Word Count. Create an extension method for the **string** class that counts the total number of words in the **string**. You can make use of the **Split** method, which works like this: **text.Split(' ');**. This returns an array of **strings**, split up into pieces using the character passed in as the split point.

For bonus points, take this a step further and split on all whitespace characters, including space (' '), the newline character ('\n'), the carriage return character ('\r'), the tab character ('\t'). For even more bonus points, ensure that words of length 0, don't get counted.

Try It Out!

Sentence and Paragraph Count. Following the example of the Word Count problem above, create additional extension methods to count the number of sentences and paragraphs in a string. You can assume that a sentence is delimited (ended/separated) by the period ('. ') symbol, and paragraphs are delimited with the carriage return symbol ('\n').

For tons of bonus points, put together a simple program that will read in a text file (see Chapter 28) and print out the number of words, sentences, and paragraphs the file contains.

Try It Out!

Lines of Code. Going even further, let's make a program that will count the number of lines of code a program has.

It is often interesting to know how big a particular program is. One way to measure this is in the total lines of code it contains. (There is some debate about how useful this really is, but it is always fun to know and watch as your program grows larger.)

Create a simple program that, given a particular file, counts the number of lines of code it has. For bonus points, ignore blank lines.

If you're up for a real big challenge, modify your program to start with a particular directory and search it for all source code files (*.cs) and add them all up to see how big an entire project is.

Lambda Expressions

In a Nutshell

- Lambda expressions are methods that appear "in line" and do not have a name.
- Lambda expressions have different syntax than normal methods, which for simple lambda expressions makes it very readable. The expression: **x => x < 5** is the equivalent of the method **bool AMethod(int x) { return x < 5; }**.
- Multiple parameters can be used: **(x, y) => x * x + y * y**
- As can zero parameters: **() => Console.WriteLine("Hello World!")**
- The C# compiler can typically infer the types of the variables in use, but if not, you can explicitly provide those types: **(int x) => x < 5**.
- If you want more than one expression, you can use a statement lambda instead, which has syntax that looks more like a method: **x => { bool lessThan5 = x < 5; return lessThan5; }**
- Lambda expressions can use variables that are in scope at the place where they are defined.
- Expression syntax can be used to define normal, named methods, properties, indexers, and operators as well: **bool AMethod(int x) => x < 5;**

The Motivation for Lambda Expressions

Lambda expressions are a relatively simple concept. The trick to understanding lambda expressions is in understanding what they're actually good for. So that's where we're going to start our discussion.

For this discussion, let's say you had the following list of numbers:

```
// Collection initializer syntax (see Chapter 24).
List<int> numbers = new List<int>(){ 1, 7, 4, 2, 5, 3, 9, 8, 6 };
```

Let's also say that somewhere in your code, you want to filter out some of them. Perhaps you want only even numbers. How do you do that?

The Basic Approach

Knowing what we learned way back in some of the earlier chapters about methods and looping, perhaps we could create something like this:

```
public static List<int> FindEvenNumbers(List<int> numbers)
{
    List<int> onlyEvens = new List<int>();

    foreach(int number in numbers)
    {
        if(number % 2 == 0) // checks if it is even using mod operator
            onlyEvens.Add(number);
    }

    return onlyEvens;
}
```

We could then simply call that method, and get back our list of even numbers. But that's a lot of work for a single method that may only ever be used once.

The Delegate Approach

Fast forward to Chapter30, where we learned about delegates. For this particular task, delegates will actually be able to go a long way towards helping us.

As it so happens, there's a method called **Where** that is a part of the **List** class (actually, it is an extension method) that uses a delegate. Using the **Where** method looks like this:

```
IEnumerable<int> evenNumbers = numbers.Where(MethodMatchingTheFuncDelegate);
```

The **Func** delegate that the **Where** method uses is generic, but in this specific case, must return the type **bool**, and have a single parameter that is the same type that the **List** contains (**int**, in this example). The **Where** method goes through each element in the array and calls the delegate for each item. If the delegate returns true for the item, it is included in the results, otherwise it isn't.

Let me show you what I mean with an example. Instead of our first approach, we could write a simple method that determines if a number is even or not:

```
public static bool IsEven(int number)
{
    return (number % 2 == 0);
}
```

This method matches the requirements of the delegate the **Where** method uses in this case (returns **bool**, with exactly one parameter of type **int**).

```
IEnumerable<int> evenNumbers = numbers.Where(IsEven);
```

That's pretty readable and fairly easy to understand, as long as you know how delegates work. But let's take another look at this.

Anonymous Methods

While what we've done with the delegate approach is a big improvement over crafting our own method to do all of the work, it has two small problems. First, a lot of times that we do something like this, the method is only ever used once. It seems like overkill to go to all of the trouble of creating a whole method to do this, especially since it starts to clutter the namespace. We can no longer use the name **IsEven** for anything else within the class. That may not be a problem, but it might.

Second, and perhaps more important, that method is located somewhere else in the source code. It may be elsewhere in the file, or even in a completely different file. This separation makes it a bit harder to truly understand what's going on when you look at the source code. It our current case, this is mostly solved by calling the method something intelligent (**IsEven**) but you don't always get so lucky.

This issue is common enough that back in C# 2.0, they added a feature called *anonymous methods* to deal with it. Anonymous methods allow you to define a method "in line," without a name.

I'm not going to go into a whole lot of detail about anonymous methods here, because lambda expressions mostly replaced them.

To accomplish what we were trying to do with an anonymous method, instead of creating a whole method named **IsEven**, we could do the following:

```
numbers.Where(delegate(int number) { return (number % 2 == 0); });
```

If you take a look at that, you can see that we're basically taking the old **IsEven** method and sticking it in here, "in line."

This solves our two problems. We no longer have a named method floating around filling up our namespace, and the code that does the work is now at the same place as the code that needs the work.

I know, I know. You're probably saying, "But that code is not very readable! Everything's just smashed together!" And you're right. Anonymous methods solved some problems, while introducing others. You would have to decide which set of problems works best for you, depending on your specific case.

But this finally brings us to lambda expressions.

Lambda Expressions

Basically, a *lambda expression* is simply a method. More specifically, it is an anonymous method that is written in a different form that (theoretically) makes it a lot more readable. Lambda expressions were new in C# 3.0.

In Depth

The Name "Lambda." The name "lambda" comes from lambda calculus, which is the mathematical basis for programming languages. It is basically the programming language people used before there were computers at all. (Which is kind of strange to think about.) "Lambda" would really be spelled with the Greek letter lambda (λ) but the keyboard doesn't have it, so we just use "lambda."

Creating a lambda expression is quite simple. Returning to the **IsEven** problem from earlier, if we want to create a lambda expression to determine if a variable was even or odd, we would write the following:

```
x => x % 2 == 0
```

The lambda operator (**=>**) is read as "goes to" or "arrow." (So, to read this line out loud, you would say "x goes to x mod 2 equals 0" or "x arrow x mod 2 equals 0.") The lambda expression is basically saying to take the input value, **x**, and mod it with 2 and check the result against 0.

You may also notice with a lambda expression, we didn't use **return**. The code on the right side of the **=>** operator must be an expression, which evaluates to a single value. That value is returned, and it's type becomes the return type of the lambda expression.

This version is the equivalent of all of the other versions of **IsEven** that we wrote earlier in this chapter. Speaking of that earlier code, this is how we might use this along with everything else:

```
IEnumerable<int> evens = numbers.Where(x => x % 2 == 0);
```

It may take a little getting used to, but generally speaking it is much easier to read and understand than the other techniques that we used earlier.

Multiple and Zero Parameters

Lambda expressions can have more than one parameter. To use more than one parameter, you simply list them in parentheses, separated by commas:

```
(x, y) => x * x + y * y
```

The parentheses are optional with one parameter, so in the earlier example, I've left them off.

This example above could have been written instead as a method like the following:

```
public int HypoteneuseSquared(int x, int y)
{
    return x * x + y * y;
}
```

Along the same lines, you can also have a lambda expression that has no parameters:

```
() => Console.WriteLine("Hello World!")
```

Type Inference and Explicit Types

The C# compiler is smart enough to look at most lambda expressions and figure out what variable types and return type you are working with. This is called *type inference*. For instance, in our first lambda expression it was smart enough to figure out that **x** was an integer and the whole expression returned a **bool**. With the expression **(x, y) => x * x + y * y** we saw that the C# compiler was smart enough to figure out that **x** and **y** were integer values and the resulting expression returned an **int** as well. And with **() => Console.WriteLine("Hello World!")**, it was smart enough to know that there were no parameters, and the lambda expression didn't return anything (**void** return type).

Type inference is actually a pretty big deal. It's not a trivial accomplishment, and I'd imagine there were a lot of smart people working on it to get it right.

Sometimes though, the compiler can't figure it out. If it can't, you'll get an error message when you compile. If this happens, you'll need to explicitly put in the type of the variable, like this:

```
(int x) => x % 2 == 0;
```

Statement Lambdas

As you've seen by now, most methods are more than one line long. While lambda expressions are particularly well suited for very short, single line methods, there will be times that you'll want a lambda expression that is more than one line long. This complicates things a little bit, because now you'll need to add in semicolons, curly braces, and a **return** statement, but it can still be done:

```
(int x) => { bool isEven = x % 2 == 0; return isEven; }
```

The form we were using earlier is called an *expression lambda*, because it had only one expression in it. This new form is called a *statement lambda*. As a statement lambda gets longer, you should probably consider pulling it out into its own method.

Scope in Lambda Expressions

From what we've seen so far, lambda expressions have basically behaved like a normal method, only embedded in the code and with a different, cleaner syntax. But now I'm going to show you something that will throw you for a loop.

Inside of a lambda expression, you can access the variables that were in scope at the location of the lambda expression. Take the following code, for example:

```
int cutoffPoint = 5;
List<int> numbers = new List<int>(){ 1, 7, 4, 2, 5, 3, 9, 8, 6 };

IEnumerable<int> numbersLessThanCutoff = numbers.Where(x => x < cutoffPoint);
```

If our lambda expression had been turned into a method, we wouldn't have access to that **cutoffPoint** variable. (Unless we supplied it as a parameter.) This actually adds a ton of power to the way lambda expressions can work, so it is good to know about.

(For what it's worth, anonymous methods have the same feature.)

Expression-Bodied Members

Lambda expressions were introduced to C# in version 3.0, and as I mentioned earlier, one of the big draws to it is that the syntax is much more concise. That's great for short methods that would otherwise require a lot of overhead to define.

C# 6.0 extends this a little, allowing you to use the same expression syntax to define normal non-lambda methods within a class. For example, consider the method below:

```
public int ComputeSquare(int value)
{
    return value * value;
}
```

Now that we know about lambda expressions and the syntax that goes with them, it makes sense to point out that this method could also be implemented with the same expression syntax:

```
public int ComputeSquare(int value) => value * value;
```

This only works if the method can be turned into a single expression. In other words, we can use the expression lambda syntax, but not the statement lambda syntax. If we need a statement lambda, we would just write a normal method.

This syntax is not just limited to methods. Any method-like member of a type can use the same syntax. So that includes indexers, operator overloads, and properties (though this only applies to read-only properties where your expression defines the getter and the property has no setter). The following simple class shows all four of these in operation:

```
public class SomeSortOfClass
{
    // These two private instance variables used by the methods below.
    private int x;
    private int[] internalNumbers = new int[] { 1, 2, 3 };

    // Property (read-only, no setter allowed)
    public int X => x;

    // Operator overload
    public static int operator +(SomeSortOfClass a, SomeSortOfClass b) => a.X + b.X;

    // Indexer
    public int this[int index] => internalNumbers[index];

    // Normal method
    public int ComputeSquare(int value) => value * value;
}
```

And of course I'll reiterate here that just because you can do this doesn't necessarily mean you should. Depending on what you're doing, this way or the original way might be more readable. Readability and understandability should always be a priority over using cool new language features.

Try It Out!

Lambda Expressions Quiz. Answer the following questions to check your understanding. When you're done, check your answers against the ones below. If you missed something, go back and review the section that talks about it.

1. **True/False.** Lambda expressions are a special type of method.
2. **True/False.** A lambda expression can be given a name.
3. What operator is used in lambda expressions?
4. Convert the following to a lambda expression: **bool IsNegative(int x) { return x < 0; }**
5. **True/False.** Lambda expressions can only have one parameter.
6. **True/False.** Lambda expressions have access to the local variables in the method they appear in.

Answers: (1) True. **(2)** False. **(3)** Lambda operator (=>). **(4)** x => x < 0. **(5)** False. **(6)** True.

36

Query Expressions

In a Nutshell

- Query expressions are a special type of statement in C# that allows you to make queries on a collection of data to get a particular chunk of that data, along with ordering it.
- A basic query expression has three parts: the **from** clause determines what you're looking through, the **where** clause determines what is filtered out and what is kept, and the **select** clause picks the actual thing that will be given back from the query. This mimics what you might see in a query language like SQL.

C# includes a feature that allows you to use query language constructs in C#. This feature is called *query expressions*. If you are familiar with SQL (Structured Query Language) for accessing databases, you'll love this. If you haven't done anything with databases, don't worry. Your task for this chapter is simply to get a basic understanding of what a query expression is and how they work.

What is a Query Expression?

Imagine you have a set of data out there somewhere. Perhaps it is in a database, or perhaps it is in a C# collection, like a **List** or **Dictionary**. One of the most common tasks that you'll do in programming is to dig through that data set and get certain parts of the data. Maybe your task is to find all the people who make more than a certain amount of money, or maybe it is to find the employee who is assigned to each task in a list of tasks.

A *query* is simply a way of choosing a specific chunk of the complete set of data, including how the results should be organized.

Query expressions are *declarative*. Most things we've done up until now have been procedural, meaning we've said step-by-step *how* to do it. With declarative code, we don't care *how* it is done. We simply state ("declare") what we want, and let the computer figure out how to get it on its own. And of course, that means we don't (and can't) know for sure how it does it, just that it does it correctly.

Anything that we might want to do with query expressions could also have been done with procedural stuff like loops and **if** statements. But there are lots of things that are just flat out more readable with query expressions, and so it is important to know about it as you learn C#.

If you're brand new to C# or programming in general, all I'd really expect you to get out of this chapter is that this feature exists, so that you aren't surprised when you see it pop up, and also to get you some practice making some really simple query expressions.

If you're familiar with SQL, much of what we're going to see here should be very familiar to you. In fact, you'll probably breeze through this chapter and think, *Awesome! I can totally use my SQL knowledge in C#!* This chapter focuses on using query expressions on C# collections, not SQL databases. But everything you see here could be used to work against SQL databases too, if you get things set up correctly. While I can't cover everything in this chapter, what you'll want to look into is called LINQ to SQL.

LINQ and Query Expressions

Query expressions are done using *Language Integrated Query* (abbreviated LINQ, pronounced "link"). In Visual Studio, you've probably seen many of the classes you've create showing a **using** directive saying **using System.Linq;**. Perhaps you've wondered what that's for. That directive brings in the world of LINQ to do query expressions!

Creating a Simple Query Expression

The purpose of a query expression is to take a set of data and retrieve a subset of that data in a new collection, with the possibility of retrieving the results in a specific order or grouping.

We'll start simple and discuss how to make more sophisticated query statements a little at a time.

Let's start with something really basic. For these examples, let's say you have a **Person** class like this:

```
public class Person
{
    public string FirstName { get; set; }
    public string LastName { get; set; }
    public int Age { get; set; }
    public int Height { get; set; }
    public int Weight { get; set; }
}
```

And of course, you have a collection (It has to implement the **IEnumerable** interface) of these; let's say they're in a **List<Person>**:

```
List<Person> allPeople = new List<Person>();
// Add a bunch of people here...
```

Using this list, you can create a query expression to retrieve all people over the age of 18, using the combination of the **from**, **where**, and **select** keywords:

```
IEnumerable<Person> adults =
        from person in allPeople
        where person.Age >= 18
        select person;
```

Ta-da! Your first LINQ query expression!

Let's look at this in a little more detail. The formatting of this, with **from**, **where**, and **select** each starting on a new line, is the most common way of writing this. It helps to keep the different parts separated, and I'd recommend following it.

The **from** part sets up where we're getting data from. In this case, the **allPeople** list. We also give a name to any particular item in the list (**person**, in this case) which we'll be able to use throughout the rest of the query.

The **where** part does our filtering. Here, we're grabbing anyone with an age of 18 or higher.

The last part, with the **select** keyword, indicates what to choose as a part of the results. In this case, the whole **Person** object.

This could have been done procedurally, with a loop and an **if** statement. It would look like this:

```
List<Person> adults = new List<Person>();
foreach (Person person in allPeople)
{
    if (person.Age >= 18)
        adults.Add(person);
}
```

But even this simple example is probably easier to understand with the query expression, and as we get more complicated, the procedural loop and **if** statement gets harder to read faster than the query expression does. So while it may take a little getting used to, queries are a very nice feature of C#.

One other important thing to keep in mind is that you are not allowed to modify the original list during a query statement. Doing so will cause the program to crash.

More Complicated where Clauses

As we move along to more sophisticated queries, one of the first things we'll want to do is be able to make the **where** clause (the part after the **where** keyword, which does the filtering) more complicated.

You are allowed to combine multiple things into one, using the standard **&&** and **||** operators:

```
// In the state of California, kids need a child restraint
// in a car, if under 5 or under 60 pounds.
IEnumerable<Person> kidsNeedingChildRestraints =
    from person in allPeople
    where person.Age <= 5 || person.Weight < 60
    select person;
```

You are also allowed to use multiple **where** clauses, which all have to be true for it to count (essentially an implicit **&&**):

```
// In the state of Utah, you need a booster seat if they are 8 or younger and also
// less than 57 inches tall. (We'll assume the height variable is in inches for now.)
IEnumerable<Person> kidsNeedingChildRestraints =
    from person in allPeople
    where person.Age <= 8
    where person.Height < 57
    select person;
```

Try It Out!

BMI. In the world of health, a number called a Body Mass Index (BMI) can be calculated for a person given their height and weight. While not perfect, this gives people an idea of how much they weigh compared to their "ideal weight." Using the **Person** class defined earlier in this chapter, and assuming that **Height** is measured in inches and **Weight** is measured in pounds, the formula for BMI is this:

$$BMI = \frac{703 * weight}{height^2}$$

1. Write a query expression to find all people in a list who are overweight, defined by a BMI higher than 25.
2. Write another query expression to find all people who are in their ideal weight, defined by a BMI > 20 and < 25.

Multiple from Clauses

One interesting thing that you can do with a query expression is combine multiple **from** clauses into one query. Multiple **from** clauses means you can pull in stuff from multiple places.

You can also use a second **from** clause to dig further into what you're looking at in the first **from** clause. For example, let's say the items in the first **from** clause have a member that is another list. A second **from** clause would allow you to look at each of the items in that list.

For instance, take the **Person** class and add another property to it, containing a **List<Person>** which lists other people who are a child of the person:

```
public class Person
{
    public string FirstName { get; set; }
    public string LastName { get; set; }
    public int Age { get; set; }
    public int Height { get; set; }
    public int Weight { get; set; }
    public List<Person> Children { get; set; }
}
```

You could create a query expression with two **from** clauses like this, which creates a list of all people who are parents of teenagers:

```
// Finds all parents of teenagers.
IEnumerable<Person> parentsOfTeenagers =
    from person in allPeople
    from child in person.Children
    where child.Age >= 13 && child.Age < 20
    select person;
```

The second **from** clause does not need to be pulled from a member of the object retrieved from the first **from** clause. It can be entirely different. This might look something like the code below, though this particular case won't do anything amazing:

```
IEnumerable<Person[]> allPairings =
    from employee in allEmployees
```

```
    from customer in allCustomers
    select new Person[] { employee, customer };
```

> **Try It Out!**
> **Back to Procedural Programming.** Take the query statement to find all parents of teenagers just presented and rewrite it without using query syntax. This will probably require a couple of loops and an **if** statement. Compare which of the two is easier to understand at first glance.

Ordering Results

One very powerful feature of query statements is the ability to dictate how the results should be given back to you. To accomplish this, we'll add in a new clause to our query: the **orderby** clause. The query below orders the people from shortest to tallest:

```
IEnumerable<Person> shortestToTallest =
    from person in allPeople
    orderby person.Height
    select person;
```

There's another part to this too. You have the ability to sort in ascending or descending order. The default is ascending. To choose between the two you simply stick the **ascending** or **descending** keyword at the end of the **orderby** clause:

```
IEnumerable<Person> tallestToShortest =
    from person in allPeople
    orderby person.Height descending
    select person;
```

Retrieving a Different Type of Data

In all of the examples we've done so far, we've always created a resulting list of the same type as the original. In all of our examples, our final result has always been an **IEnumerable** of the type **Person**, which is the type we started with. This doesn't need to be the case though.

We can return a list of a different type, based on a *part* of what we were looking at:

```
IEnumerable<string> fullNameOfAdults =
    from Person person in allPeople
    where person.Age >= 18
    select person.FirstName + " " + person.LastName;
```

In this example, we end up with a list of **string**s, not a list of **Person**.

Method Call Syntax

We've seen a lot in this chapter. And now let's do it all over again! Everything we've seen with query syntax could alternatively be done with a variation called *method call syntax*. All of the things we've discussed with query syntax can also be done by calling methods that have similar names. Method call syntax is often less readable, but it allows you to do a few other things that query syntax does not.

Let's go back a variation on the first query we discussed, which looked like this:

```
IEnumerable<Person> adults =
    from person in allPeople
    where person.Age >= 18
    orderby person.Age descending
    select person;
```

This could alternatively be done with method call syntax, which would look like this:

```
IEnumerable<Person> allAdults = allPeople.
    Where(person => person.Age > 18).
    OrderByDescending(person => person.Age);
```

This uses lambda expressions, which we talked about in the previous chapter. This is another example of why lambda expressions are so powerful.

We're Just Scratching the Surface

Query expressions are one of those things that we just can't cover completely in this book. There are entire books dedicated to query expressions, and Internet forums are peppered with questions about how to accomplish this task and that task with a query expression.

We've covered quite a bit. And it is definitely enough to get going with it. As you start to feel like you've really got the hang of the basics of query statements that we've discussed here, feel free to explore. It's a very powerful tool.

37

Threads

> **In a Nutshell**
>
> - Threading allows you to run sections of code simultaneously.
> - Starting a new thread is relatively simple in C#: **Thread thread = new Thread(MethodNameHere); thread.Start();**
> - Starting a thread this way runs the method indicated. It must have a return type of **void** and must not have any parameters.
> - You can also use **ParameterizedThreadStart**, which will allow you to pass a parameter to the method the thread is running.
> - You can wait for a thread to finish with the **Join** method: **thread.Join();**
> - If you need to worry about thread safety (preventing problems when multiple threads are modifying the same data), you can use the **lock** keyword: **lock(aPrivateObject) { /* code in here is thread safe */ }**. Typically, you create a private instance variable if the critical sections are all within the same class.

Back in the day, all computers had a processor. Nowadays, they usually have a lot more than one. (More specifically, this is usually done by having multiple cores that are all a part of the same processor chip.) I'm currently working on a computer with four processors, each of which is "hyper-threaded," making it appear that I have a total of eight processors. And this computer isn't even a particularly fancy computer. There are machines out there that have 16 or 32 processors, or even hundreds or thousands, all working together. (Turns out, they're kind of expensive.) Beyond just that, it is possible to give work to *other* computers to do over the network, giving you the possibility of having an unlimited number of processors at your disposal.

Computers have a lot of power, but unless you intentionally structure your code to run on multiple processors, it will only ever use one. Think of all of that raw computing power going to waste!

In this chapter, we're going to take a look at threading. The basic process of threading is that we can take chunks of code that are independent of each other and make them run in separate *threads*. A thread is almost like its own program, in that the computer will run multiple threads all at the same time on different processors. (For the record, a thread is *not* its own program. It still shares memory with the program/process that created it.)

When many threads are running, the computer will let one thread run on each processor for a little while, and then it will suddenly switch it out for a different one. The computer gets to decide when it is time to make the switch and which thread to switch to, but it does a pretty good job, so that's one less thing we need to worry about. This switching is called a *context switch*.

All threads will be treated fairly equally, but they *will* get switched around from time to time. It's something that you need to be aware of, and even write your code in a way that can handle it. If you fail to do it correctly, you can end up with strange intermittent errors that only appear to happen on Tuesdays where you ate tacos for lunch and the moon is waxing. We'll talk more about dealing with this in the section about Thread Safety.

There's a lot that goes into threading, and we simply don't have enough time to discuss all of the ins and outs of it here. So we won't. Instead, we'll take a look at the basics of threading, and I'll allow you to dig further into threading as you need it.

Threading Code Basics

We'll start out with a very simple example of how to start a new thread. To get started, let's say we have some work we want to do on a separate thread. This can really be anything, but let's say it is this:

```
public static void CountTo100()
{
    for (int index = 0; index < 100; index++)
        Console.WriteLine(index + 1);
}
```

To run this in a separate thread, you will need to create a new thread, tell it what method to run, and then start it. The following code does this:

```
Thread thread = new Thread(CountTo100);
thread.Start();
```

You'll also need to add a new **using** directive to get access to the **Thread** class (**using System.Threading;**) like we discussed back in Chapter 26.

The first line creates a new **Thread** object, but take a look at that constructor. We've passed in the method that we want it to run as a parameter. This is using a delegate. (Another good use of delegates!) This happens to be the **ThreadStart** delegate, which has a **void** return type and no parameters, so the method we use needs to match that.

After we call the **Start** method, a new thread will be created and will start executing the method we told it to run (**CountTo100**). At that point, we'd have two threads: the original thread, and the new one that is off running in **CountTo100**.

To have the original thread wait at some point for this new thread to terminate, we use the **Join** method:

```
thread.Join();
```

When a thread runs into this statement, it freezes and waits there until the other thread finishes up, effectively joining the execution of the two threads.

You can create as many threads as you want. As I mentioned earlier, they'll all get their fair share of the total processor time. (Though you don't want to have *too* many threads, because it takes time to create them all, and more threads means more frequent context switches, which take time.)

For example, here is some code that runs two threads and waits for them to finish:

```
using System;
using System.Collections.Generic;
using System.Linq;
using System.Text;
using System.Threading;
using System.Threading.Tasks;

namespace Threading
{
    class Program
    {
        static void Main(string[] args)
        {
            Thread thread1 = new Thread(CountTo100);
            thread1.Start();

            Thread thread2 = new Thread(CountTo100);
            thread2.Start();

            thread1.Join();
            thread2.Join();

            Console.ReadKey();
        }

        public static void CountTo100()
        {
            for (int index = 0; index < 100; index++)
                Console.WriteLine(index + 1);
        }
    }
}
```

Try running this code. There are two threads that are both going to print out the numbers 1 through 100. Typically, you'll see one thread active for a little while, and then the other active. So you might see the numbers 1-24 printed out, and then that thread will stop and the other thread will print out, say, 1-52, and then the first one will print out 25-87, and so on, until they both finish up.

You'll never get the exact same output the second time around. Things will be a little bit different because the operating system is doing context switches as it sees fit. That's an important point to remember about threads: the timing and ordering is unpredictable. This can be a big pain when you're trying to debug a problem in a multi-threaded program, and it is worth considering before you decide to run things on separate threads.

One other thing that I should point out is that you can make the current thread "sleep" for a certain amount of time. This is useful when you know one thread needs to wait a while for another thread to do something. To make the current thread sleep, you can call the static **Sleep** method in the **Thread** class:

```
Thread.Sleep(1000);
```

The parameter that you pass in is the amount of time to sleep for, measured in milliseconds. So 1000 is 1 second. The code above will cause the current thread to stop execution for one second, while other threads have a chance to run.

Using ParameterizedThreadStart

In the code we were using above, we used the **ThreadStart** delegate. This meant the method we ran needed to have no parameters, which meant that we couldn't pass any information to the method we were calling, which is somewhat limited.

There is an alternative that allows us to use methods that take a single parameter instead. To do this, we simply use the **ParameterizedThreadStart** delegate instead. This delegate has a return type of **void** and a single parameter of type **object** (the base class for everything). Because the parameter is of type **object**, we can basically use it for anything, as long as we cast it correctly to the right type.

For example, consider this variation of our **CountTo100** method, which counts to any number you want:

```
public static void CountToNumber(object input)
{
    int n = (int)input;
    for (int index = 0; index < n; index++)
        Console.WriteLine(index + 1);
}
```

This has a parameter of the type **object**, which we then cast to an **int**, which we use through the rest of our method.

Because this method matches the **ParameterizedThreadStart** delegate, we can create a new thread, with a reference to this method, and start it with a parameter, which gets passed along into our method:

```
Thread thread = new Thread(CountToNumber);
thread.Start(50);
```

It is really important to point out that while it may seem limiting to *only* have the option of passing in an **object**, this can be used to work for nearly anything. Whatever type you want it to be, you can cast to it. Furthermore, while the **ParameterizedThreadStart** doesn't allow you to return information, you could easily construct an object that has a property that can store what would have been returned.

To illustrate, below is a simple little program that will do division on a separate thread (yeah, it's overkill):

```
using System;
using System.Collections.Generic;
using System.Linq;
using System.Text;
using System.Threading;
```

```
using System.Threading.Tasks;

namespace Threading
{
    public class DivisionProblem
    {
        public double Dividend { get; set; } // the top
        public double Divisor { get; set; }  // the bottom
        public double Quotient { get; set; } // the result (normally would be returned)
    }

    class Program
    {
        static void Main(string[] args)
        {
            Thread thread = new Thread(Divide);

            DivisionProblem problem = new DivisionProblem();
            problem.Dividend = 8;
            problem.Divisor = 2;

            thread.Start(problem);
            thread.Join();

            Console.WriteLine("Result: " + problem.Quotient);

            Console.ReadKey();
        }

        public static void Divide(object input)
        {
            DivisionProblem problem = (DivisionProblem)input;
            problem.Quotient = problem.Dividend / problem.Divisor;
        }
    }
}
```

Try It Out!

Frog Racing. Let's make a little simulator for a frog race. The idea is that there are multiple frogs that are lined up and competing against each other to jump across a finish line. In order to finish the race, a frog will need to jump a total of 10 times. Each frog will run on its own thread, and we'll use three frogs total.

To do this, create a method that follows the **ParameterizedThreadStart** delegate. As input, the object that is passed in will be the frog's number. Inside of that method, use a loop to print out "Frog #X jumped" and then use **Thread.Sleep** and a **Random** object to have the frog/thread sleep for a random amount of time between 0 and 1 seconds. When the frog/thread has jumped ten times total, and the loop ends, print out "Frog #X finished." (See Chapter 16 for information about generating random numbers.)

Start the frog race by creating three separate threads and starting them all with different numbers. Wait for each thread to finish using the **Join** method.

Thread Safety

We've covered the basics of threading, but it is worth looking at another, more advanced threading topic. Not all situations need to be worried about this, but some do, so it is worth knowing a little about.

In the examples that we've done so far, each thread has been working with its own data. But what if there was something that they had to share? Remember that threads get swapped out whenever the operating system decides, and a thread may actually be in the middle of something important when that context switch hits.

To help explain the concept of thread safety, I'm going to draw on a real-world example. Imagine you're in your car, driving down a road that has three lanes. You're on the outside lane (the black car), and another car is on the inside lane (the white car):

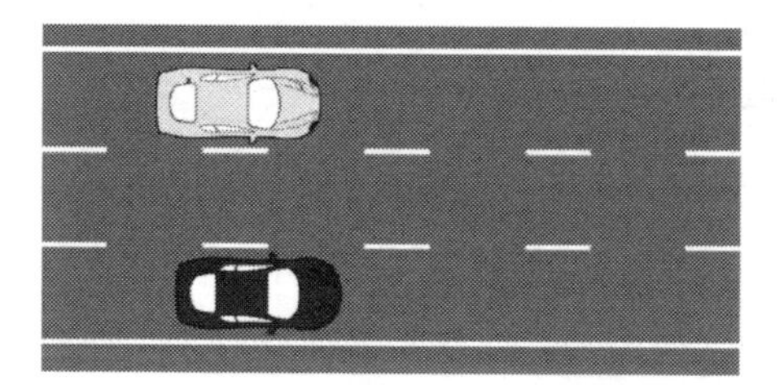

You want to change lanes to the middle. The normal process is that you look over into that other lane, make sure it is clear, and if so, you move over. But what if the white car was doing the same thing, without you knowing it? The other driver looked as well, saw that the lane was clear, and moved over. Unless one of you realizes the problem, you're headed for a crash.

Like with the cars, when you have multiple threads that are using any of the same resources, if you aren't careful, two threads can run into each other and cause problems.

Imagine even the simple problem of adding one to a variable. To do this, the computer will do the following steps:

- Read the current value from the variable.
- Do the math to figure out the result of adding 1 to the value.
- Store the new value back in the variable.

Now, imagine that two separate threads are both given the task of adding 1 to the variable. In a normal "happy" scenario, this variable should get incremented twice.

But let's see how things could go bad with multiple threads. Let's say our variable starts with a value of 1. In theory, both threads should increment the variable, and it should end up with a value of 3. *But* the following could happen as well:

- Thread #1 reads the current value from the variable (1).
- Thread #1 does the math to figure out the result of adding 1 to the value (2).
- Thread #2 reads the current value from the variable (which is still set to 1).
- Thread #2 does the math to figure out the result of adding 1 to the value (getting 2 also).
- Thread #2 stores the new value back into the variable (2 gets assigned back).
- Thread #1 stores the new value back into the variable (once again, 2 is assigned back).

Both threads did the work they were supposed to do, but because the two are running at the same time (or because of a context switch) things didn't turn out the way they should, and the variable is left with a value of 2, instead of 3 like it should be.

This example is kind of a toy problem, but the reality is that any time you have more than one thread modifying some specific data, you're likely to run into problems like this.

We want to be able to address this problem and make certain sections of code that we know might be problematic be *thread safe*, meaning that the code can prevent bad things like this from happening when multiple threads want to use it.

If we have a certain block of code that modifies data that could be accessed by other threads, we call that section a *critical section*. We will want to make it so that only one thread can get inside it at a time. This principle of only allowing one thread in at a time is called *mutual exclusion* and the mechanism that is used to enforce this is often called a *mutex*.

To enter a critical section, we require that the thread that is entering the section acquire the mutual exclusion lock for a specific object. When the thread has this lock, it can enter the critical section and execute the code. When it's done, it releases the lock, freeing it up for the next thread. If a thread shows up and tries to grab the lock that is already taken, the thread will be suspended until the lock is released.

Doing this in code is pretty simple. The first step is to create some sort of object that is accessible to anything that needs access to the critical section, which can act as the lock. Often, this is best done with a private instance variable in the same object or class as the data that is being modified:

```
private object threadLock = new object(); // Nothing special needed. Any object will do.
```

You could alternatively use a static variable instead of an instance variable if you need thread safety across all instances of the class.

To actually make a block of code thread safe, you simply add a block surrounding the critical section with the **lock** keyword:

```
lock(threadLock)
{
    // Code in here is now thread safe. Only one thread at a time can be in here.
    // Everyone else will have to wait at the "lock" statement.
}
```

That's the basics of threading in C#. There is a lot that goes on in multi-threaded applications, and there are entire books dedicated to the subject. We can only cover the basics here, but it should give you an idea of the fundamentals.

Asynchronous Programming

In a Nutshell

- Asynchronous programming is where you take a particular task or chunk of code and run it on a separate thread, outside of the main flow of execution.
- C# has used many ways of achieving asynchronous programming in the past.
- The Task-based Asynchronous Pattern (TAP) uses the **Task** and **Task<TResult>** class to represent a chunk of work that is running asynchronously.
- Tasks can be awaited with the **await** keyword (if the method is marked with **async**), allowing you simple syntax for scheduling additional work that happens once the task has been completed: **HighScores = await highScoreManager.LookupScores();**

In this chapter, we're going to continue what we started in the previous chapter, and introduce the best way to do most asynchronous things in C#. This uses the **async** and **await** keywords that were introduced into the C# language in C# 5.0.

Asynchronous programming is hard. There's not enough space in this chapter to cover every single detail and strange corner case. Rather, the goal of this chapter is to introduce you to the syntax surrounding calling code asynchronously, and get you to a point where it's not a deep, dark mystery, but something you feel comfortable with doing, while leaving some of the finer points of asynchronous programming for a later point in time.

In this chapter, we'll start by defining what asynchronous programming is and why you might use it. Then we'll take a look at how asynchronous programming has evolved in C# over time. We'll finally introduce the Task-based Asynchronous Pattern and the **async** and **await** keywords, which serve as the basis of modern asynchronous programming in C#.

What is Asynchronous Programming?

Asynchronous programming means taking a piece of code and running it on a separate thread, freeing up the original thread to continue on and do other things while the task completes.

The typical use case is when you start something that you know is going to take a while to happen. "A while" is measured in computer time, so we may be talking only a few milliseconds (or not). Some common scenarios include making a request to a server or a website, making requests to the file system, while running a particularly large or long running algorithm that you're using, or any other time where you know a particular piece of code is going to take some work to get done.

Modern computers have a lot of computational power, and users are expecting responsive user interfaces. Those two combined mean asynchronous programming is becoming more and more necessary and important to know.

The opposite of asynchronous programming (sometimes called "asynchrony") is synchronous programming, and it's what we've been doing up until now.

Approaches from the Early Days

There have been lots of approaches to doing asynchronous programming in C#'s past. While no longer recommended or preferred, those older approaches are worth mention because they illustrate the complexity of the problem, the beauty of the final solution, and the decade long journey it took to get there.

Throughout this section of this chapter, we'll stick with a recurring example of getting a list of high scores for a game from a web server. This will help us see the trade-offs of the various approaches that we'll see.

A Synchronous Approach

Let's start by looking at how we might do this task (looking up high scores) using a traditional synchronous approach. That means blocking and waiting for the results to come back before moving on. Obviously, this defeats the purpose of what this chapter is talking about, but it is a worthwhile exercise anyway. This biggest thing we'll gain from this is an example of what "ideal" code looks like here. Because a synchronous request is so straightforward, the methods we call are clean and simple.

To do our high score lookup in a synchronous fashion, we might use code like this:

```
Score[] highScores = highScoreManager.LookupScores();
```

That's relatively straightforward. We just call the method directly, which returns the results upon completion, and we store them in an array.

I'm skipping the actual implementation of the **LookupScores** method. I know that is going to be a disappointment to some people, but it's irrelevant to the current discussion. The point is, it doesn't matter what the work is, just that it's some sort of long running task that we really don't want to sit around waiting for. (But if you're dying of curiosity, one possible way to do this might be to do an HTTP request using C#'s **WebClient** class, which will return the results in a text stream of some sort, which could then be parsed and turned into a list of scores.)

While that code is short and easy to understand, it's not asynchronous.

Creating Threads Directly

Having seen a synchronous version of our high scores lookup problem, let's turn our attention to doing this in an asynchronous way.

Our first option is to create a thread like we did in the last chapter. It's using the most basic building block (starting a thread directly) but it certainly fits the definition of asynchronous programming.

Based on what we learned before, this might look something like this:

```
Thread thread = new Thread(() =>
{
    Score[] highScores = highScoreManager.LookupScores();
     // Do something with the results.
    HighScores = highScores;
});

thread.Start();
```

That code isn't so terrible. It gets the job done for sure. But as we'll soon see, there are better ways still of structuring this code.

Using the ThreadPool

While directly creating a thread was our approach in the last chapter, it has long been discouraged. One obvious problem with this approach has been that it's pretty easy to end up with so many threads that you drown the CPU. You spend far too much of your time switching between threads, and far too little doing real work.

C# has the concept of a thread "pool", wrapped up in the **ThreadPool** class. The idea here is that the .NET Framework and the Base Class Library can provide a collection of threads that already exist (no wasted time in creating them yourself) that are reused for various tasks. The **ThreadPool** class has the smarts to maintain the optimal number of threads.

When you have a chunk of work that you need done asynchronously, you can just hand it off to the thread pool to do. This is done by using the static **QueueUserWorkItem** method:

```
ThreadPool.QueueUserWorkItem(data =>
{
    Score[] highScores = highScoreManager.LookupScores();

    // Do something with the results.
    HighScores = highScores;
}, null);
```

This is pretty close to what we had in the previous iteration. This code uses a lambda statement. We could have used a named method instead, but this type of thing is a perfect use for lambda statements, so I went with that approach here in the sample code as well.

This gets us over the problem of not knowing when we should create new threads and when we should reuse an old thread, and we were also able to ditch the **Thread.Start** call.

You'll notice with this approach that the method required a parameter (**data**), which I filled in with **null**. This is more along the lines of the **ParameterizedThreadStart** that we talked about in the last chapter.

At any rate, this is a bit simpler than what we had before. Using the thread pool is preferable to not using it, but it's also still not our final solution. So let's keep looking.

Event-Based Asynchrony

The idea with the event-based asynchronous approach is that you call a method that is known to take a long time and it returns immediately. When the method completes asynchronously, an event on the class is raised.

Using Event-based asynchrony like this, we have to subscribe to the right event, and in most cases, the asynchronous method will end with the word **Async**:

```
HighScoreManager highScoreManager = new HighScoreManager();
highScoreManager.HighScoresLoaded += OnHighScoresLoaded;
highScoreManager.LookupScoresAsync();
```

And elsewhere we define our event handler method:

```
private void OnHighScoresLoaded(Score[] highScores)
{
    // Do something with the results.
    HighScores = highScores;
}
```

In the past, this has been a rather common approach to long running tasks. As you explore the Base Class Library C# offers, you'll inevitably see methods that end with **Async** like this, which expect you to subscribe to a related event, and put your code for dealing with the results in the event handler.

While it works, it's definitely a little annoying that you've now separated your code into two separate pieces. You have the part that sets up and calls the method, and then in a different location, you have the method for dealing with the results. You'll see this same problem in the next few iterations as well.

A second problem we introduce with this is the event subscription. The second time we make this request, it's easy to forget that we may already have subscribed to the event (with the **+=**) and we may accidentally subscribe again. We can work around that but it does create more things you have to remember to do as a developer.

The AsyncResult Pattern

Moving along in our tour of asynchrony, our next option is the AsyncResult pattern. This approach is similar to various fast food restaurants and other places where you make your order or request, and you're given a ticket of some sort. When it's done, you can return the ticket to get your order (the results). Or you can take your ticket and stand at the counter waiting impatiently.

Using this pattern, you typically start with a synchronous method (let's call it **X**). You then add a **BeginX** method and **EndX** method to the mix. The **Begin** method starts the synchronous method running in an asynchronous way and returns your "ticket" in the form of an object that implements the **IAsyncResult** interface. You can use this object to periodically check if the asynchronous task is completed or not, or you can call the **End** method, passing in the **IAsyncResult** that you got from calling the **Begin** method, which will block, and cause the thread to do nothing else until the results come back.

This sometimes shows up in the form of three distinct named methods (**X**, **BeginX**, and **EndX**) but the easiest way to implement this is by using some functionality on the **Delegate** class:

```
HighScoreManager highScoreManager = new HighScoreManager();
Func<Score[]> scoreLookupDelegate = highScoreManager.LookupScores; // Synchronous version.

IAsyncResult asyncResult = scoreLookupDelegate.BeginInvoke(null, null);

HighScores = scoreLookupDelegate.EndInvoke(asyncResult);
```

In that second line, we store the **LookupScores** method in a delegate type. Remember from the chapter on delegates (Chapter 30) that we can usually get away with some variation of **Action** or **Func**, and that's what we've done in this case. (**Func<Score[]>** is a delegate for any method that has no parameters and returns an array of **Score**.)

Once stored in the delegate, we can start off the process by calling **BeginInvoke**. The two **null** values are for a callback method that should be called when the asynchronous task is completed, and a parameter that can be passed into the callback method.

If the method we were trying to make asynchronous had any parameters, we'd be able to list them in **BeginInvoke** before those two parameters.

In the last line of that code, we stop and wait for the task to complete with **EndInvoke**. That obviously defeats the purpose of making it asynchronous in the first place, but in a more realistic scenario, you wouldn't call **EndInvoke** immediately. Instead, you'd continue on to other things, and you could use the **IAsyncResult** that comes back to keep track of the asynchronous task's progress.

The best part about this delegate approach is that you can use it on any method. There are no limitations on it. It's a pattern that's infinitely reusable.

Unlike the event-based version, you don't have to worry about subscribing and unsubscribing correctly. There's no way to make that mistake with this approach.

At this point, you might be starting to feel like this async stuff just seems to be getting uglier and uglier, and more and more complicated. I definitely think that's the case so far. But let's keep going. It gets better from here.

Callback Methods

One of the cleanest ways we could structure an asynchronous call is to have our method run asynchronously, but include a parameter that allows the programmer to supply a method that will be called on completion. The method you pass in (in the form of a delegate) is called a callback method.

```
public void LookupScores(Action<Score[]> callbackMethod)
```

This can be called like this, using a lambda statement:

```
highScoreManager.LookupScores(scores => { HighScores = scores; });
```

We used a lambda here, but it could have been a normal named method as well.

This code is actually fairly readable. It's not quite as good as the original synchronous version, but it's the cleanest version we've seen so far. It probably seems strange that **LookupScores** has a parameter that's a delegate, until you realize that this is a callback method. We're definitely making progress though.

The Task-based Asynchronous Pattern

As you can see, asynchronous programming has been a mess in C# in the past. And it's not just C#. Every language and every programmer has struggled with this. Obviously we need a better solution that what we've seen so far.

C# 4.0 introduced yet another pattern for solving this problem: the Task-based Asynchronous Pattern (TAP). TAP doesn't magically solve our code clarity problems, but it lays the foundation for what came in C# 5.0, which does (for the most part).

TAP introduced two new classes: **Task** and **Task<TResult>**. **Task** represents a long running asynchronous chunk of work, while **Task<TResult>** is a generic variant that additionally serves as a promise of a resulting value when the task is completed. **Task** is used when the asynchronous task doesn't return a value or when you don't care to do anything with the returned value. When you care about the result, you would use the generic version, which has the option of grabbing the result from the task upon completion and doing something more with it.

Let's look at how we'd implement our high score lookup using the TAP approach:

```
public Task<Score[]> LookupScores()
{
    return Task.Run(() => {
        // Do the real work here, in a synchronous fashion.
        return new Score[10];
    });
}
```

Instead of directly returning the score array, we return a **Task** that contains it. More accurately, we return a task that promises to give us an array of scores when it completes. In this case, because we want to do something with the result, we want to use the generic version of **Task**, which promises a resulting value upon completion.

Outside of this class, we can handle a **Task** return value in a few different ways. The first way would be to just ignore it. If it were an asynchronous task that is "fire and forget," we could just ignore the return value.

The second option is to take the task and call **Wait()** on it:

```
Task<Score[]> scoresTask = highScoreManager.LookupScores();
scoresTask.Wait();
HighScores = scoresTask.Result;
```

Wait causes the current thread to block, doing nothing else until the task finishes. This isn't the desired effect because it forces the main thread to suspend. It's not happening asynchronously anymore.

Which leads us to the third option: **ContinueWith**.

```
Task<Score[]> scoresTask = highScoreManager.LookupScores();
scoresTask.ContinueWith(task => HighScores = task.Result);
```

ContinueWith specifies a second method (another lambda expression in this case) that should execute when the task finishes. That code doesn't happen when the flow of execution reaches that line in the code. Rather, the continuation is only scheduled then, and executed when the task actually completes at some later point in time.

The 'async' and 'await' Keywords

We're now ready for the latest and greatest in asynchronous programming in C#. Building off of the TAP pattern, C# 5.0 made this far simpler and easy to read by integrating it into the language itself. It does this by adding two keywords to the language: **async** and **await**.

The **Task** and **Task<TResult>** classes still function as the core building block of asynchronous programming with these new keywords. You'll still create a new **Task** or **Task<TResult>** using **Task.Run**, like in the previous section. The real difference happens when you get a **Task** returned to you from a method, where you can do something when the task completes or do something with the result the task gives you upon completion:

```
Score[] highScores = await highScoreManager.LookupScores();
```

Before looking at what that code actually does, let's compare it to the synchronous version, right at the start of this chapter. Remember this guy?

```
Score[] highScores = highScoreManager.LookupScores();
```

It's *surprisingly* similar. And yet, it is done asynchronously. While previous approaches to asynchronous programming have been convoluted, using **Task** or **Task<TResult>**, combined with the new **await** keyword, everything seems to fall in place and we get a clean solution.

So what does that **await** keyword do exactly?

Well let's start by rounding that piece of code out a little bit. I'm going to put some **Console.WriteLine** statements in there and wrap it in a method:

```
public async void GrabHighScores()
{
    // Preliminary work...
    Console.WriteLine("Initializing asynchronous lookup of high scores.");

    // Start something asynchronously.
    Score[] highScores = await highScoreManager.LookupScores();

    // Work to be done after completion of the asynchronous task.
    Console.WriteLine("Completed asynchronous lookup of high scores.");
}
```

Let's start there in the middle, with the **await** keyword, since it's the central point of this whole discussion. That **LookupScores** method returns a **Task<Score[]>**—an uncompleted task with the promise of having a **Score[]** when finished. Any time a method returns a task, you can optionally "await" it with the **await** keyword, which creates a situation similar to the **ContinueWith** that we saw in the previous section.

The preliminary code before the **await** happens immediately, by the thread that called the method.

Once an **await** is reached, the thread that called the method jumps out of the method and continues merrily on its way to something else. The asynchronous code will either happen on a separate thread (if you use **Task.Run**) or perhaps by no thread at all (if it's waiting for the network or file system to complete some work).

But everything after the **await** is scheduled to happen when the task finishes. The **await** effectively slices the method in two while still using syntax that is easy to understand.

You'll notice from the previous code sample that the **await** keyword also extracts the value out of the **Task<Score[]>**. We don't have to call **Task.Result** to get it. If a method returns a generic **Task<TResult>**, then the **await** keyword will do that for you. In other words, it extracts the promised value out of the task upon completion.

If a method returns just a **Task**, the non-generic version which doesn't have a promise value, you can still await it. It just has to be structured as a statement, rather than an assignment.

```
Console.WriteLine("Before await.");
await SomeMethodThatReturnsTask();
Console.WriteLine("After await.");
```

The **await** keyword can't be used just anywhere. It can only be called inside a method that is marked with the **async** keyword. You can see that I added that in the earlier code:

```
public async void GrabHighScores()
```

The **async** keyword can be attached to a method that has a return type of **void**, **Task**, or **Task<TResult>**.

- An **async** method with a **void** return type implies a fire-and-forget nature; you don't care when it finishes, just that it happens asynchronously.
- An **async** method with a **Task** for the return type implies you care about when it finishes, including scheduling stuff after the task completes (with **await**) but that the task itself doesn't return any meaningful data.
- An **async** method that returns the generic **Task<TResult>** implies the task returns a value (it promises a value upon completion) and allows you to grab it when it's done.

Asynchronous programming can be tough. This chapter has hopefully demystified it to a large extent, but there are lots of additional corner cases that we haven't even begun to consider. (For example, how do you best handle exceptions when dealing with asynchronous programming?) Becoming an expert in asynchronous programming takes time and practice. But hopefully, we've gotten that particular adventure off on the right foot.

39

Other Features in C#

In a Nutshell

- This chapter covers a large collection of random advanced topics. The purpose of this chapter is to ensure that you know they exist.
- **yield return** produces an enumerator without creating a container like a **List**.
- **const** and **readonly** define compile-time and runtime constants that can't be changed.
- The **unsafe** keyword lets you work with pointers and interoperate with non-.NET code.
- Attributes let you apply metadata to types and their members.
- The **nameof** operator gets a string representation of a type or member.
- The bit shift operators let you play around with individual bits in your data.
- Reflection allows you to inspect code while your program is running.
- **IDisposable** lets you run custom logic on an object before being garbage collected.
- C# defines a variety of preprocessor directives to give instructions to the compiler.
- C# can often automatically deduce the type of local variables (**var x = 1;**)
- Nullable types allow value types to take on a value of **null**.
- You can read in command line arguments from the command prompt.
- You can make your own user-defined conversions between types.

In this chapter, we're going to cruise through a variety of features and tricks that we haven't talked about yet. Some of these are very useful but aren't big enough for their own chapter. Others are less useful, so we'll just cover their basics, and let you to explore it in depth if you feel the need.

There is a lot of stuff in this chapter and I don't expect you to become a master of it all overnight. Some of these topics are so big that it you could write books about it. Many of these topics will only apply to some people, but not everyone.

There are two real purposes to this chapter. One is to open your eyes to the possibilities with C#. The other is to make it so that when you see these things, instead of being completely blindsided by it, you'll at least be able to say, "Hey, I remember reading something about this!"

You don't necessarily *need* the stuff in this chapter to write C# programs. Everything we've covered up until now will go a very long way. This chapter will help tie together some of the loose ends.

Iterators and the Yield Keyword

Way back in Chapter 13, we first introduced **foreach** loops. In Chapter 24 we introduced the **IEnumerable<T>** interface, and stated that anything that implements **IEnumerable<T>** (or the non-generic variant **IEnumerable**) can be used in a **foreach** loop. Remember, an **IEnumerable** is simply anything where you can examine multiple values in a container or collection, one at a time. This capability makes types that implement this interface *iterators*.

In addition to a straight up simple **IEnumerable** implementation, you can additionally define an iterator using the **yield** keyword. This would look something like this:

```
class IteratorExample : IEnumerable<int>
{
    public IEnumerator<int> GetEnumerator()
    {
        for (int index = 0; index < 10; index++)
            yield return index;
    }

    IEnumerator IEnumerable.GetEnumerator()
    {
        return GetEnumerator();
    }
}
```

Much of this code should make sense by now. We're simply implementing the **IEnumerable<int>** interface, which requires us to have the **GetEnumerator** method that you see. And since the **IEnumerable<T>** interface is derived from the non-generic **IEnumerable** interface, we also need to define that second method, which just simply calls the first method.

The part that will likely seem strange to you is the **yield return**. As we're iterating, the method will be called multiple times, and each time, it will return the next item that is "yielded." Unlike a normal **return** statement, when we use **yield return**, we're saying, "pause what you're doing here and return this value, but when you come back, start up again here." That is very powerful. For example:

```
public IEnumerator<int> GetEnumerator()
{
    yield return 0;
    yield return 1;
    yield return 4;
    yield return 6;
}
```

Interestingly, the compiler isn't simply turning all yielded values into a list. It is happening one at a time. This means you could theoretically create an iterator that never ends:

```
public IEnumerator<int> GetEnumerator()
{
    int number = 0;
    while(true)
    {
        yield return number;
        number++;
```

```
    }
}
```

The calling code could keep calling the iterator forever, or stop when it has completed its job.

The **yield return** syntax cannot be used anywhere, just inside of methods that **return IEnumerator<T>** or (as we'll see in a second) **IEnumerable<T>** and their non-generic counterparts.

Named Iterators

There's a second way to do iterators. Instead of implementing the **IEnumerable<T>** interface and adding a **GetEnumerator** method, you can create a method that returns an **IEnumerable<T>.** You can then use this method in a **foreach** loop whenever you want. This approach lets you have multiple iterators for a single data collection class, and it lets you have parameters for those methods:

```
class Counter
{
    public IEnumerable<int> CountUp(int start, int end)
    {
        int current = start;
        while (current < end)
        {
            yield return current;
            current++;
        }
    }

    public IEnumerable<int> CountDown(int start, int end)
    {
        int current = start;
        while (current > end)
        {
            yield return current;
            current--;
        }
    }
}
```

You then call these iterators like this:

```
Counter counter = new Counter();

foreach (int number in counter.CountUp(5, 50))
    Console.WriteLine(number);

foreach (int number in counter.CountDown(100, 90))
    Console.WriteLine(number);
```

Constants

We've spent a lot of time talking about variables. True to their name, the contents of a variable can... well... vary. But there are times where we want to assign a value to something and prevent it from ever changing again. C# provides two ways of addressing this: the **const** and **readonly** keywords.

If you mark a variable with either of these, then you will not be allowed to change the value of the variable later on. There's a subtle difference between the two that is worth outlining here. Let's start by taking a look at the **const** keyword. In this book, we've used **Math.PI** several times. If we were to write our own **Math.PI** variable, we'd use the **const** keyword, and it would look something like this:

```
public const double PI = 3.1415926;
```

Adding in **const** tells the compiler that we're assigning it a value and it will never change. In fact, if we try to assign a value to it at some point in our program, we'll get a compiler error.

If you have a value that you know will never change, it is a good idea to make it into a constant by adding the **const** keyword to it. That way, no one will mess it up on accident.

It is worth pointing out that anything marked with **const** is automatically treated as though it were **static**. Because this is assumed (and required) you don't need to (and can't) make it **static** yourself.

The **readonly** keyword is similar to this, but has an important difference. Anything that is marked **const** must be assigned a value right up front, when the program is being compiled. As such, they are often called *compile-time constants*. Things that are marked with **readonly** can be assigned to in a constructor or at the same place it is declared, and not after. These are called *runtime constants*.

You can assign a value to a **readonly** field either when you declare the variable (like with compile-time constants) or also, in a constructor of the class. Which brings us to the key distinction: run-time constants can be different in every instance of a class.

For example, we could create a **Point** class that stores an x- and y-coordinate in such a way that once it has been created, it is impossible to change the values.

```
public class Point
{
    private readonly double x;
    private readonly double y;

    public double X { get { return x; } } // We could have also just used a readonly property
    public double Y { get { return y; } } // for these, but this illustrates the point well.

    public Point(double x, double y)
    {
        this.x = x;
        this.y = y;
    }
}
```

We can then use this class, creating two point with different values, but both are unchangeable:

```
Point p1 = new Point(5, -2);
Point p2 = new Point(-3, 7);
```

This, by the way, is an excellent way of building immutable types like we discussed in Chapter 20.

Unsafe Code

In C#, all memory is managed for us, which as we've discussed in other chapters is a huge help. Earlier languages like C and C++ didn't have this feature. As nice as it is though, there are a few very rare cases where we want to jump outside of the safety bounds and work in an unmanaged arena.

I'm not going to go into a lot of detail on this, since it opens up a whole can of worms that can't effectively be addressed here. It is enough to know that if you want to do this, and free yourself of both the benefits and constraints of managed code, you can do so. This allows you to work with pointers directly, using a C/C++ inspired syntax. Starting a block of code that is "unsafe" is simple:

```
public void DoSomethingUnsafe()
{
    unsafe
    {
        // You can now do unsafe stuff here.
    }
}
```

Alternatively, if an entire method is going to be unsafe, you can simply add the **unsafe** keyword to the method declaration:

```
public unsafe void DoSomethingUnsafe()
{
}
```

The name "unsafe" is not the best name. Unsafe code doesn't mean the code is dangerous, necessarily, it just means that the code is *unverifiable*—the compiler can't guarantee safety.

In an unsafe context (and only in an unsafe context) you can also create *pointer types*, in addition to value and reference types. Pointer types are a little like reference types, except that they store the actual address of the variable. They aren't managed references. You can create a pointer to any of the numeric types, **bool**, enumerations, any pointer types (pointers to pointers) and any structs that you've made, as long as they don't contain references. Pointer types are not managed, and they are not garbage collected. They exist almost solely for interoperability with code that is running outside of the .NET Framework. You declare a pointer type with a ***** by the type:

```
int* p; // A pointer to an integer.
```

There's actually quite a bit more involved with pointer types than just this, and if you're going to use them, you'll need to do some additional learning. I bring them up only so that you know they exist. In reality, it will be pretty rare to need or even want unsafe code and pointer types, it is worth knowing it's an option. This brings us to the final version of our C# type hierarchy chart:

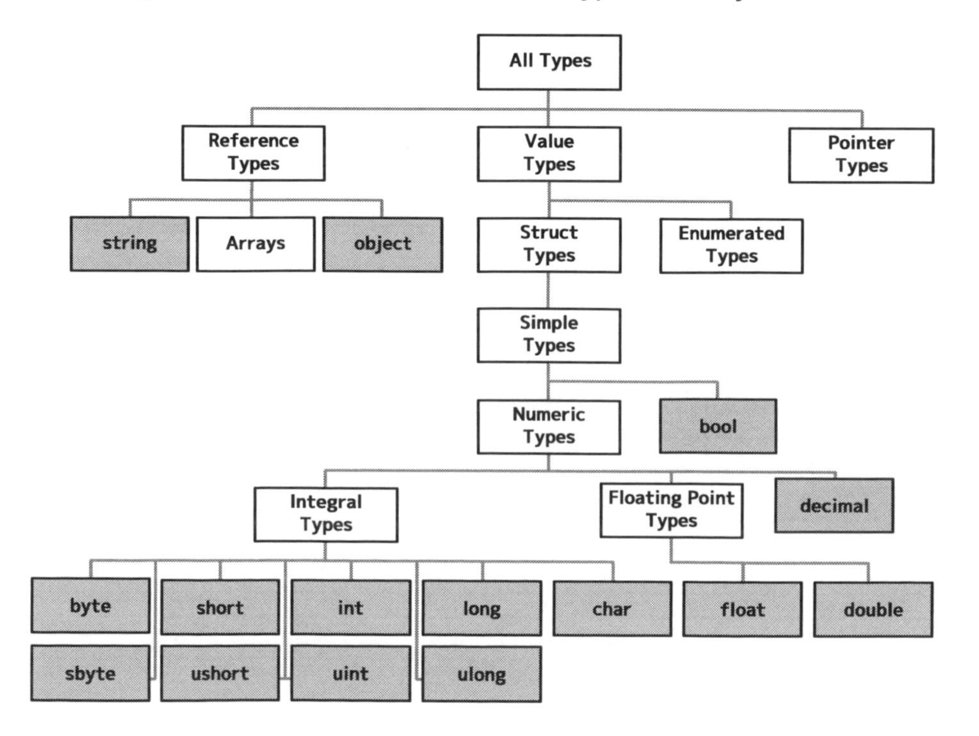

Attributes

Attributes in C# provide a way for you to add metadata to an element of your code. They can be added to types that you create or their members. They can even be applied to a method's parameters and return types.

These attributes can be detected by the compiler, external code tools like unit testing frameworks, or even detected while your code is running, using reflection. (We'll talk about reflection in a second.)

There are hundreds of types of attributes that come with the .NET Framework, and we can't cover them all. In the various places where you may need to use them, you can learn about the specific attributes that apply as you go.

But to illustrate the basics of how attributes are used, we'll look at one specific but simple attribute called the **Obsolete** attribute. You can use this to mark classes and methods that are out of date, and should no longer be used. To apply an attribute to a method, you simply put the attribute name in square brackets (**[** and **]**) above the element:

```
[Obsolete]
private void OldDeadMethod()
{
}
```

Depending on what namespace the attribute lives in, they may require an additional **using** directive.

The compiler uses this attribute to detect calls to obsolete methods and warns the programmer about it. If you use a method with this attribute on it, you'll see a warning in the Error List telling you so.

Many attributes have certain parameters that you can set as well, including the **Obsolete** attribute:

```
[Obsolete("Use NewDeadMethod instead.", true)]
private void OldDeadMethod()
{
}
```

In this case, the first parameter is text that will be displayed in the warning message, and the second indicates whether the compiler should treat the problem as an error rather than a warning.

Attributes are simply classes that are derived from the **Attribute** class, and you can make your own. When we use an attribute, we're essentially calling a constructor for that class.

Multiple attributes can be applied to an element:

```
[Attribute1]
[Attribute2]
[Attribute3]
public class MagicClass
{
}
```

In most cases, when creating an attribute, people put **Attribute** at the end of the name for their attribute (e.g. **ObsoleteAttribute**). If you do this, when you use the attribute you can use either **[Obsolete]** or the longer **[ObsoleteAttribute]**.

For specific instructions on how to create your own attributes, see: **http://msdn.microsoft.com/en-us/library/84c42s56.aspx**.

The 'nameof' Operator

It's quite common to want to do something with the name of a property, method, type, or other bit of code. Among other things, this is useful in debugging and when you're doing data binding in a UI framework like WPF or Windows Forms, where property names are frequently used.

Let's consider a really simple example related to debugging, where we want to print out an object's properties and their values.

Let's say we've got a **Book** class similar to what we created in Chapter 18, modified slightly to include the properties we introduced in Chapter 19:

```
using System;
using System.Collections.Generic;
using System.Linq;
using System.Text;
using System.Threading.Tasks;

namespace CreatingClasses
{
    class Book
    {
        public string Title { get; set; }
        private string Author { get; set; }
        private int Pages { get; set; }
        private int WordCount { get; set; }

        public Book(string title)
        {
            this.Title = title;
        }
    }
}
```

Let's say we want to override the **ToString** method to display the values of each of these properties:

```
public override string ToString()
{
    return $"[Book Title={Title} Author={Author} Pages={Pages} WordCount={WordCount}]";
}
```

This lets us see the details of a book either by writing it to the console window or in Visual Studio's debugger, revealing the values of all of the properties. It will appear like this:

```
[Book Title=There and Back Again Author=Bilbo Baggins Pages=118 WordCount=54386]
```

That's all fine and good, but what happens when we decide we want to change the name of something? For example, let's say we decide to rename **WordCount** to **Words** to mirror the **Pages** property. It's easy to refactor a property name like this (Ctrl + R, Ctrl + R), but in this case, our text remains unchanged. We'll still print out the same text, which is now misnamed.

This is where the **nameof** operator comes in. The **nameof** operator allows you to refer to a variable, type, or type member (like a method or property) and have the compiler change it into a string that matches the name.

For example, consider the following uses of the **nameof** operator:

```
Book book = new Book("There and Back Again") {
            Author = "Bilbo Baggins", Pages = 118, WordCount = 54386 };
Console.WriteLine(nameof(Book));
Console.WriteLine(nameof(book));
Console.WriteLine(nameof(Book.Pages));
```

This results in the following output:

```
Book
book
Pages
```

Look carefully at how that works. Microsoft has gone out of its way to make sure it works in an intuitive way. For a type, it gives you the name of the type without the namespace. For the name of a variable, it gives you the name of the variable. When you access a property like **Book.Pages**, you get the name of the property. That syntax is special because you normally can't access a class's properties like that unless they're static properties (which this isn't).

If we update our **ToString** override from earlier in this section to use the **nameof** operator, we can make it much more resilient to changes in our code:

```
public override string ToString()
{
    return $"[{nameof(Book)} {nameof(Title)}={Title} {nameof(Author)}={Author} " +
             $"{nameof(Pages)}={Pages} {nameof(WordCount)}={WordCount}]";
}
```

Now any changes to variable, property, or class names will be reflected correctly in this code.

Bit Fields

We usually work with variables at a relatively high level, thinking of them as containers for numbers, strings, or complicated classes. Behind the scenes, though, they are essentially little containers to store a pile of bits. Every type of data that we've talked about is represented as bits.

To illustrate, let's briefly look at how a **byte** stores its values. Remember, a **byte** can hold the values 0 to 255 in a single byte, with 8 bits. These values are stored using binary counting. So the number 0 is stored with the bits **00000000**, the number 1 is stored with the bits **00000001**, the number 2 is stored with the bits **00000010**, and so on.

There's a lot to know about counting in binary, and if you haven't had any exposure to it before, it is probably worth taking some time to learn how it works.

There are a few cases where we actually *want* to sort of fall back down to that low of a level, and work with individual bits and bytes. In fact, you may use this extensively in certain types of projects.

One of the popular ways of using the raw bits of data is a bit field. Let's start our discussion with a brief explanation of why people are even interested in bit fields. Let's say you're creating an online discussion board or a forum. You'll have lots of users, each with different permissions and abilities. Some people may be able to delete posts, edit posts, add comments, etc., while others may not.

Knowing what we've discussed before, we can easily imagine storing each of these values as a **bool**. And that gets the job done. But each **bool** variable takes up 1 full byte, and if you have dozens of settings per user, and millions of users, that adds up very quickly.

One popular solution to this is to have multiple Boolean values be packed into a single **byte**, or a single **int**. In one single byte, you can store eight Boolean values, with one in each bit. For each bit, a 1 represents a true value, and a 0 represents a false value. Something like this:

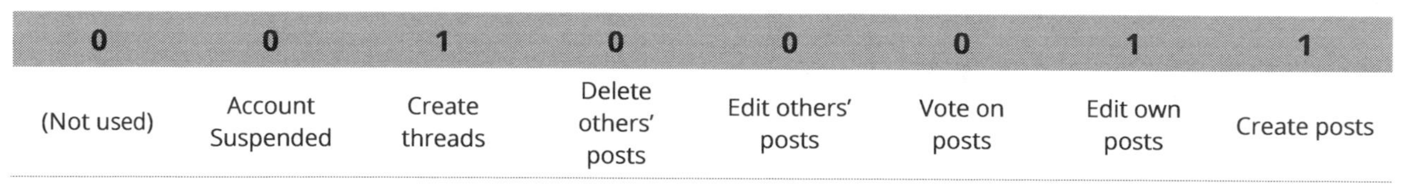

0	0	1	0	0	0	1	1
(Not used)	Account Suspended	Create threads	Delete others' posts	Edit others' posts	Vote on posts	Edit own posts	Create posts

Doing this kind of trick is actually very widespread, especially if you're talking about things like very large data sets, operating systems, and sending things across the network.

Even though we're treating our byte as a pile of Boolean values here, it is worth pointing out that as you're working with this byte, C# is going to want to present it to you as the **byte** type, or the **int** type with a specific, single value for the entire collection. The bit field above will be shown to you as the number 35. Of course, since we don't care about it as a complete number, but as a collection of Boolean values or Boolean flags, we'll need some tools to help us work with the raw bits in a bit field. Fortunately, C# has those tools built in for us, which we'll now discuss briefly.

Bit Shift Operators

There are two operators that will take the bits in a **byte** or other integral type, and move them over a certain number of slots. These two operators are called bit shift operators. The left bit shift operator is two less than signs together (**<<**) while the right bit shift operator is two greater than signs together (**>>**). These operators look like little arrows that tell you which way the bits will be shifted.

To use the bit shift operators, you would do something like this:

```
int result = 5 >> 2; // You'd usually use the contents of a variable, instead of '5'.
```

This takes the number five, which when written in binary, looks like **00000101**, and shifts all of the bits to the right by two. This results in **00000001**, which is 1. Notice that as things are shifted, some bits drop off one end, which just go away, and 0's are added to fill in empty spots on the other end.

The left bit shift operator works the same way:

```
int result = 5 << 2; // 00000101 turns into 00010100, or 20.
```

Bitwise Logical Operators

Think back to when we first discussed the logical **&&** and **||** operators. Remember that **&&** evaluates to true if and only if both sides are true, and **||** evaluates to true if either side is true. There are a couple of related operators that work on bits themselves, and do a similar thing. As we look at this, think of 1 as being true, and 0 as being false.

The bitwise *and* operator is a single ampersand (**&**) and the bitwise *or* operator is a single vertical bar (**|**). These operators go down the bits, one at a time, and do an *and* or *or* operation on each bit. This will probably make more sense with an example. Imagine the two binary numbers:

```
01010101
11111111
```

If you use the bitwise *and* operator, the program will look at the first bit in each number. The top number has a 0, while the bottom one has a 1. That's like having a false and a true, which if they are combined together with *and* result in false, or a 0.

```
01010101
11111111
0
```

Then you look at the next value in each number. They are both 1, which is like having a true and a true, which when combined with *and*, would result in true, or a 1.

```
01010101
11111111
01
```

You continue down the entire sequence, doing this for each bit:

```
01010101
11111111
01010101
```

The bitwise *or* operator does the same thing, but combines the two numbers with *or* instead.

In code, these two operators look like this:

```
int a = 5; // 00000101
int b = 3; // 00000011
int combinedWithAnd = a & b; // results in 00000001, or 1.
int combinedWithOr = a | b;  // results in 00000111, or 7.
```

There's a third bitwise operator that is the *exclusive or* operator. (This is sometimes called *xor*, pronounced "ex-or.") The "normal" *or* operator returns true if either of the two parts are true. This still applies where both are true. The *exclusive or* operator is only true if *exactly one* is true. If they are both true, then it evaluates to false. The exclusive or operator is the caret symbol (^).

Continuing on with the last code snippet, this would look like this:

```
int combinedWithXor = a ^ b; // results in 00000110, or 6.
```

One final, similar operator is the bitwise complement operator (~), which is a little like the **!** operator we saw when we first looked at logical operators. This takes each bit and changes it to the opposite:

```
int number = 200; // 11001000
int bitwiseComplement = ~number; // 00110111, or 55.
```

There is a compound assignment operator that you can use for all of these:

```
int number = 8;
number <<= 1;    // equivalent of number = number << 1;
number >>= 1;    // number = number >> 1;
number &= 32;    // number = number & 32;
number |= 32;    // number = number | 32;
number ^= 32;    // number = number ^ 32;
```

Enumeration Flags

While it is possible to work with the **byte**, **int**, or another type to accomplish what we've seen in this section, this kind of stuff is typically done with an enumeration in C#, to make the whole thing more readable. (Enumerations were introduced in Chapter 14.) Before we can use an enumeration for this, we must add the **Flags** attribute and assign the right values to each member of the enumeration:

```
[Flags]                          // Don't forget to add the Flags attribute.
public enum ForumPrivileges
```

```
{
    CreatePosts =       1 << 0, //  1 or 00000001
    EditOwnPosts =      1 << 1, //  2 or 00000010
    VoteOnPosts =       1 << 2, //  4 or 00000100
    EditOthersPosts =   1 << 3, //  8 or 00001000
    DeletePosts =       1 << 4, // 16 or 00010000
    CreateThreads =     1 << 5, // 32 or 00100000
    Suspended =         1 << 6, // 64 or 01000000

    // Note we can also add in "shortcuts" here:
    None = 0,
    BasicUser = CreatePosts | EditOwnPosts | VoteOnPosts,
    Administrator = BasicUser | EditOthersPosts | DeletePosts | CreateThreads
}
```

A Practical Example

Before you start thinking, "Well that's completely useless," let me show you why these are so useful. The **|** operator can be used to "turn on" a bit in the bit field:

```
ForumPrivileges privileges = ForumPrivileges.BasicUser;
privileges |= ForumPrivileges.Suspended; // Turn on the 'suspended' field
```

The **&** operator can be used to check if a particular flag is set, using some simple trickery:

```
bool isSuspended = (privileges & ForumPrivileges.Suspended) == ForumPrivileges.Suspended;
```

The trick that is going on here that if we do a bitwise and operation with something that only has one bit set to true, it will either become all 0's, which indicates that the field was not set, or if the field was set, we'll get back the value of the field we were checking for, turning all other fields to 0.

Using a combination of the **&** and the ~ operators, we can turn off a particular field:

```
privileges &= ~ForumPrivileges.Suspended;
```

You can also toggle a field using the ^ operator:

```
privileges ^= ForumPrivileges.DeletePosts;
```

Reflection

C# has the ability to inspect elements of executable code, and explore what types are in an assembly. They can see the methods, properties, and variables it contains, even while your program is running. The ability to have code analyze the structure of other code is called *reflection*.

The biggest use for reflection is when you want to look at an unknown assembly or object. Not every program has a use for this. Most won't. Doing this is always slower than directly accessing the code.

Having said that, there are still times where it is useful. To name a few examples, a unit testing framework might want to dig through an assembly to find any method that ends with the word "Test" or have a **Test** attribute applied to them, or a plugin system might want to find all classes in a DLL that implements a specific interface. Reflection can also be used to bend the rules, like calling a private method from outside the type it belongs to (a great way to shoot yourself in the foot).

The core class used in reflection is the **Type** class. The **Type** class represents a compiled type on the .NET Framework, including classes, structs, or enumerations.

For any given type, you can use the **typeof** keyword to get the **Type** object that represents it:

```
Type type = typeof(int);
Type typeOfClass = typeof(MyClass);
```

You can alternatively get a type from an object:

```
MyClass myObject = new MyClass();
Type type = myObject.GetType();
```

With a given type, you can figure out what constructors, methods, properties, or other things the type has defined. For example:

```
ConstructorInfo[] contructors = type.GetConstructors();
MethodInfo[] methods = type.GetMethods();
```

If you're looking for a member of the type with a particular name and parameter list, you can do that too:

```
ConstructorInfo constructor = type.GetConstructor(new Type[] { typeof(int) });
MethodInfo method = type.GetMethod("MethodName", new Type[] { typeof(int) });
```

In all of these cases, if what you're looking for doesn't exist, an empty array or **null** will be returned.

Once you've got the constructor, method, or whatever thing you asked for, you can execute that code with the **Invoke** method that they have. Doing so with constructors will return an object that was created by the constructor, and doing so with a method or property will return the result of the method (or **null** if the method's return type is **void**):

```
object newObject = constructor.Invoke(new object[] { 17 });
method.Invoke(newObject, new object[] { 4 });
```

Using Statements and the IDisposable Interface

Back in Chapter 16 when we first talked about the heap, we talked about garbage collection. One of the big things that C# and the .NET Framework do for us is get rid of memory that we're no longer using. Our program's memory is managed for us. Occasionally though, the task we're trying to accomplish requires using unmanaged memory and resources. When we do this, we can no longer count on garbage collection to clean up those resources and memory.

While there may be times where we create our own unmanaged memory, it is much more common to use a pre-existing type that uses unmanaged memory. As an example, when we open a file like we did back in Chapter 28, we need to access unmanaged memory to get to the file system. Back in that chapter, we simply used the **Close** method when we were done with a file, which freed any unmanaged memory or resources that were in use, but there's a better way.

Typically, types that access unmanaged memory will implement the **IDisposable** interface. These classes have a **Dispose** method which cleans up any unmanaged memory the object is using. We could directly call this method (in fact, the **Close** method we used with files does that) but there are some tricky issues that come up. For instance, if an exception is thrown while the file is open, the **Close** method may not get called, leaving the file open. Using what we already know, it is possible to write code to handle this, but C# provides a simpler way to handle this: the **using** statement.

A **using** statement (not to be confused with a **using** directive) will look something like this:

```
using (FileStream fileStream = File.OpenWrite("filename"))
{
```

```
    // Do work here...
}
```

When the ending curly brace is reached, the object inside of the parentheses in the using statement is disposed (the **Dispose** method is called). This holds true even if an exception is thrown in the middle of the block.

As you write code, be on the lookout for types that implement the **IDisposable** interface and dispose of them correctly when you're done using them. A **using** statement like this is a very readable and simple solution for doing this.

Preprocessor Directives

Many languages, including C#, give you the ability to include instructions for the compiler within your code. These instructions are called preprocessor directives. (The C# compiler doesn't have a preprocessor like some languages, but it treats these preprocessor directives as though there is.)

These preprocessor directives all start with the **#** symbol, which tips you off to the fact that something is a preprocessor directive.

#warning and #error

Two simple preprocessor directives are the **#warning** and **#error** directives, which give you the ability to make the compiler emit a warning or error with a specific message. While the compiler typically only generates errors or warnings when it detects a problem, this allows you to force it to happen.

You're probably wondering why you'd ever do that, and that's a great question. As an example, I recently made a change to a program, but had to leave a part of the old approach in place for short term backwards compatibility. I added in a **#warning** directive, which now shows up every time I compile. In a few weeks, I'll be able to complete the change (which I'll remember because of the **#warning** directive) and be able to strip out the **#warning** directive.

To add in a **#warning** or **#error** directive, you'd simply add something like this to your code:

```
#warning Enter whatever message you want after.
#error This text will show up in the Errors list if you try to compile.
```

#region and #endregion

The **#region** and **#endregion** directives allow you to add little regions to your code, which Visual Studio then allows you to collapse and expand. This is one potential way to group related methods or instance variables (or anything else) within a type, and allow you to collapse and hide each group on the fly.

To add a region, you'd do something like this:

```
#region Any region name you want here
public class AwesomeClass
{
    // ...
}
#endregion
```

With a region defined, you can collapse and expand the entire region with the outlining feature in Visual Studio. This can be done by clicking the little '+' and '-' icons on the left hand side of your code:

```
namespace CreatingClasses
{
    #region Any region name you want here
    class Program
    {
        static void Main(string[] args)
        {
        }
    }
    #endregion
}
```

You can nest regions.

#if, #else, #elif, #endif, #define, and #undef

There is a whole collection of other useful compiler directives, but before I get into them, I need to discuss compilation symbols briefly. Compilation symbols are special names that can be defined (turned on) or undefined (turned off—the default). For example, there's a **DEBUG** symbol that is defined when you compile in debug mode, but not when you compile in release mode. You can define your own symbols, which I'll explain in a second.

The **#if**, **#else**, **#elif**, and **#endif** directives work very much like an **if** statement, except they're instructions for the compiler to follow, and don't end up in your final program. So for instance, you could add the following to your program, and the parts inside of the **#if #endif** block would only be included in your program and compiled if **DEBUG** is defined:

```
public static void Main(string[] args)
{
#if DEBUG
    Console.WriteLine("Running in debug mode.");
#else
    Console.WriteLine("Running in release mode.");
#endif
}
```

If you're running in debug mode, the compiled program would be equivalent to this:

```
public static void Main(string[] args)
{
    Console.WriteLine("Running in debug mode.");
}
```

And of course, if you're running in release mode, it's this:

```
public static void Main(string[] args)
{
    Console.WriteLine("Running in release mode.");
}
```

Note that despite this example, it is always a little dangerous to have things run differently in the final release version than it does in testing, because you may not discover all of the bugs it has until the program has been released.

#elif is short for "else if" and can be used to chain together a sequence of conditional blocks.

You can use the **&&** and || operators with symbols as well.

You can also define a symbol with **#define**, and undefine it with **#undef**, right at the top of a file:

```
#undef DEBUG
#define MAGIC
```

This little block would undefine or unset the **DEBUG** symbol, and define your own special **MAGIC** symbol. It is important to point out that since you can't determine the ordering of files, using **#undef** and **#define** only apply to a single file.

Alternatively, you can define a symbol for the entire compilation process (much like how **DEBUG** will be set in every file, unless you turn it off with **#undef**). To do this, right-click on your project in the Solution Explorer and choose Properties. Select the Build tab, and at the top, under Conditional Compilation Symbols, enter the list of symbols you wish to define for all files, separated by spaces.

Implicitly Typed Local Variables and Anonymous Types

Up until now, every time we've created a variable, we've indicated the type of the variable. C# provides a way to infer the type of a variable at compile time. In the past, we've always done this:

```
int a = 3;
```

You can also create implicitly typed local variables. This is done by using the **var** keyword, instead of any specific type:

```
var a = 3;
```

It is very important to understand that the type in use is determined at compile time, based on what is assigned to it. This is not the same as having no type, or *any* type. There are some languages out there that are called *weakly typed languages*, where a variable can contain any kind of information you put in it. The **var** keyword does not do this. When the compiler looks at this, it will turn it into the **int** type using type inference.

Type inference only applies for local variables. Instance variables or static class variables must always have an explicit type.

Even though this is an option, in most cases I opt to use an explicit type like we've done all along. I do this for the sake of clarity. Just because the compiler *can* deduce the type doesn't mean you should force humans to do so as well.

Different programmers have different feelings about the **var** keyword. Some people use it extensively. Other people never use it. I'll let you judge for yourself when you want to use it, but I believe it is useful to illustrate a few cases to help you understand the tradeoffs. At the simple end:

```
var number = 0;
```

Because of the variable's name and the integer literal, it is easy for a human to deduce the type of that **number** variable (**int**).

Here's another one:

```
var names = new List<string>();
```

The type is also pretty intuitive here. It's a generic **List** class, where the type parameter is **string** (a list of strings). This is another case where it's easy to figure out the type; you even save a few keystrokes.

Now consider this:

```
var results = SendDataToServer(42);
```

Good luck on that one. There's no way to know what type **results** is without either knowing the return type of **SendDataToServer**, which is located somewhere else in the file at best, or in a completely different file in other cases, possibly not even your own code. Moving the mouse over the **var** keyword in Visual Studio will reveal the type. That doesn't help if you happen to not be using Visual Studio at the moment, or you're only given a part of the code to look at (like viewing somebody else's code on the web). In that case, it becomes impossible to figure out the type.

I'm not trying to convince you to go one way or another. It's just important to understand the tradeoffs of using **var** so that you can understand why some programmers really like it while others really hate it. And with this information, you can now make your own decision about when to use it and when not to.

Having said all of that, there is at least one time where using an implicit type is useful, and that is with anonymous types. Back in Chapter 30, we talked briefly about anonymous methods. In addition to anonymous methods, we can also create anonymous types. This means you can define a type without actually creating a class for it. There are, of course, limitations to what we can do with anonymous types, which we'll discuss in a minute. To keep track of an anonymous type, we *have to* use the **var** keyword.

This works like this:

```
var anonymous = new { Name = "Steve", Age = 34 };
Console.WriteLine($"{anonymous.Name} is {anonymous.Age} years old.");
```

Basically, without defining a specific type for it, we've created a new type that has two properties: one for **Name** and one for **Age**. On the second line, you can see that we can then use that type and access the properties that it has. When using anonymous types like this, we *can't* use an explicit type, because we've never created a class for it.

Situations where this comes up are rare, but do occur. For example, some LINQ queries put this feature to good use.

Nullable Types

Value types are not allowed to be assigned a value of **null**. This is a fundamental feature of value types, but there are times where you wish you could bend the rules a little. For example, **bool** is a value type and can only store **true** and **false**. But sometimes, we're in a situation where something may be true, false, or unknown.

The concept of nullable types addresses this. A *nullable type* is simply one that uses the **Nullable<T>** generic type:

```
Nullable<bool> nullableBool = true;    // We can assign true or false...
nullableBool = null;                   // but also null.
```

Even better, C# provides a shorthand way of doing this:

```
bool? nullableBool = true;
```

You can use the **HasValue** property to figure out whether the nullable type is **null** or contains a real value, and you can use the **Value** property to grab the value, assuming that **HasValue** is true:

```
if(nullableBool.HasValue)
    bool actualValue = nullableBool.Value;
```

You can also use the *null-coalescing operator* (**??**) to assign a default value if a nullable type is **null**:

```
bool actualValue = nullableBool ?? false;
```

The **actualValu**e variable will contain the value of **nullableBool** if it has a value, or if it's **null**, it will be false.

(The null-coalescing operator can also be used in every other situation where null might show up. It's not just useful with nullable types.)

Nullable types can't be created from reference types, but that's OK because they already support **null**.

It is interesting to note that while **Nullable<T>** nominally makes value types be able to have a value of **null**, **Nullable<T>** is still a value type, and can't even contain a **null** reference itself. Instead, it basically associates an additional **bool** value with the rest of the data. If this is extra value is true, then the normal value is assumed to be good. If it's false, then **Nullable<T>** will throw exceptions when trying to access the data. The ability to treat it as though it can actually take on a value of **null** is magic that the compiler takes care of for you.

Simple Null Checks: Null Propagation Operators

C# 6.0 introduced another new feature called *null propagation*, *succinct null checking*, or the *null propagator operator* that is quite powerful. One of the greatest annoyances of programming in many programming languages, especially object-oriented languages like C# is null checking.

Let's say we have a **HighScoreManager** class that has a **Scores** property that contains an array of items in order from highest to lowest. Each score is represented with an instance of the **HighScore** class, which has a reference to the player that got the score, which has the player's name. (I know, that's kind of a mouthful.) To get the player's name that has the highest score, your code might look like this:

```
private string GetTopPlayerName()
{
    return highScoreManager.Scores[0].Player.Name;
}
```

That's all fine and good until you realize that there are five things in this statement that might be null. If any of those pieces is null, you're looking at a null reference exception which could bring your program to a crashing halt and veiled threats from angry users about their friends in the KGB, mafia, and/or something about them sending you on a "vacation" to the Spice Mines of Kessel. (Some people take their **NullReferenceException**s very seriously.)

The point is, checking for null in your code is a good idea. If there is any chance that something could be null, checking to make sure it's not is good defensive coding practice.

So what does that do to our code?

```
private string GetTopPlayerName()
{
    if(highScoreManager == null) { return; }
    if(highScoreManager.Scores == null) { return; }
```

```
    if(highScoreManager.Scores[0] == null) { return; } // Could still fail if empty.
    if(highScoreManager.Scores[0].Player == null) { return; }

    return highScoreManager.Scores[0].Player.Name;
}
```

Ouch. That's a ton of code just to make things safe for use. That's four whole lines of null checking!

C# 6.0 introduced a new operator called the *null propagator operator* (or simply, the *null propagator*) that makes it easy to do these null checks without causing your code to get ugly. It comes in two forms, depending on what comes after it: **?.** and **?[]**.

Let's look at a simple example that uses this for the first null check:

```
return highScoreManager?.Scores[0].Player.Name;
```

What this operator does is checks if the thing before it was **null**. If it was, it simply evaluates to a **null** result. If not, it continues on and evaluates whatever is beyond the operator. You could say that this line of code turns into the following:

```
HighScoreManager manager = highScoreManager;
if(manager == null) { return null; }
return manager.Scores[0].Player.Name;
```

So that's one null check down. Four more to go. We can of course string these together in a line to get what we want very quickly:

```
return highScoreManager?.Scores?[0]?.Player?.Name;
```

This shows us the alternative form (with the question mark, followed by the square brackets) in addition to the entire series of null checks.

With all of these null propagators, if any of the objects involved (immediately before a null propagator) turn out to be null, the expression will short circuit and evaluate to null. Otherwise, it will keep going down the chain and evaluate the next part.

This has three interesting caveats that are worth looking into.

First, it's important to point out that when a null propagator is evaluated, the result is stored for use later. The previous code illustrates that, by copying **highScoreManager** into a local variable (**manager**) which is then used throughout the rest of the evaluation. That means that if something causes the original **highScoreManager** variable to change (perhaps to a null value, or perhaps to just a different non-null value) our evaluation doesn't change.

Basically, we've protected ourselves against things that are about to become null while we perform this statement by creating a local/cached copy of the object. That's good defensive programming, so it's good that null propagators work like that.

Second, it's important to understand that these evaluate to a null value. This will cause a small problem for us if the final value that we want was a value type instead of a reference type. Our earlier code worked because the final return value was of the type **string**, which can deal with a value of **null**. It's a reference type.

What happens if we were trying to get the high score (perhaps an **int** instead of a **string**)?

```
int? score = highScoreManager?.Scores?[0]?.Score;
```

Because the null propagators must be able to return null, this expression doesn't evaluate to a plain **int**, but rather a nullable **int**. (We discussed nullable types in the previous section.)

If we want to make sure it gets turned back into a non-null value, we would simply use the null-coalescing operator (**??**) from the previous section to turn nulls into some other specific value:

```
int score = highScoreManager?.Scores?[0]?.Score ?? 0;
```

In this case, we'll get our (non-null) score back, or if we bump into any nulls during the evaluation process, we'll get the default value of 0.

The final caveat to null propagators is that you can't directly call a delegate (including an event) immediately after it. The following is *not* valid syntax:

```
Func<string, string> delegateMethod = null;
string resultOfMethod = delegateMethod?("3");          // Invalid...
```

Instead of this, you would have to call the delegate's **Invoke** method instead of the shorthand notation from above:

```
Func<string, string> delegateMethod = null;
string resultOfMethod = delegateMethod?.Invoke("3");  // Valid...
```

In this particular case, we know that the **delegateMethod** variable is null, so the null propagator will kick in here and our result will be a null value.

This might be a little confusing, so another example is probably in order, this time with events.

In the events chapter, we talked about how it is a good idea to store a copy of the event in a local variable, just in case the event gets unsubscribed while you're attempting to call it. With a null propagator, this is now done for you. The following is now preferred syntax for raising an event:

```
public class SimpleClassWithAnEvent
{
    public event EventHandler<EventArgs> SomethingHappened;
    public void OnSomethingHappened()
    {
        SomethingHappened?.Invoke(this, EventArgs.Empty);
    }
}
```

Command Line Arguments

Not all programs have a user sitting in front of them typing commands or pushing buttons. It is worthwhile to cover a very popular alternative to manually typing in commands, and that is to use command line arguments. This allows you to put your program in scripts, and run it as a part of a larger automated task.

Let's say you have a program that adds two numbers together, and when the code is compiled, your program is called Add.exe. Instead of asking the user to type in two numbers, as an alternative, the user can specify those numbers on the command line when they start the Add.exe program. From a command prompt, this might look like this:

```
C:\Users\RB\Documents>Add.exe 3 5
```

These numbers on the end get pulled into your program as command line arguments. If you look at your **Main** method, you'll see that we have an array of **string**s as a parameter to the method. Command line arguments appear in this array:

```
static void Main(string[] args)
{
    int a = Convert.ToInt32(args[0]);
    int b = Convert.ToInt32(args[1]);

    Console.WriteLine(a + b);
}
```

User-Defined Conversions

Way back in Chapter 9, we looked at typecasting, which allows you to do things like convert a **float** into a **double**, or a **double** into a **float**, even though they're different types.

We briefly discussed implicit casting and explicit casting. Some conversions will happen automatically. These are called implicit conversions, and generally happen when the conversion is a widening conversion, with the idea being that the "wider" data type can always hold any value of the narrower type. We see this if we assign an **int** to a **double**. A **double** can hold any **int** value.

Casting works great for the built-in types. It also works well in an inheritance hierarchy.

But perhaps you've ran into the situation where you want to convert one of your own types to a different one. Let's say you've got a simple class like the **MagicNumber** class below, where you have basically a number with an extra property:

```
public class MagicNumber
{
    public int Number { get; set; }
    public bool IsMagic { get; set; }
}
```

If you want to convert this to and from an **int**, you could always add methods to do this like this:

```
public int ToInt()
{
    return Number;
}

public static MagicNumber FromInt(int number)
{
    return new MagicNumber { Number = number, IsMagic = false };
}
```

Simple enough, but it doesn't let you do implicit or explicit conversions with the casting operator. However, C# *does* allow you to define your own implicit or explicit casts, and it is pretty easy to do:

```
public static implicit operator MagicNumber(int value)
{
    return new MagicNumber() { Number = value, IsMagic = false };
}
```

From looking at this, you'll see that this is defined as an operator, and the syntax looks like what we saw in Chapter 32. Like with all other operators, this must be **static** and **public**.

We then specify whether it is **implicit** or **explicit**. In this case, we've chosen **implicit**. We then put down the type that is being converted to (**MagicNumber**, in this case), and then in parentheses, what is being converted from. Inside of the body of the conversion, we write code to actually convert from one to the other.

Now, with this added to our **MagicNumber** class, we can simply convert from **int**s to **MagicNumber**s:

```
int aNumber = 3;
MagicNumber magicNumber = aNumber;
```

We can also create an explicit cast in the same way. Here, we'll define a conversion going in the opposite way, from **MagicNumber** to **int**. It is typically a good idea to make our cast/conversion explicit whenever you're losing information in the conversion process. (A "narrowing" conversion.)

To make an explicit cast, we would add something like the following to our class:

```
static public explicit operator int(MagicNumber magicNumber)
{
    return magicNumber.Number;
}
```

Now, elsewhere in our code, we can use this conversion. But this time, we'll have to explicitly state that we want to convert from one type to another (which is why it's called an explicit conversion):

```
MagicNumber magicNumber = new MagicNumber() { Number = 3, IsMagic = true };
int aNumber = (int)magicNumber;
```

There are some strange problems that can come up with user-defined conversions. Because of this, you should think long and hard before adding in a user-defined conversion.

Of greatest importance is the fact that when you create a user defined conversion, you're going to be creating an entirely new object. You can see that we use the **new** keyword, and that should raise a flag in your mind. Imagine the scenario where you have two classes, and each can be converted to another using user-defined classes:

```
TypeA a = new TypeA();
TypeB b = a; // Converted with an implicit user-defined conversion

b.DoSomething();
```

When we do this user-defined conversion, our **b** variable is holding a completely different object than **a**. When we call the **DoSomething** method, which may change the data in **b**, **a** is completely unaffected, which seems kind of counterintuitive here. If we had used the following code instead, it would have been much clearer that **a** and **b** were separate objects:

```
TypeA a = new TypeA();
TypeB b = a.ToTypeB();
b.DoSomething();
```

So just because you *can* create a user defined conversion, doesn't mean it is always a good idea.

Instead of adding user-defined conversions, consider adding in a new constructor, or a **ToWhatever** or **FromWhatever** method to your type, as it will be much more obvious what's happening.

Part 5

Mastering the Tools

In order to master C#, you need to thoroughly understand the development tools that you use. We've spent the bulk of this book looking at how to write C# code, but in order to truly understand how a program works, you'll need to learn the details of how C# and the .NET Framework function, and how to use Visual Studio effectively. In this section, we'll look in depth at how these things work, and we'll pay particular attention to how to debug your code and fix common problems.

We'll cover:

- The .NET Framework in more depth (Chapter 40)
- A detailed look at Visual Studio and useful features it has (Chapter 41)
- How to work with multiple projects, including using code that other people or businesses created and made available to you (Chapter 42)
- Dealing with common compiler errors (Chapter 43)
- How to debug your code (Chapter 44)
- A behind-the-scenes guide through the way Visual Studio organizes and manages your code and other assets in a project or solution (Chapter 45)

C# and the .NET Framework

In a Nutshell

- Outlines the basic process of compiling in the classical sense.
- Describes what a virtual machine is and why the .NET Framework has one (the CLR).
- Explains how the CLR and IL work.
- Points out the advantages and disadvantages of using a virtual machine like the CLR.
- Describes the Base Class and Framework Class Libraries, which have lots of reusable code.

Binary, Assembly, Languages, and Compiling

I've mentioned several times throughout this book that computers only understand binary—1's and 0's. All of the data that computers store and use is ultimately represented with a pile of bits, and so are the instructions that they execute. For instance, check out the following binary code:

```
00100100 00001000 00000000 00000010
00100100 00001001 00000000 00000010
00000001 00001001 01010000 00100000
```

Can you tell what that does? Turns out, it adds 2 and 2, and stores the result. While it is easy for a computer to understand this, humans can't easily make sense of it. So we go up a step or two.

Instead of working with raw 1's and 0's, there's a low-level language called *assembly language*, or *assembler*, which is a human-readable form of binary. The binary code might look like this in assembly:

```
li $t0, 2
li $t1, 2
add $t2, $t0, $t1
```

We can now see the 2's in there, and we can even identify the **add** instruction. Humans can make some sense of this now, and typos are much easier to catch. But these three lines only do a single addition operation. Imagine if you were trying to write a *real* program.

For the most part, assembly directly mirrors binary. Every line of assembly will result in a single instruction in the binary code. (There are a few shortcuts in many versions of assembly language that are exceptions to this.) This means that the actual instructions we have available to us in assembly are the exact same as the instructions that the physical computer has available. If our computer isn't able to do addition, then there won't be an **add** instruction in our version of assembly.

Now that we've introduced assembly, we can more formally define the word *compiling*. Compiling code is simply translating from a higher level language to a lower level language. We can make a simple compiler to turn assembly code into machine code. It would be pretty simple and straightforward, because every line of assembly code becomes a single line of machine code.

But even assembly is not very easy to work with. The **if** statements, and loops that we learned about in Chapters 10 and 12 don't exist in assembly. We need to do a lot more work to make that kind of stuff happen. Working in assembly still isn't much fun.

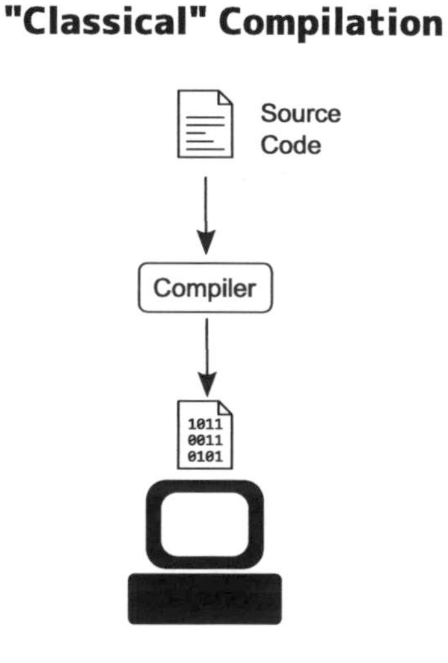

This is why all of these programming languages were invented. Programming languages are built in a way that mimics a human language (usually English, though that's not a requirement) but has a formal structure to it that allows for a compiler to figure out, step-by-step, what binary instructions it needs to use to make it happen.

This is the "classical" compilation process, but as we learned in Chapter 1, with C#, it's not quite this simple. Instead, C# relies on a virtual machine called the Common Language Runtime.

Virtual Machines and the CLR

Since C# utilizes a virtual machine to execute, it is worth taking the time to explain what a virtual machine is, and why it is so helpful. So we're going to take a brief detour and look into this.

Binary instructions are directly tied to the hardware that they are running on. This means that the same set of instructions won't run on all computers. Additionally, every operating system that you use is going to present the hardware to your program (as well as various operating system tools) in different ways. So even if you have the same architecture, if you have a different operating system, you'll potentially need a different compiler once again.

What does that mean for compilers?

Well, you'd need a different compiler (or at least different configuration settings) for each hardware architecture that you want to have your code run on.

Let's complicate things even further and say there are multiple programming languages, because... well... there are. With the "classic" approach, we'll need a different set of compilers for *each language*.

If you have four architectures and three languages, that's a whopping 12 compilers! Needless to say, this gets out of hand very quickly.

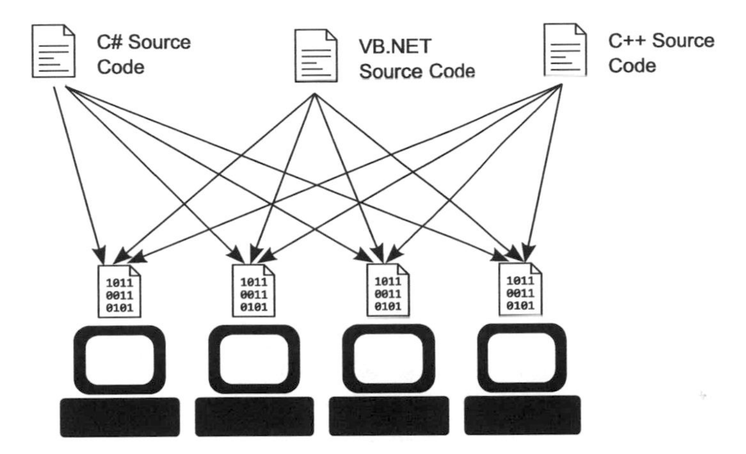

Fortunately, there's a better way to deal with this. Instead of trying to make compilers for each language and computer architecture, let's just make one very special architecture: a *virtual machine*.

A virtual machine is a software program that mimics a full computer. You can give it computer programs to run, just like their physical counterparts. More than that, virtual machines typically provide controlled access to the hardware of the computer that the virtual machine is running on.

In effect, part of its role is to be a bridge between the software it is running, and the operating system and hardware of the physical computer that it is running on. Since our goal with a virtual machine is to provide a single architecture for our code to run on, regardless of what physical hardware or operating system is being used, the virtual machine can be thought of as an abstraction, hiding away all of the details that make the different physical machines different from each other.

The *Common Language Runtime*, or *CLR*, is the .NET Framework's virtual machine, and it is the keystone of the .NET Framework. For languages that want to run on the CLR, rather than compiling directly to binary instructions, the compiler emits code in a special type of assembly language called Common Intermediate Language, which the CLR reads and ultimately compile to binary instructions.

As you can see, this simplifies the process substantially.

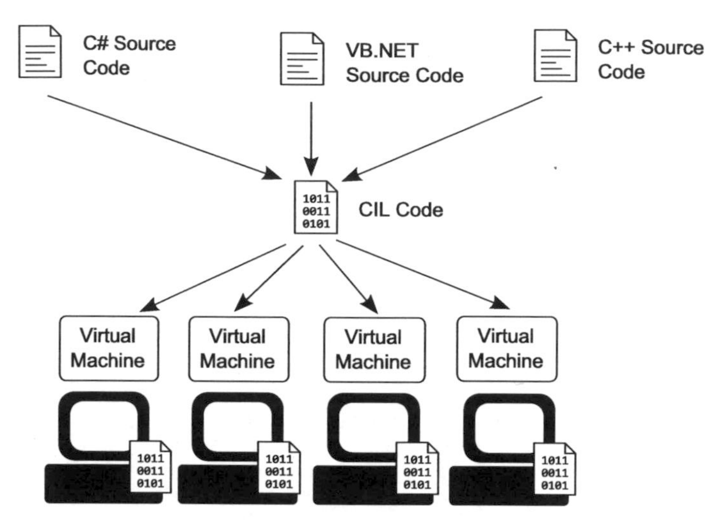

If you want to add another language to the .NET Framework, all you need to do is add another compiler. If you want to add another operating system or architecture, all you need to do is add another CLR, custom made for the system you want to run on.

The Common Language Runtime

The Common Language Runtime (CLR) is the virtual machine that is used within the .NET Framework. The CLR is a complicated piece of software, and we simply can't cover it all here, but we can outline the basics.

The basic process that the CLR follows is pretty simple. When it is time to load an assembly, like an EXE or DLL, it figures out what types it includes and what methods each one has, and builds a table for them. When it comes time to run a method, if it has not been run before, the CLR will load the IL code for that method and compile it to binary code that the current physical computer can execute. This is called *Just-in-Time compiling*, or *JIT compiling*. The next time this method runs, it sees that it was already compiled and just re-runs it. JIT compiling only needs to happen once for any given method.

Common Intermediate Language

Because C# is targeting the CLR, rather than compiling to binary code, the C# compiler will turn your C# code into a special language that is designed for the CLR. This language is called *Common Intermediate Language* (often abbreviated *CIL* or just *IL*). IL is very much like the assembly language I showed you earlier. However, IL is actually higher level than most assembly languages. You can think of IL as an object-oriented version of assembly. IL is aware of things like exceptions (Chapter 29), working with data types, and creating and initializing objects.

I think it's worth showing you what IL code looks like, so to start, look at the following code:

```
static void Main(string[] args)
{
    int x = 2;
    int y = 2;
    Console.WriteLine(x + y);
}
```

The above method gets turned into something that looks like this in IL:

```
.method private hidebysig static void  Main(string[] args) cil managed
{
  .entrypoint
  // Code size       13 (0xd)
  .maxstack  2
  .locals init ([0] int32 x,
           [1] int32 y)
  IL_0000:  ldc.i4.2
  IL_0001:  stloc.0
  IL_0002:  ldc.i4.2
  IL_0003:  stloc.1
  IL_0004:  ldloc.0
  IL_0005:  ldloc.1
  IL_0006:  add
  IL_0007:  call       void [mscorlib]System.Console::WriteLine(int32)
  IL_000c:  ret
} // end of method Program::Main
```

Microsoft has a tool that allows you to take any compiled program and open it up and see the IL it contains. This tool is called the IL Disassembler (ildasm.exe) and is installed with Visual Studio.

IL is independent of any particular hardware, operating system, or architecture, which is its primary purpose. IL is the language that the CLR knows and uses as its input, compiling it to actual, machine-dependent executable code as it is running it. It is called "Common" because any language that wants to use the .NET Framework can simply provide a compiler that generates IL. This includes C#, VB.NET, and Visual C++ (a version of C++ designed for the CLR).

Advantages of the CLR

Using a virtual machine like the CLR provides you with a lot of interesting advantages. For instance, as you write code, you'll very rarely need to worry about the differences in hardware architectures or operating systems. The CLR takes care of all of those details for you.

The fact that compilers for any given language emit IL code instead of machine code means that if you're making a compiler for a language, you don't need to worry about hardware and architectures. You just need to know enough about the language you're compiling to turn it into the appropriate IL.

Memory Management

One big thing that the CLR does for you is memory management. C# and other .NET languages do not require you allocate memory for your objects, nor do you have to clean it up when you're done (in most cases). The CLR will manage what memory is used, and what isn't, and move things around to keep things organized. Memory management is like having a personal assistant whose job is to take care of these things for you, freeing you up to work on the interesting parts of your program.

I should point out, as I've done several times throughout this book, that there are a few times where you go beyond the bounds of the CLR's managed memory, and when you do, you'll need to take extra care to clean things up correctly yourself. (See the sections named *Unsafe Code* and *Using Statements and the IDisposable Interface* in Chapter 39 for more information about this.)

Security

Because C# code is running on a virtual machine, it has a high level of control over what code can access the hard drive, the network, and other hardware. And because it is running inside a virtual machine, a program can't gain access to the memory of other programs. This prevents code from doing a whole slew of dangerous, virus-like behaviors.

Multiple Languages

The .NET Framework supports a wide variety of different languages, including C#, Visual Basic .NET, C++ (Visual C++), F#, Python (IronPython), Ruby (IronRuby), and many more. Each of these use their own compiler to generate IL, which is then executed by the same CLR.

Interestingly, the side benefit of this is that code written in one language can be used by code written in any of the other languages. This also applies to the BCL and FCL, the massive class library that C# has available for use. Any of these languages can take advantage of this library, along with any other code you write. This also means that you can choose the language that is best suited for the task at hand, which is a very powerful proposition.

The Drawbacks of Virtual Machines

Virtual machines are great inventions, but they don't come without a few strings attached. In many cases, the advantages outweigh the problems, but it is still important to understand these tradeoffs, when it comes time to pick a language to use.

Performance

The primary drawback of a virtual machine like the CLR is performance. Code running on a virtual machine is generally considered to be somewhat slower than code running without one.

But this isn't as bad as it seems.

For starters, in many applications, speed isn't an issue. Think about a GUI application, like a word processor or a web browser. Computers are fast enough that despite everything that these programs do, the vast majority of the time, the computer is sitting there waiting for the user to press the next key, or click the next button. The holdup is the user, not the computer. If you're in this kind of a situation, running on a virtual machine is not going to hurt anything.

Additionally, it is entirely possible that code running on a virtual machine is actually faster. This may seem kind of strange at first, knowing the overhead that the virtual machine needs, but it is possible.

Without a virtual machine, the compiler makes the final decision about what machine instructions are actually going to be used. That happens on the developer's machine.

With a virtual machine, the final compilation step happens within the virtual machine. It knows a lot more about the computer it is running on than the compiler does. It can choose the instructions that make the most sense at the time the program begins running, rather than long in advance on a completely different computer. These optimizations by the JIT compiler and the CLR can make up enough ground to mostly offset the cost of running on a virtual machine in the first place.

Finally, when performance is a critical issue, C# has the ability to inter-operate with unmanaged code as needed. There's some overhead to this of course, and covering this in-depth is beyond what we can cover here, but it can be done. So if you have a section of code where performance is a critical issue, you can write that code with unmanaged C++ or something else.

Bad for Low-Level Coding

Since virtual machines need to run on top of an existing physical machine and operating system (the "host" operating system) there are some things that just don't work inside of a virtual machine. You can't really write an operating system with C#, nor can you make device drivers with it. The CLR itself couldn't have been written in C# either. These things are just too low level. If this is what you're trying to do, you'll need to pick something that doesn't use a virtual machine. C++ is by far the most popular choice for things like this. (Though I'll be honest with you, after working in C# for so long, it makes me a little queasy to work with C++. It's just not nearly as pretty, elegant, or fun.)

On the other hand, if you're doing anything else—web development, application development, etc.—using C# and the CLR is an excellent choice.

Your Code is More Visible

Sometimes, people are concerned that with C# and the .NET Framework, your code will be accessible by others. To be clear, this is true to some extent. No one will have access to your C# source code, but they will have your EXE file, containing IL code. IL is essentially a high-level version of assembly, so this is more information than raw bits and bytes of binary instructions, but by no means is it easy to just sit down and figure out what's going on.

If you're concerned about this, you can always apply a code obfuscator to your IL before making it available to the world. A code obfuscator will go through and rename variables, methods, classes, and the like, making it even tougher for someone to reverse engineer your software.

On the other hand, (a) most people won't care, (b) most people won't even know where to begin, and (c) the ones who both care and have the know-how, won't be stopped by you choosing a strictly compiled language like C or C++.

If you're writing code to launch nuclear missiles or something else where it is absolutely critical that no one discovers how your code works, don't use C# or any .NET language. (Though note that in those cases, there's far more to consider than just using a non .NET language.)

The BCL and FCL Class Libraries

In addition to the CLR, the .NET Framework also includes a very large collection of previously built code and types. This collection is absolutely massive, and contains more than the "standard library" of most languages, including C++. The standard libraries in Java are similar in size. This large collection is called the *Base Class Library*, or *BCL*.

The BCL contains all of the built-in types, arrays, exceptions, math libraries, basic File I/O, security, collections, reflection, networking, string manipulation, threading, and more. While not a perfect guide, a general rule is that any namespace that start with **System** is a part of the BCL.

Beyond the BCL, there are many more classes that Microsoft ships with the .NET Framework. In general, these additional things cover broad functional areas, such as database access or graphical user interfaces (Windows Forms or WPF). This entire collection, including the BCL, is called the *Framework Class Library*, or *FCL*. In casual discussion, sometimes people use FCL and BCL interchangeably, which isn't strictly correct, but it is perhaps good enough for most things.

In this book, we looked at some of the most common parts of the BCL, but no book can cover all of it. If you are about to do something that you think others have done before, it's worth a quick web search to see if something in the BCL or FCL already does it, because there probably is.

In fact, after going through this book, you'll know nearly everything there is to know about C# itself. Much of the rest of your C# learning will involve learning how a specific type or collection of types work to accomplish a particular task.

Try It Out!

.NET Framework Quiz. Answer the following questions to check your understanding. When you're done, check your answers against the ones below. If you missed something, go back and review the section that talks about it.

1. **True/False.** C# is compiled to binary that is immediately executable by the computer.
2. Is a virtual machine software or hardware?
3. What is the name of the .NET Framework's virtual machine?
4. Name the two key parts of the .NET Framework.
5. **True/False.** A slightly different version of the CLR is required for every platform you want to run C# code on.
6. **True/False.** The CLR understands C# code.
7. Name two other languages that can be used to write code for the .NET Framework.
8. **True/False.** Virtual machines will always be slower than compiling directly to binary.
9. **True/False.** IL code is easier to reverse engineer than compiled code.

Answers: (1) False. **(2)** Software. **(3)** Common Language Runtime (CLR). **(4)** the CLR and the BCL/FCL. **(5)** True. **(6)** False. **(7)** C++, VB.NET, IronPython, IronRuby, ... **(8)** False. **(9)** True.

41

Getting the Most from Visual Studio

In a Nutshell

- You can exclude files that you do not want in a project without deleting them. You can also include previously excluded files.
- You can show line numbers in your source code, which helps in debugging.
- Describes how to use IntelliSense.
- Outlines some basic refactoring techniques.
- Points out some useful and interesting keyboard shortcuts.

Visual Studio is a very sophisticated program. It is impossible to fit everything that is worth knowing about Visual Studio into a single chapter in a book. Like many other topics that we've talked about, you could write books about this. In this chapter, we'll cover some of the most important and most useful features of Visual Studio. There is a lot more to learn beyond what I describe in this chapter, but these things will get you started.

Windows

The Visual Studio user interface is essentially composed of a collection of windows or views that allow you to interact with your code in different ways. Each window has its own specific task, and there are lots of windows to choose from.

The Code Window

The main code window, where you have been typing in all of your code for all of your projects, is the most important window. This window has a lot of settings that you can customize, which I'll outline in the section below, about the Options Dialog.

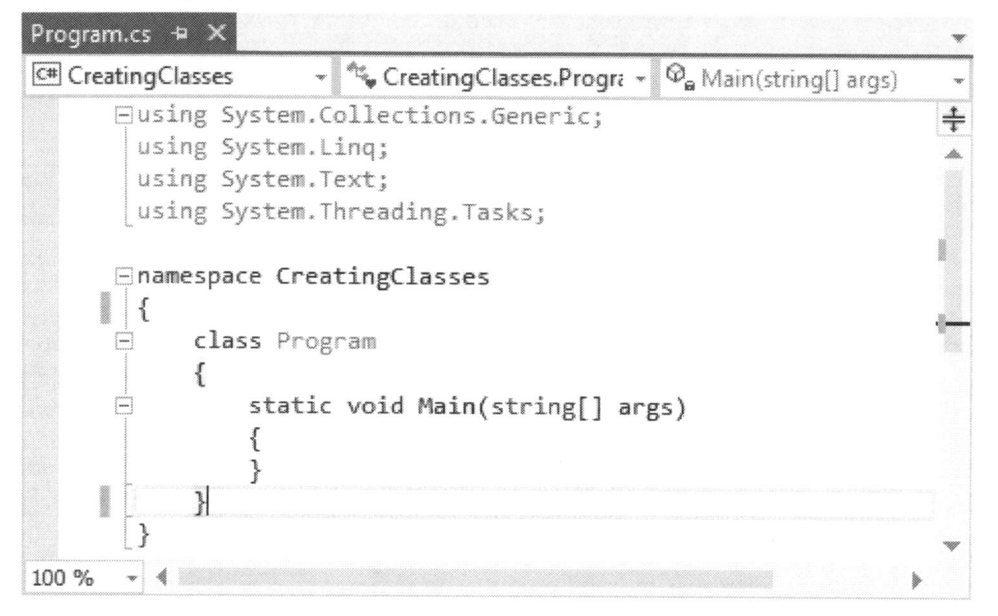

While this window is pretty straightforward, I want to point out that at the top of the window, just under the tabs to select which file to view, there are two drop down boxes that you can use to quickly jump to a specific type or member of a type in the current file. The left drop down box lets you jump to a specific type within the file (you'll usually only have one) and the second one lets you jump to a member of the type, like a method or instance variable.

Speaking of jumping to a type or member, there are a few keyboard shortcuts that are well worth knowing here, as you work on code. If you select something and press **F12**, you'll automatically jump to where that thing is defined. If that happens to be in another file, that file will be opened.

Going the other way, if you press **Shift + F12**, Visual Studio will find all of the places in the code where the current selection is used. This works for any type, method, variable, or nearly anything else. These two shortcuts are convenient for quickly navigating through your code.

The Solution Explorer

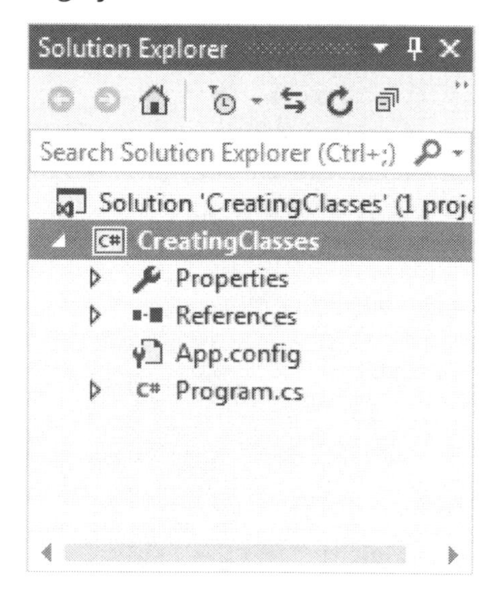

This shows a high level view of how your solution and projects are organized. The Code Window and the Solution Explorer are the two most commonly used windows.

At the top of the Solution Explorer, you will see an item for your entire solution. A solution can contain multiple projects, but by default, when you create a new project you will get a solution with one project, and both will have the same name. You can always add more projects to a solution if you need.

Under each project, you'll see a Properties node, which you can use to modify a variety of properties about your project.

You'll also see that each project has its own References node, which you can use to manage the other projects or DLLs that the project needs access to. (This is described in detail in Chapter 42.)

Your code is typically organized into namespaces, which are usually placed in separate folders under your project, and each new type that you create is usually in its own .cs file.

The Properties Window

You can right-click on any item in the Solution Explorer and choose Properties to view properties of that file, project, solution, etc. Doing so will open up the Properties window.

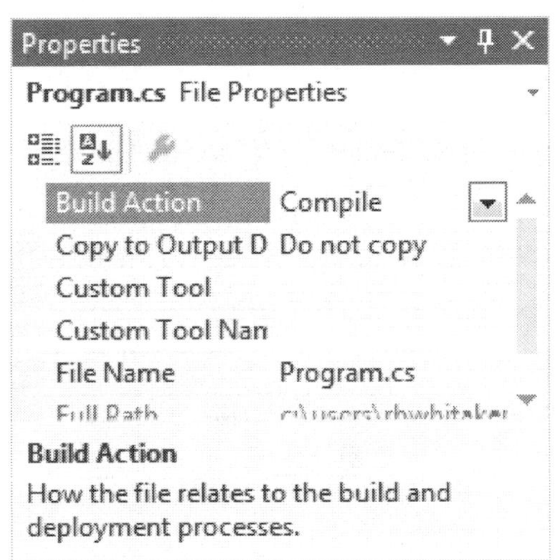

The Properties window shows you various properties about what you have selected. As you select different things, the properties window will update to reflect the current selection.

This window becomes even more useful as you start to build GUI applications, because you'll be able to use a designer to lay out your user interface, and by selecting buttons, checkboxes, or other UI controls, you'll be able to modify various properties and settings that those items have.

The Error List

The Error List shows you the problems that occurred when you last compiled your program, making it easy to track down the problems and fix them.

Error List — 2 Errors | 0 Warnings | 0 Messages | Search Error List

Code	Description	Project	File	Line
CS0103	The name 'x' does not exist in the current context	HelloWorld	Program.cs	13
CS0103	The name 'y' does not exist in the current context	HelloWorld	Program.cs	13

You can double click on any item in the list, and the Code Window will open up to the place where the problem occurred in your code. You can show or hide all errors, warnings, or messages at the top by clicking on the appropriate button.

Other Windows

Visual Studio has many other windows as well. To see what other views you can access, look for them under the View menu. Even the windows that are described above can be opened from here.

The Options Dialog

Visual Studio has tons of settings that you can modify. There are far too many to try to cover here, other than a few of the most popular ones. To get to these settings, on the menu, click on **Tools > Options**, which will bring up the Options Dialog. Like many other programs, these settings are organized into various pages, and the pages are organized by category in the tree on the left. Selecting different items in the tree will present different options to configure.

Including and Excluding Files

Sometimes you have a file that is not being used in your project. You may want to remove it from your project so that you don't keep seeing it in your Solution Explorer. One option is to simply delete the file (right-click on it and choose **Delete**) but sometimes, you don't actually want to delete the file, just stop it from being included in your solution.

To exclude a file from your project, simply right-click and choose **Exclude From Project**. When you do this, the file will disappear from your Solution Explorer, but it is not permanently deleted.

You can also add a file back in to your solution. This is also helpful if you create a file outside of Visual Studio and save it in your project directory. To do this, right click on your project (not the solution at the very top) and choose **Add > Existing Item...**. Browse to find the file you want to add and press Add. The file will now be included in your project.

Showing Line Numbers

Being able to see line numbers is very important. A lot of times, when something goes wrong, the error message will tell you what file and line number the problem occurred, but if you have to manually count every line to get down to line 937, that information would be pretty useless.

Visual Studio has a way to turn on the display of line numbers. To do this, on the menu, choose **Tools > Options**. This opens the Options dialog box:

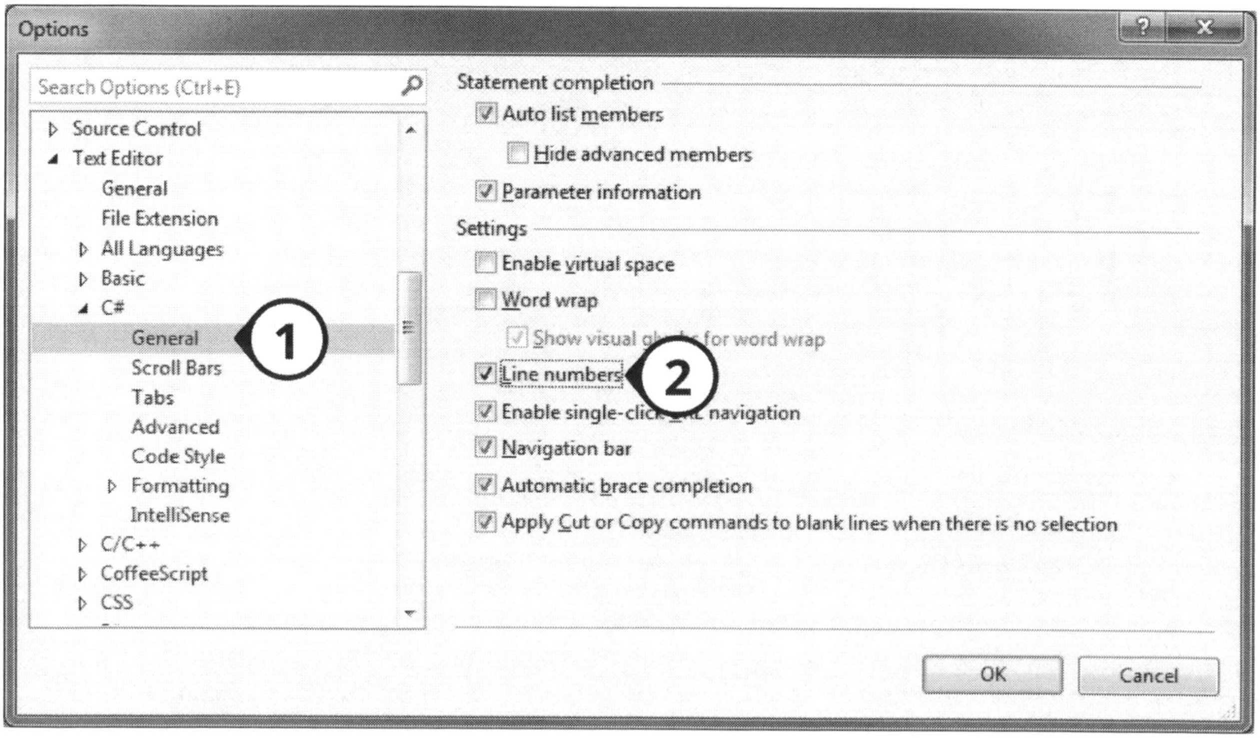

In the panel on the left, click on the node that is under **Text Editor > C#**. When you click on this, the panel on the right will show several options. At the bottom under Settings, check the box that says "Line numbers." Once you have done this, you will see line numbers on the left side of your code, as shown below.

```
 1  using System;
 2  using System.Collections.Generic;
 3  using System.Linq;
 4  using System.Text;
 5  using System.Threading.Tasks;
 6
 7  namespace HelloWorld
 8  {
 9      class Program
10      {
11          static void Main(string[] args)
12          {
13          }
14      }
15  }
```

In my opinion, it is unfortunate that showing line numbers isn't the default, but it's not.

IntelliSense

Visual Studio has an incredibly powerful tool called IntelliSense (also called AutoComplete in other programs or IDEs) that makes typing what you want much faster, as well as providing you with quick and easy access to documentation about code that you are using. You have probably seen this by now. It usually pops up when you start typing stuff, like this:

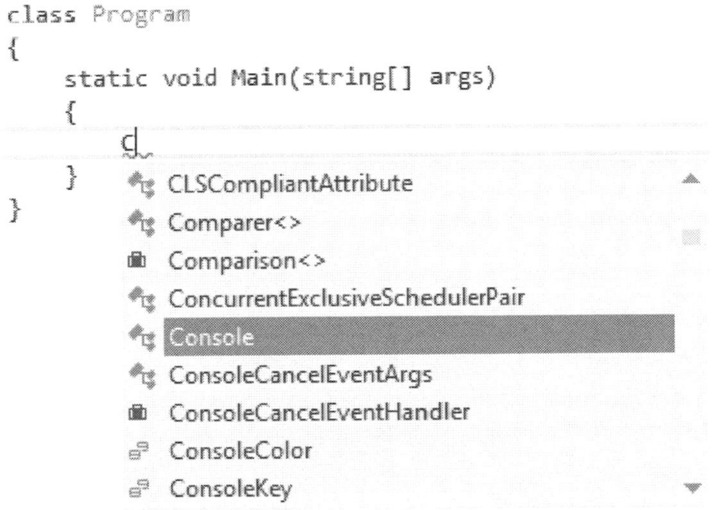

```
class Program
{
    static void Main(string[] args)
    {
        c
    }
}
        CLSCompliantAttribute
        Comparer<>
        Comparison<>
        ConcurrentExclusiveSchedulerPair
        Console
        ConsoleCancelEventArgs
        ConsoleCancelEventHandler
        ConsoleColor
        ConsoleKey
```

IntelliSense will highlight the item in the list that is most recently used, making it so that you can simply press **<Enter>** to choose it. This means that even if you give a variable a long name, you may only need to type the first few letters and press **<Enter>**. This feature makes it easy to give things descriptive names without needing to worry about how hard it will be to type it all in later.

There are several things that cause IntelliSense to pop up automatically. This includes when you first start typing a word, when you type a "." (the member access operator) or parentheses. You can get IntelliSense to go away by pressing **<Esc>**, and you can bring up IntelliSense whenever you want it by pressing **<Ctrl> + <Space>**.

I also want to point out that IntelliSense shows you the comments that have been provided for any type or its members. This includes your own comments, so it is very helpful to write XML documentation comments for your code as we discussed in Chapter 15.

Basic Refactoring

Refactoring is the process of changing the way code is organized without changing how it functions. The idea with refactoring is to make it so that your code is better organized and cleaner, making it easier to add new features in the future. There are books written on the best way to refactor code, but it is worth pointing out that Visual Studio provides a small set of refactoring tools.

If you really want refactoring power, you should consider a Visual Studio add-on called ReSharper (http://www.jetbrains.com/resharper/). It's not cheap, but provides a massive amount of refactoring support among many other useful features.

Visual Studio has a few basic refactoring tools that are worth pointing out, though. For starters, if you have something that you want to rename, rather than manually typing the new name everywhere, simply select it in the Code Window, right-click, and choose **Rename** (F2). Also, if you have a block of code that you want to pull out into its own method, you can select the code, right-click, and choose **Quick Actions**, then **Extract Method**.

Keyboard Shortcuts

Before finishing up here, there are several keyboard shortcuts that are worth pointing out.

- **F5:** Compile and run your program in debug mode.
- **Ctrl + F5:** Compile and run your program in release mode.
- **Ctrl + Shift + B:** Compile your project without attempting to run it.
- **Ctrl + Space:** Bring up IntelliSense.
- **Ctrl + . :** Show Quick Actions.
- **Ctrl + G:** Go to a specific line number.
- **Ctrl +]:** Go to the matching other curly brace (**{** or **}**).
- **F12:** Go to the declaration of a variable, method or class.
- **Shift + F12:** Locates all places where something is referenced, throughout the project.
- **Ctrl + R then Ctrl + R:** Rename an element of code.
- **Ctrl + R then Ctrl + M:** Extract selected code into its own method.
- **Ctrl + -:** Move back to the last place you were at.
- **Ctrl + Shift + -:** Move forward to the place you were at before moving back.
- **Ctrl + F:** Find in the current document.
- **Ctrl + Shift + F:** Find in the entire solution.
- **Ctrl + H:** Find and replace in the current document.
- **Ctrl + Shift + H:** Find and replace in the entire solution.

42

Referencing Other Projects

In a Nutshell

- A project can reference other projects that you have made, DLLs, or other parts of the .NET Framework. This gives you access to large piles of previously created code.
- To add a reference for one project to another, right-click on the References node in the Solution Explorer and choose Add Reference, then click on the Projects tab and select the project to add a reference to.
- To add a component that is on the .NET Framework, right-click and choose Add Reference, and click on the Assemblies tab and find the assembly that you are interested in.
- To add a reference to a project that has already been compiled into a DLL, use Add Reference, click on the Browse tab, and hit the Browse button to find the DLL.

Many of the projects that we've done have been small. But with the skills that you've learned in this book, it won't be long before you're off making a new program to save the world. Or take over it. Whichever.

But you won't be able to do that all by yourself. At least, not very quickly. You can't and shouldn't write code to do anything and everything. There are some piles of code out there that have been written by smart people, are well tested, and do really great things. You can take advantage of these other libraries and DLLs to get a quicker start on your own project. You can even reuse code that you made previously in another project.

In this chapter, I'll show you how. We'll look at three different ways to access previously written code, including how to reference another project, how to reference another assembly on the .NET Framework, and how to access code that others have put in a special code library called a DLL.

Adding a reference to pre-existing code will typically follow the following three steps:

1. Tell your solution about the other source code you want to use (if you have the source code).
2. Add a reference to your project to tell the project about the code.
3. Add appropriate **using** directives to your code to make it easy to start working with it.

Referencing Existing Projects

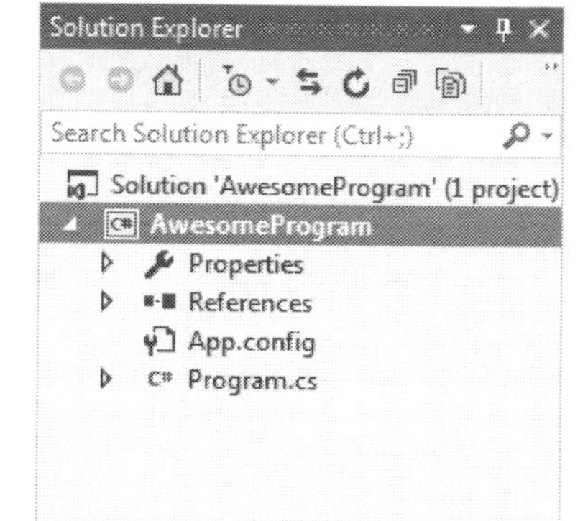

The simplest approach to reusing another external collection of code is by referencing another project. Perhaps it is another project that you wrote yourself. Or it might be a project from elsewhere that you have the source code for. (This happens a lot if you are taking advantage of open source software.)

If you have another project that contains the code you want to use, it is a pretty easy process to get access to that code.

Let's say you've got your main project, which we'll call **AwesomeProgram**, for the moment. And let's say that at some point in the past, you created another project called **HelperCode**. To get going quickly, you want to allow your **AwesomeProgram** solution to know about and use the **HelperCode** project that you made earlier. When starting, your Solution Explorer might look like the image to the right.

Step 1. To add the **HelperCode** project, right-click on the Solution node at the top and choose **Add > Existing Project**. Browse your file system to find the .csproj file for the project that you want to add. In this particular case, mine is called HelperCode.csproj.

> **Side Note**
>
> **Adding a New Project Instead.** Instead of adding an existing project, you could also add a completely new project. The files for the new project will be contained within the same folder as the rest of the solution. To add a new project, instead of choosing **Add > Existing Project**, you'd simply choose **Add > New Project**. The rest of the steps for referencing this second, new project are the same as adding a reference to an existing project.

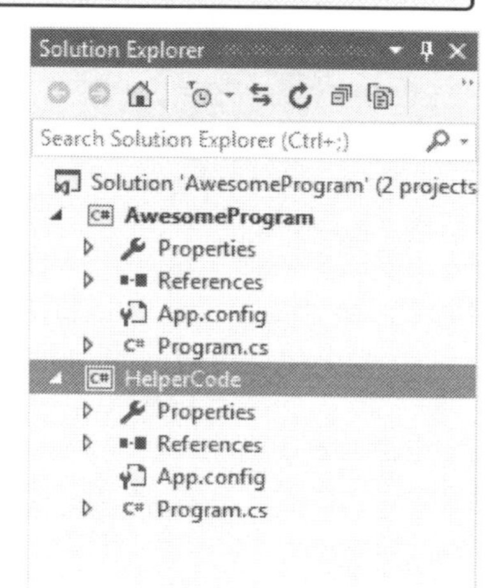

Once you've chosen the project file, Visual Studio will add that project and all of its contents to your solution, which will now look something like this in the Solution Explorer:

You'll see that now your solution contains the two projects! But we're not quite done yet. Our solution knows about the two different projects, but our first project, the **AwesomeProgram** project, doesn't know it is allowed to use the code in the second project yet.

Step 2. Under the **AwesomeProgram** project, right-click on **References**, and choose **Add > Reference....** This will open up a dialog that will allow you to choose a reference for the other project. It looks like the image below.

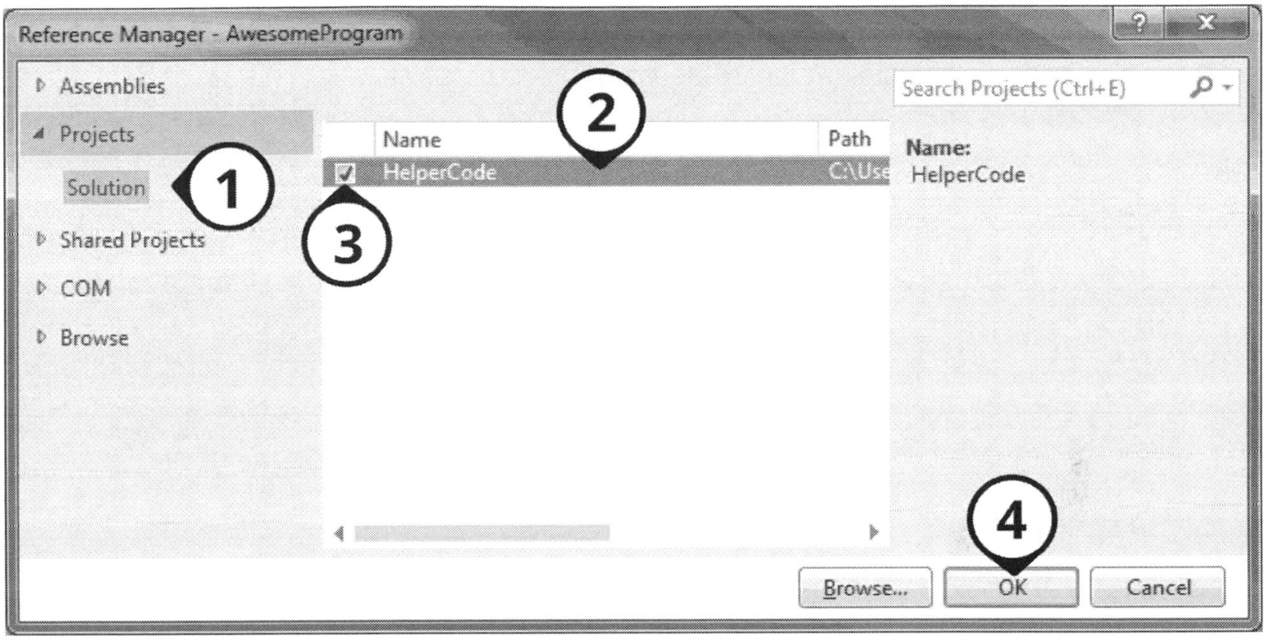

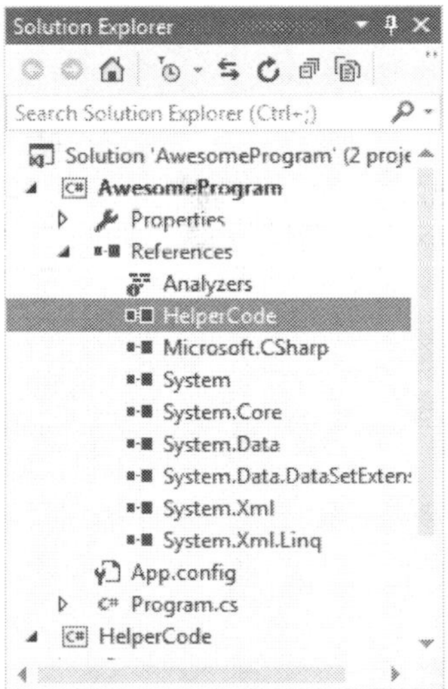

Select the **Projects > Solution** tab on the left, and check the box for the project you want to reference—**HelperCode**, in this case. Your project will now include a reference to the **HelperCode** project that you added, visible under the project's **References** node in the Solution Explorer.

It is important to keep in mind that the newly imported project in the Solution Explorer is still pointing to the original location for the **HelperCode** project. This is a double-edged sword. Making changes here affect the original as well. Any other solution or project that also uses it will get those changes as well. That can be good, because improvements will automatically be added to anything that uses it, but if you make any changes that aren't backwards compatible, you may break your other projects. If this is problematic, you can consider making a copy of the project to work with instead.

It is also worth pointing out that when doing this, if you share your code with others, things may not work out perfectly. Since you're picking out a project that is elsewhere on *your* computer, Visual Studio tries to use a relative path to the other project. If the person you are sharing your code with doesn't have that project in the same place relative to the original project, there's a really good chance that Visual Studio won't be able to load the added project.

Step 3. At this point, once you've got your second project added and referenced, you'll still need to add **using** directives if you want to use the simpler, unqualified name for classes or other types. For more information about **using** directives, see Chapter 26.

Adding a Reference to .NET Framework Assemblies

The .NET Framework is absolutely huge. It contains a lot of code to do some really cool things. It's so large though, that Microsoft has chosen to not make all of it immediately accessible to a C# project by default. This helps keep your project light weight. Instead, you're provided with access to the most common parts, and you can reference other pieces as the need arises.

Unfortunately, the times that you realize that you're missing a reference is usually when you're following someone else's example code or tutorial, and they'll just start using a class. You may try to add a **using** directive, only to discover that the compiler seems to be unable to find the type you're looking for.

So you go do a Google search for the type name, leading you to the **msdn.microsoft.com** site, where you see that the class *does* in fact exist, and the person wasn't making stuff up. But you have no clue where it actually is. In most cases, it is in one of these extra .NET assemblies, and you simply need to add a reference to the right thing.

Of course, the first step is to find out which .NET assembly you actually need to reference. If you're looking at the type's page on **msdn.microsoft.com**, you can usually see near the top, something that lists the type's assembly. That's the one we want to add.

To add a reference to one of these existing .NET Framework components, you'll follow similar steps to what you did to add a reference to any other project.

Step 1. In the Solution Explorer, go to the project that's missing the assembly and find its **References** node inside of it. Right-click on that node and click **Add Reference...**.

The Reference Manager dialog will appear. Switch to the **Assemblies** tab on the left and look for the correct assembly under either **Framework** or **Extensions**.

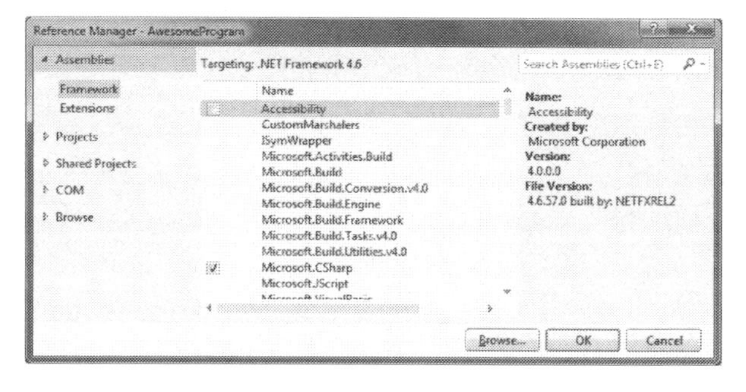

Select the assembly that you want to add and press **OK**. Once this is done, you'll now see the new assembly added to your project under the **References** node of your project in the Solution Explorer.

Step 2. Like when we added a project before, you'll still need to add the necessary **using** directives to your code if you want to be able to use the common, unqualified name of various types.

Adding a DLL

It is really common to get executable code, packaged in a DLL. They are versatile and widespread, and so the chances of eventually needing to bring in a DLL to your project is fairly high.

Adding a reference to an existing DLL is pretty easy to do. It follows steps that are similar to what we've seen in adding a reference to an existing project or a component of the .NET Framework.

Step 1. We start by going to the Solution Explorer, and under our project node, right-click on the References node and choose **Add Reference...**.

This will bring up the Add Reference dialog like we've seen with adding other types of references. Click on the Browse button at the bottom, and then browse to find the DLL that you want to add a reference to. When you've selected it, press OK to close the dialog.

You should now see the DLL under the **References** node of your project in the Solution Explorer.

Step 2. Like with the other approaches, you'll still probably want to add **using** directives to your code to make it easy to work with the new code.

Handling Common Compiler Errors

In a Nutshell

- Outlines common compiler errors, what causes them, and how to fix them.
- Compiler errors are shown in the Error List window.
- If you don't know what a compiler error means, take the time to understand what it is telling you, fix what errors you do understand, and go to the web for help if all else fails.

As you are writing code, you're bound to write some that just doesn't compile. When this happens, C# will point out the problems that came up so you can fix them. There are hundreds of different types of errors, each with a variety of possible causes, so we clearly can't cover everything here (or anywhere). But it is worth taking some time to look at the most common errors. Additionally, we'll take a look at a few general guidelines for fixing these compiler errors in general.

Understanding Compiler Errors

There's an important step that occurs before you can run your program called *compiling*. This is where a special type of program (the compiler) takes the human-readable source code that you've made and turns it into machine-readable binary code. If there are any mistakes in what you've written, the compiler can't turn it into something executable. This is called a *compiler error*, or a *compile-time error*.

Fortunately, this type of problem is relatively easy to solve. The compiler can tell you exactly what went wrong, and can often even point out how you might be able to fix it.

Compiler errors are shown to you in the Error List, which I described in Chapter 41. The Error List can be opened at any time by choosing **View > Error List** from the main menu.

Compiler Warnings

Instead of an error, sometimes you'll get a more minor problem called a warning. But warnings can sometimes be more dangerous than an error.

Errors mean that the compiler was completely unable to make sense of what you wrote, so it didn't finish compiling. When you get a warning, it means is that the compiler noticed something odd, but it still found a way to compile your source code. Sometimes, warnings are harmless. A lot of times they come up because you're only halfway done with a piece of code, and so things are naturally a little out of place.

But occasionally, a warning means that you've done something wrong in your code, like you forgot to assign a value to a variable. The compiler doesn't feel sure enough of the problem that it's willing to stop compiling. It doesn't like what you've done and points it out, but you still end up with executable code.

These warnings usually mean that something is wrong in your program, yet you'll still be able to run it anyway. Because of this problem, it is best to always try to treat compiler warnings as errors, and eliminate them as soon as you can. Don't let dozens (or hundreds) of warnings pile up. Fixing warnings up front will save you a great deal of trouble down the road.

Common Compiler Errors

We'll now take some time to look through some of the most common compiler errors, see what they mean, and look at how to fix them. Of course, we won't be able to cover *all* errors, so when we're through, I'll give you some basic principles for fixing other errors that we haven't been able to discuss.

"The name 'x' doesn't exist in the current context"

Sometimes, you'll get an error that says that a variable name doesn't exist in a particular context, meaning it either can't find the name anywhere, or if it can find it, it's in a different place, making it unusable in the spot it is currently at. One very common time that this comes up is if you accidentally misspell a variable name. If so, that's an easy fix.

If it's not a spelling problem, then what this usually means is that you forgot to declare your variable. You'll see this if you do something like this:

```
static void Main(string[] args)
{
    int b = x + 7;
}
```

You've used the variable **x** without ever declaring it. This can be fixed by simply declare it first:

```
static void Main(string[] args)
{
    int x = 3;
    int b = x + 7;
}
```

There are times when you may think you've already declared a variable. In fact, you can even *see* the declaration! This is where that "in the current context" part comes into play. You may have declared it, but not in the context that you're using it.

This gets right down to the heart of variable scope, which I described in Chapter 18. If you're sure the variable has been declared and you're still getting this error, you'll want to make sure that the variable in question is still in scope at the place that you're trying to use it. This may mean moving the variable in question to a bigger scope.

One example in particular that I think is worth looking at is one where you have a variable that is declared at block scope, but you try to use it after the block (but still within the method) like this:

```
static void Main(string[] args)
{
    for(int index = 0; index < 10; index++)
    {
        // ...
    }

    index = 10; // Can't use this here. It has block scope, and doesn't exist after
                // the loop.
}
```

Here, the variable **index** can't be used beyond the scope of the **for** loop. If you want to use this variable outside of the loop, you also need to declare it outside of the loop:

```
static void Main(string[] args)
{
    int index;

    for(index = 0; index < 10; index++)
    {
        // ...
    }

    index = 10; // You can use it here, now.
}
```

") expected", "} expected", or "; expected"

It is very common to see errors that say a parenthesis, closing curly brace, or semicolon is expected.

Interestingly, the solution isn't always to add **)**, **}**, or **;** at the spot that the compiler is complaining about. (Sometimes, but not always.) It just means that the compiler was unable to figure out where the end of a statement or block was.

In fact, the compiler sometimes thinks the error is in a location that is very different from where the problem actually lies! This is because the compiler only realizes there's a problem when it eventually runs into a place where it no longer makes sense to still think you're still in the same block or statement.

It's kind of like driving down the road and missing your turn, only to discover the error ten minutes later when you see you're leaving the city limits. You don't know when you missed the turn specifically, just that at some point along the line, you went too far. That's exactly what the compiler does, and so you potentially get the error much later than when it actually occurred.

Take this code, for instance:

```
namespace Example
{
    class Program
    {
```

```
        static void Main(string[] args)
        {
            for(int index = 0; index < 10; index++)
            {
              // missing curly brace here...
        }
    }
} // error shows up here...
```

We're missing our closing curly brace for the **for** loop, which is pretty obvious when we've formatted the code this way. But the compiler doesn't care about whitespace, so it tries to match the next curly brace it sees with the end of the **for** loop, the one after that with the end of the **Main** method, and the last one for the end of the **Program** class. But then it reaches the end of the file, and it knows it should have come across one more curly brace. Adding the right missing curly brace, bracket, parenthesis, or semicolon will fix this problem, though sometimes you need to study your code a bit to find out where you went wrong.

It is also worth pointing out that sometimes, missing a single parenthesis, semicolon, or curly brace will cause a whole pile of errors to show up in the Errors List, because the compiler can't figure out where things begin and end. Simply fixing the one problem will often fix lots of errors.

Cannot convert type 'x' to 'y'

There's a category of data type conversion errors that you're bound to see at some point or another. This can come in one of several flavors, like these, below:

- Cannot implicitly convert type 'x' to 'y'.
- Cannot convert type 'x' to 'y'.
- Cannot implicitly convert type 'x' to 'y'. An explicit conversion exists (are you missing a cast?)

What this error means is that you are trying to take one type of data and turn it into another, and the compiler doesn't know what to do with it. If you fit in the category of that third error, and you are sure you want to change from one type to another, it is an easy solution. Just put in a cast to the correct type:

```
int a = 4;
short b = (short)a; // The cast tells the compiler you know what's happening here.
```

Whenever you're required to use an explicit cast, it means there's the potential to lose data, so you really should be sure of the explicit cast before doing it.

On the other hand, if you're running into one of the other errors, it means the C# compiler doesn't know of any way to get from the type you have to the type you want. If your intention was truly to convert between types, there are usually easy ways around that. In most cases, you can simply write a method that will convert from one type to another, passing in one type as a parameter, and returning the converted result from it.

Other times this means you made an entirely different mistake in your code. If you weren't intending to convert from one type to another, then this error means there was something else wrong here. For example, it may just mean you didn't finish typing all of the code you needed. For instance, if you have a **Point** class, with **X** and **Y** properties, casting is probably not what you wanted in this case:

```
Point p = new Point(4, 3);
int x = p;    // Fails because the compiler has no clue how to convert Point to int.
```

Instead, you want to just change your code to get the X property of the point:

```
Point p = new Point(4, 3);
int x = p.X;
```

So the true error may not be that the compiler can't convert from one type to another, but that you accidentally forgot a part of the code that left the compiler thinking you were trying to convert types.

"not all code paths return a value"

If you see an error that says something along the lines of **[Namespace].[Type]. MethodName(): not all code paths return a value**, it simply means that it is possible for the program to go through your code in a way that reaches the end of the method without ever returning anything.

The code below is perhaps overly simple, but it gets to the heart of what's going on:

```
public int DoSomething(int a)
{
    if (a < 10)
        return 0;
}
```

The method is supposed to return an **int**. If the value of **a** is less than 10, a value is returned (0). But if **a** is 10 or higher, it skips that **return** statement and gets to the end of the method without returning anything.

To solve this problem, you need to analyze your code to find what paths through your code are failing to return a value, and add it in. One possible way to fix the problem from the code above is this:

```
public int DoSomething(int a)
{
    if (a < 10)
        return 0;

    return a;
}
```

"The type or namespace name 'x' could not be found"

As soon as you start trying to use someone else's code or code in other projects that you've created, or even just putting things in different folders or namespaces, you're going to run into this error.

You'll see this error with something like this:

```
using System;
using System.Collections.Generic;
using System.Linq;
using System.Text;
using System.Threading.Tasks;

namespace Example
{
    class Program
    {
        static void Main(string[] args)
        {
            MissingClass c = new MissingClass(); // Error here
        }
```

```
    }
}
```

This means is that the compiler is running into a type name (**MissingClass**, in this case) that it can't find.

It's possible that you just misspelled something. Easy fix.

It is also possible that it is spelled correctly, and that the problem is a little deeper. When you get this error, you'll see that it also says "(are you missing a using directive or an assembly reference?)". The compiler is pointing you at the two most common causes of this problem.

It is either a missing **using** directive or a missing assembly/DLL reference. To tell the two apart, think about where the code is that you are trying to refer to. Is it something you wrote? Something in the project that you're currently working on? If so, then it is probably just missing **using** directive. The details about what's going on with missing **using** directives and how to fix them are covered in Chapter 26, but in short, all you need to do is figure out what namespace the missing type is in and add a new **using** directive at the top of your file:

```
using System;
using System.Collections.Generic;
using System.Linq;
using System.Text;
using System.Threading.Tasks;
using BlackHole.CodeEatingGravity; // Add the appropriate namespace here

namespace Example
{
    class Program
    {
        static void Main(string[] args)
        {
            MissingClass c = new MissingClass();
        }
    }
}
```

If you don't think the compiler knows where the code is, because it is in a different project, DLL, or library that you haven't told it about, you'll need to add a reference to the missing assembly, which is discussed in Chapter 42. Even after you've added a reference to the project, you'll also need to add in a **using** directive to access the type in most cases.

"No overload for method 'X' takes N arguments"

If you see this error, it means that you're not calling a method with the correct number of arguments. Keep in mind that a method could be overloaded, meaning that there are multiple methods with the same name. You'll want to make sure you get the one you want. Double-check to see what parameters are needed for the method you are trying to call.

In some cases, you may also see this error if you've got parentheses in the wrong spot or if they're missing, so if you think you've supplied the right number of parameters, double-check your parentheses.

"The best overloaded method match for 'X' has some invalid arguments"

Like the previous error, this means you're not calling a method correctly. Unlike the previous error, this one means that you've got the right *number* of parameters, but they aren't the right *types*. Go back and check that the types of all of your parameters match.

Sometimes, to fix this problem, all you need to do is add in a cast to the correct type.

General Tips for Handling Errors

We can't cover all possible compiler errors here. There are just too many of them. Microsoft lists over 800 different compiler errors that could come up! And that doesn't even begin to look at the many root causes and possible fixes for those problems.

One of the key parts of making software is knowing how to find and fix your own problems. Think about it; you're making software that no one else has ever made before! That's cool, but that means you're going to run into problems that no one else has ran into before. So it is critical that you know how to deal with any and every error that comes up in your code.

So in this section, I'll outline a few guiding principles that should keep you going, even when you have no clue how to fix the problem.

Ensure that You Understand What the Error Says

Look at the error message. Does each word in the error make sense, or is there something there that you don't understand? Do you understand each of the phrases in the error? For instance, take the following error message:

```
Access modifiers are not allowed on static constructors
```

When you see this, stop and think: do you know what an access modifier is? (If not, see Chapter 18) Do you know what a static constructor is? (Also Chapter 18.) Once you know what all of the pieces mean, the solution is often pretty straightforward as well. This error is saying you need to remove the **public** keyword from this:

```
public static MyClass()
{
    // initialize the class here...
}
```

Fix the Errors You Do Understand

There are times that a single *actual* problem causes lots of errors to show up in the list. If you get a large pile of errors, look through the list and fix one or two that you understand and recompile. Doing so might get rid of *all* of the errors.

Compiling happens in parts. If the compiler can't get through one part because of an error and you fix it, it is also possible that when the compiler advances to the next part, additional errors are might come up. Because of this, don't worry if by fixing one error, extra errors appear to pop up. The error count shown to you doesn't necessarily show the actual number of problems in your code.

The Error May be in a Different Spot

Just because the error list takes you to a particular location, doesn't mean that the error is actually right there. When you double-click on an error in the Error List, it takes you to the place that it realized there's a problem. That does not necessarily mean that the error is actually on that line. Look around for other things that look out of place if nothing stands out to you.

Use Microsoft's Documentation

It used to be that documentation for programming languages and their code libraries were really cryptic and poorly done. But Microsoft has done an excellent job describing everything that they've done with C# and the .NET Framework. They're an excellent source for figuring out what your error means. Check it out here: **http://msdn.microsoft.com/en-us/library**

Use Other Programmers for Help

Programmers are usually willing to help each other when they run into problems. Programmers, by nature, like finding problems and fixing things. They're usually willing and interested in helping beginners (or pros) to learn, grow, and solve the problems they're running into.

I can't help but recommend **stackoverflow.com** as one of the best sites for overcoming software development problems. It's extremely well put together, and it is designed specifically for programming. In fact, in 99% of all cases, you won't even need to ask a question, because the question has already been asked. (Just be sure to read the FAQs before asking questions, because they sometimes throw a fit when you ask a question that isn't a "good fit.")

If all else fails, there's always the good old-fashioned Google search. Whatever problem you're running into, someone else out there has probably come across a similar problem before, and they've probably posted about it on the Internet. (Another thing programmers love doing.)

Debugging Your Code

In a Nutshell

- Debugging your code allows you to take a detailed look at how your program is executing. This gives you an easy way to see how things are going wrong.
- Describes the difference between running in debug mode and running in release mode.
- Describes how to analyze exceptions that occur while your code is running.
- Shows how to modify your code while you the program is running.
- Describes how to set breakpoints in your code, which suspend the program when reached.
- Outlines how to step through your code to see how things change.

Once you get through any compiler errors like we discussed in the previous chapter, you can run your program. Even though we've gotten rid of all of the compiler errors, sometimes things go horribly wrong while your program is running. This could be a crash, or just that your program isn't doing what you thought it would. These are called runtime errors, and we'll talk about how to deal with them here.

Visual Studio gives us the ability to dig into our program *while it is running* to see what's going on, so that we can fix it. This feature is called *debugging*.

Debugging is extremely powerful and useful when things are going wrong. Your program is unique, so no one else can tell you what's going wrong specifically, but I can show you how to use the debugging tools so that you can find the problems yourself.

Launching Your Program in Debug Mode

The first step is to make sure you start your program in debug mode. When the program is compiled and ran in debug mode (as opposed to "release" mode) it includes a whole lot of extra information in the EXE file. This extra information allows Visual Studio can keep track of what is currently being executed. This extra information slows down your program and makes the EXE file significantly larger, but it makes finding problems in your code much easier. Of course, when you're done constructing your program and you know it is bug free, you can build it in release mode and sell it!

Your program has to be compiled in debug mode or you won't be able to easily debug it.

When we first made a program back in Chapter 3, we discussed how you could start your program by pushing **F5** or **Ctrl + F5**. If you want to be able to debug your program, it will be important to use **F5**, or choose **Debug > Start Debugging...** from the menu. If you start it in release mode (**Ctrl + F5**) you won't be able to debug.

Viewing Exceptions

When your program is running in debug mode, if an unhandled exception is thrown, rather than dying, your program will pause and the computer will switch back to Visual Studio for you to take a look at it. You've probably seen this by now, but below is a screenshot that shows what this looks like:

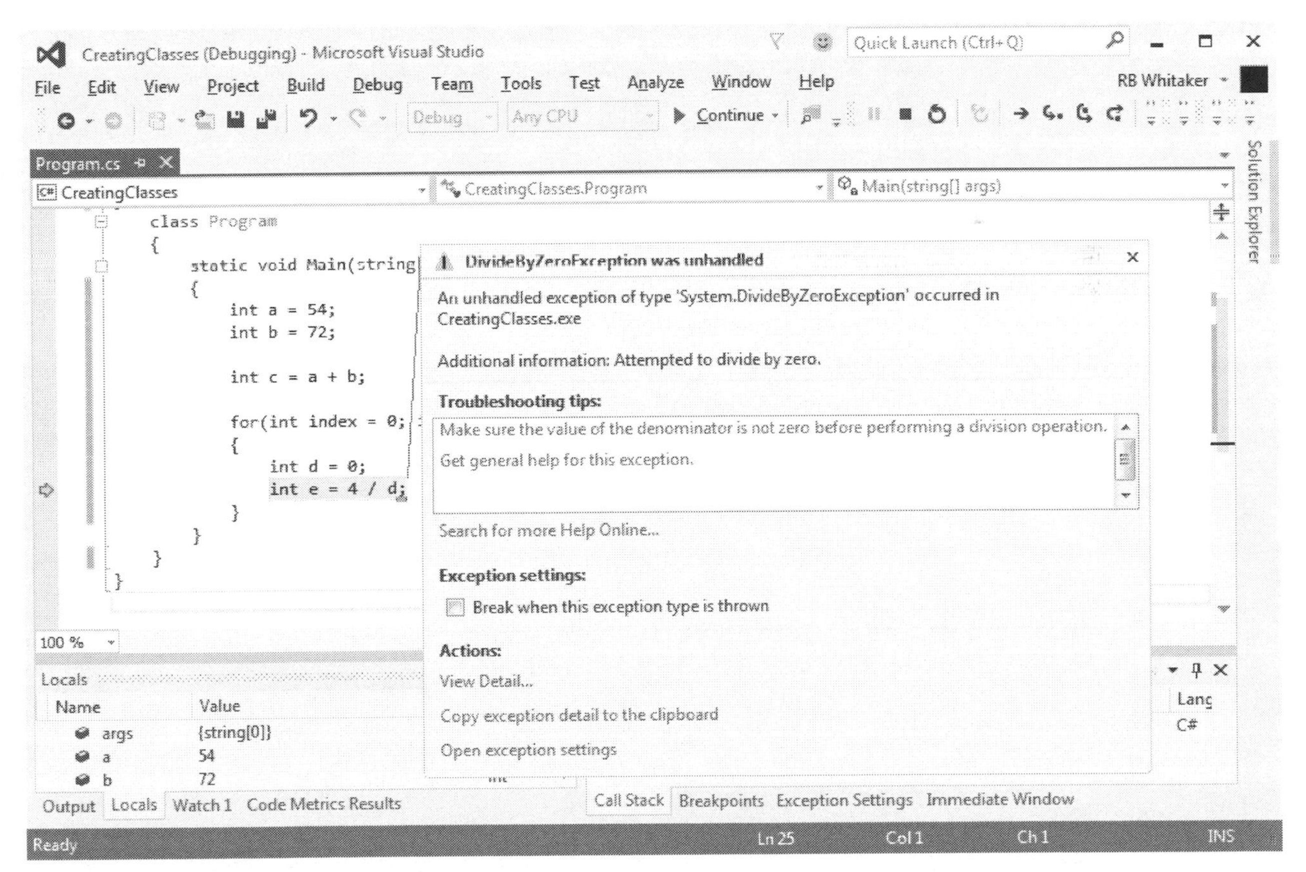

Visual Studio will mark the line that the exception or error occurred at in yellow, and it will bring up the Exception popup. On this popup, you'll see the specific name of the exception that occurred (**DivideByZeroException**, in the screenshot above) along with the text description of the problem. Ideally, this text would have a lot of useful details, but unfortunately, it's often kind of vague.

The dialog also brings up some troubleshooting tips, including "Get general help for this exception" and "Search for more Help Online...." Both of these are so generic that they're usually not very helpful. Also on this popup is the **View Detail...** link, which brings up all of this same information, and more (that we'll still talk about in a minute) in a dialog.

When the Exception dialog comes up, Visual Studio is giving you a chance to take a detailed look at the problem so that you can figure out what went wrong and fix it. All of the tools that are available in debug mode are there to assist you in doing this. Sometimes, you'll be able to edit your code while the program is still running, fix the problem, and continue on. In other cases though, once you reach this point, the computer won't be able to continue on and you'll need to restart to check your fixes.

Looking at Local Variable Values

When the program is stopped, you'll have the chance to see what state your program is currently in. One of the most useful features available is that you can hover over any variable and see its current value.

If the variable is an object, you'll be able to hover over it, and then in the little popup, you'll be able to dig down and see all of the details and current state of the object, including its private variables. If a particular class is a derived class, you'll also have the ability to dig down to the base class, so everything you want to know about an object's state will be visible.

Note that this information is also available in the Locals Window. In many cases, this window will already be open for you, down towards the bottom, where the Error List is. If you don't see it anywhere in Visual Studio, you can open it up by going to **Debug > Windows > Locals**.

The Call Stack

The call stack is one of the most useful tools that you'll have to help you figure out what happened. The call stack tells you what method called the current method. And yes, this is directly related to the stack that we described in Chapter 16.

The problem is sometimes caused, not by the method that you are inside of, but by the calling method instead. You can use the Call Stack Window to see what methods have been called to get you to there.

Like the Locals Window, the Call Stack Window is usually already visible when you're debugging, but if it is not, it can be opened by choosing **Debug > Windows > Call Stack** from the menu. The Call Stack Window looks like this:

Call Stack

Name	Lang
CreatingClasses.exe!CreatingClasses.Program.DoSomethingElse() Line 22	C#
CreatingClasses.exe!CreatingClasses.Program.DoSomething() Line 18	C#
CreatingClasses.exe!CreatingClasses.Program.Main(string[] args) Line 13	C#
[External Code]	

The method that you're currently in is at the top. (**DoSomethingElse**, in this case.) Each line below that shows the method that called it. (**DoSomethingElse** was called by **DoSomething**, which was called by **Main**.) You can click on one of those other lines and Visual Studio will jump over to that method, allowing you to see what state your variables are in over there.

Editing Your Code While Debugging

In some cases, you will have the ability to edit your code while it is running and then continue running the program, exactly where you left off, with the edited code. I can't overstate how cool this is. This is like being able to change the tires of a car as you drive it down the road without even stopping.

But this doesn't work in all cases. For starters, if the problem occurred outside of the current method (on the top of the stack trace) you won't be able to edit that method and continue. You can only continue from changes made to the top method on the call stack.

Also, if you're doing a LINQ query when the exception occurs, you won't be able to edit.

In many other cases, you'll be able to edit your code while it's stopped. To edit your code, once the program has stopped and switched back to Visual Studio, just start editing your code like you normally would. If, in the process of editing, you introduce an error that prevents it from compiling, you won't be able to resume until you fix the compiler errors.

There are some types of edits that you won't be able to do. You won't be able to add a new **using** directive. And that means that the normal steps you may do to add a **using** directive just flat out won't show up in the menu. You won't be able to structurally modify a type, so no adding, deleting, renaming methods, properties, instance variables, or class variables. The way the debugger swaps out code just doesn't allow you to swap out structural changes like that, just changes to functionality. If you make a structural change, the change you made will be underlined in blue, and Visual Studio will tell you it needs to recompile and restart to allow the changes to take effect.

Breakpoints

Up until now, we've only about debugging code after your program runs into a problem. But you'll of course want to be able to occasionally analyze running code *before* there's a crash.

Whenever you want, you can set a breakpoint on a particular line of code in Visual Studio. A breakpoint is a way to mark a line of code so that when the program reaches that line during execution, the program will halt, allowing you to take a look at the program's current state.

To set a breakpoint, you simply locate the line you want to set the breakpoint at, and on the left side, click in the gray area, marked in the image on the left below:

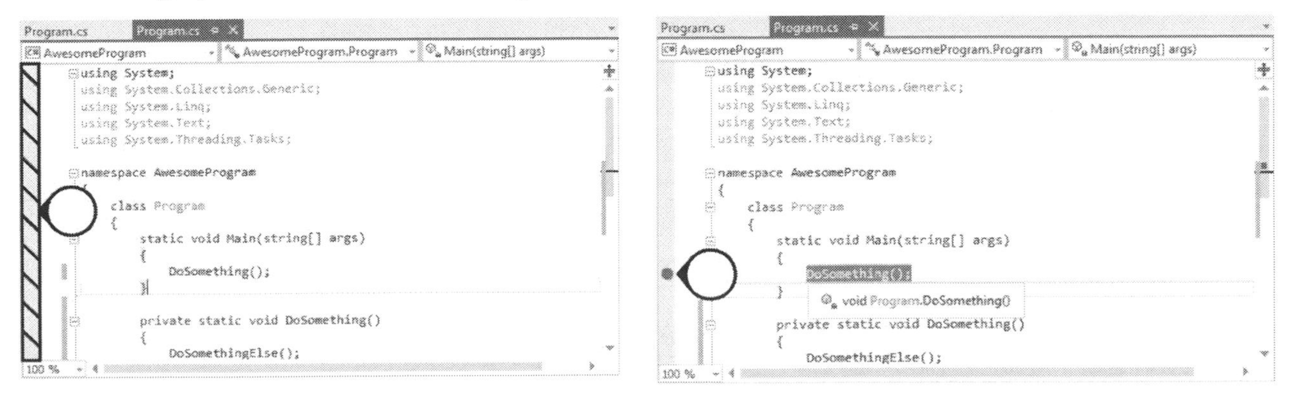

When you do this, a red circle will appear in the left margin on the line you have selected, and the text on that line will also be marked in red. This looks like the image above on the right side.

You can set as many breakpoints as you want in your program. You can also add and remove breakpoints whenever you want, even if the program is already running. To remove a breakpoint, simply click on the red circle and it will go away.

Whenever your program is stopped at a breakpoint, you can press the Continue button (located on the Standard toolbar, taking the place of the Start button while your program is running) which will leave the breakpoint and continue on until another breakpoint is hit, or until the program ends and closes.

▶ Continue

Stepping Through Your Program

While stopping your program in the middle of execution allows you to do everything you could before (looking at local variables, the call stack, etc.) it can do a whole lot more. After all, unlike before, the program hasn't run into a problem. You've stopped it before that happened.

To use these options, while your program is running, find the Debug toolbar among the various toolbars at the top of Visual Studio. The Debug toolbar looks similar to this:

If you don't see it, go to **View > Toolbars > Debug** to open it. This toolbar contains a collection of very useful tools for stepping through your code.

The debug toolbar may look a little different, depending on your settings and the specific version of Visual Studio that you have, but the items on the toolbar that we'll discuss should be available in all of them. If you feel like you're missing something useful, this toolbar (and all other toolbars) can be customized by clicking on the little down arrow at the end of the toolbar and choosing the items you want displayed on it.

Items on the Debug Toolbar

The first button on the Debug toolbar is the Break All button, represented by a pause icon. While your program is executing, you can press the Break All, which will halt execution of your program immediately, and you can see where it's at. The computer runs your program so fast that this isn't useful for fine-grained control. (We'll see a better tool for that in a second.) Instead, this feature is usually more useful to figure out why your program is taking so long to do something. Usually, it is stuck in some infinite loop somewhere, and by hitting the pause button, you can track down where the problem is.

The second button is the Stop Debugging button. This terminates your program that is running. It is important to remember that as soon as you push this, you'll lose any debugging information you were looking at, so be sure you've got the information you need before hitting it.

The third button is the Restart button. This closes the current execution and starts over from the beginning. This is very convenient for the times that you've made a change and you want to restart to see if the changes are working.

There are three other buttons on the Debug toolbar which make it easy to slowly step through your code. These are the Step Into, Step Over, and Step Out buttons. The first two allow you to advance through your code one line at a time, giving you the ability to study what your code is doing. Step Into and Step Over both advance by one line, but there's a subtle difference between the two. The difference is most obvious when you are on a line of code that calls a method. If you use Step Into, the debugger will jump into the method that you're looking at, and the next line you'll see will be at the start of that method. You can then continue stepping through the code there. If you use Step Over, the debugger will "step over" the method call, running it entirely, and only pausing again when the flow of execution returns back to the current method.

If you are looking at a line of code that doesn't call a method, the two are identical.

On the other hand, if you're in a method and you know you're ready to just move back up to the method that called it, you can use the Step Out button. This will jump forward as many lines as it needs to until it is back up to the calling method.

There are two other tools that you may find handy for debugging as well, though these are buried away in a context menu. If you know you want to skip ahead to another specific line, you can click on the line you want to jump to, right-click, and choose **Run To Cursor**. This causes the program to run without pausing until it hits the line that you indicated. This is like setting a breakpoint on that line temporarily and hitting Continue. This makes it so you don't need to keep pressing the Step Over button repeatedly.

There's one other tool that is very powerful, but can also be a bit dangerous. Use it wisely. I'll point out the pitfalls that come with it in a second. But first, let me show you what the feature is.

At any point when you're debugging, you can right-click on a line elsewhere in the method that you're debugging and choose **Set Next Statement**. This is a very handy feature. To illustrate my point, take a look at the simple example code below, which simulates rolling a six-sided die:

```
public int RollDie(Random random)
{
    return random.Next(6);
}
```

In most cases, this will probably work fine. But if the object passed in for **random** was **null**, you'll get a null reference exception on that line when it tries to generate the number.

If you're debugging, the debugger will halt on that line of code and point out the problem to you. As I said earlier, in some cases, you can make edits to your code and continue. Perhaps you want to fix this by checking to ensure that **random** is not **null** before calling the **Next** method. It's simple to make this change:

```
public int RollDie(Random random)
{
    if(random != null)
        return random.Next(6);

    throw new ArgumentNullException(nameof(random));
}
```

But after you've made these edits, if you try to continue executing, you'll still be on that line (now inside of the **if** statement). You could just restart the program, or you could use the Set Next Statement tool by selecting the **if(random != null)** line, then right-click and choose **Set Next Statement**, and then press the Continue button on the Debug toolbar. This will start your program running at the chosen line. Your program will check for **null** like you want, and you'll be able to move forward without needing to restart.

It's an extremely powerful tool, but there's a dark side to it as well. It is incredibly easy to misuse it and put your program in a weird state that it can't get into naturally. Take the code below, for instance:

```
static void Main(string[] args)
{
    int b = 2;

    if (b > 2)
        Console.WriteLine("How can this be?");

    b *= 2;
}
```

In the normal course of execution, we'd never be able to get inside of that **if** block. But, if you run this code and set a breakpoint at the closing curly brace of the method, then use Set Next Statement to move back up to the **if** statement, **b** will be 4, and the code inside of the **if** block will get executed.

It is very easy to accidentally get your program into an inconsistent state like this when you use the Set Next Statement tool, so it is important to keep this in mind as you use the feature. Even if you think you've fixed a problem, you'll still want to rerun the program from scratch again, just to be sure everything is working like it should.

Try It Out!

Debugging Quiz. Answer the following questions to check your understanding. When you're done, check your answers against the ones below. If you missed something, go back and review the section that talks about it.

1. **True/False.** You can easily debug a program that was compiled in release mode.
2. **True/False.** Debug mode creates executables that are larger and slower than release mode.
3. **True/False.** If an exception occurs in debug mode, your program will be suspended, allowing you to debug it, even if you haven't set a breakpoint.
4. **True/False.** In release mode, Visual Studio will stop at breakpoints.
5. **True/False.** In some cases, you can edit your source code while execution is paused and continue with the changes.

Answers: (1) False. **(2)** True. **(3)** True. **(4)** False. **(5)** True.

How Your Project Files are Organized

In a Nutshell

- Your solution structure looks like this:
  Solution Directory
  - **.sln** file: Describes the contents of your solution.
  - **.suo** file: User specific settings for a particular solution.
  - Project Directory (possibly multiple).
    - **.csproj** file: Describes the contents of a project.
    - **.csproj.user** file: User specific settings for a project.
    - Code files and directories, matching various namespaces.
    - **Properties** folder.
    - **obj** directory: a staging area for compiling code.
    - **bin** directory: finalized compiled code.

I once had a professor who hated Visual Studio. He explained that the reason why is because it's making programmers dumb. Students were turning in their assignments and they had no clue what they were even submitting, and they didn't know what to do when things went wrong with their submission.

Visual Studio has a bit of a bad habit of spewing lots of files all over the place. Some of these files are your **.cs** files, which contain your code. Others might be images, DLLs, or other resource files that you have. Those are all OK. But in addition, Visual Studio loves configuration files.

I, personally, don't think that this is a good reason to hate Visual Studio. In fact, this is evidence that Visual Studio is doing its job. (Though I do wish the contents of these files were simpler.) It hides all of the information needed to compile, run, and publish your program, and it does it while juggling flaming swords and singing a rock ballad about doing the Fandango. And the fact that people can get away with not knowing how it works means it's doing its job right.

Still though, I think it is worth digging into a project and seeing how things are organized and what everything is for. In this chapter, we'll take a look at the directory structure that Visual Studio creates and look at what is contained in each of the files we discover there.

Try It Out!
Demystifying Project Files. Follow along with this chapter by opening up any project of yours and digging around until you understand what every file you discover means.

In Depth
Version Control. Throughout this chapter, I'll be pointing out parts of your project's directory structure that should be placed in version control and parts that should not. While a full explanation of version control software is well beyond the scope of this book, it is worth a brief introduction. Version control software is a category of software that serves two primary purposes: the ability to share source code among team members, and the ability to track changes to that code over time, keeping a history of your software's development. SVN, Git, and Mercurial are all popular and free version control systems that you can use.

The general rule for what goes into version control are that user-specific project files don't belong in version control (everyone should have their own copy) and anything that can be rebuilt from other things (like compiled executable files) should be skipped as well. Everything else can go in your version control system.

Visual Studio's Projects Directory

Visual Studio will place each solution or project you create in its own directory. These project directories are all in the same place by default, though you can pick a different location when you create a project.

By default, these projects are all stored in **[Documents Directory]/Visual Studio 2015/Projects**, where [Documents Directory] is your 'My Documents' or 'Documents' directory.

If you've been putting all of your projects in the default location, you should be able to open that directory and see a folder for each of the projects that you've created. All projects have the same overall directory structure, so once you've figured one out, the others should make sense as well.

The Solution Directory

Solutions vs. Projects

The top level folder corresponds to a solution. Remember that a solution can be thought of as a collection of projects that are all related, and used together to accomplish a specific task.

You may remember that in the past, when you've been ready to start on something new, that you've chosen **File > New Project** from the menu. What you get is actually a new project, created inside of a new solution. In the past, when you've created a project, you've also creating a new solution to contain it.

Inside of the solution folder, you'll find three things. You'll see a **.sln** file, a folder for every project in your solution (only one if you've only got one project in your solution), and you may also see a **.suo** file. If you don't see the **.suo** file, it's probably still there, just hidden. If you don't see it, don't worry too much. It's not too critical that you can see it. It is just important to know that it is there.

The .sln File

The **.sln** file is your solution file. This file is very important. It contains information about what projects are contained in the solution, as well as solution specific settings like build configurations. The **.sln** file contains shared information that everyone will want to keep the same, and it should be included in your version control system if you have one.

The .suo File

The **.suo** file is the solution user options file. It contains various settings for the project that Visual Studio uses. Like I said earlier, it may be a hidden file, so you may not see it. This contains things like what files you had open when you last used the project, so they can get reopened when the project is reopened.

You could delete this file and nothing bad would happen.

Because this file contains user-specific information, this file shouldn't go into version control. Instead, everyone should have their own version of the file.

The Project Directories

In your solution directory, you should find that there is one folder for each project in your solution.

Let me clarify something that has confused many people in the past, including myself. By default, you'll have one project in your solution. This project is named the same thing as your solution. So if you created a new project called **LikeFacebookButBetter**, what you'll be looking at here is a folder called **LikeFacebookButBetter** inside of a folder called **LikeFacebookButBetter**. It can be confusing at first. The key to remember is that the top level one is the solution folder, and the lower level one is the project folder, and they represent two different things.

The Project Directory

Inside of a project directory, you'll see even more files. For starters, you will see a file ending with **.csproj**. You will also likely see another file ending with **.csproj.user**. You should also see a Properties folder. And you'll also probably find a pile of **.cs** files, or other folders than contain **.cs** files and other resources. You may also see a **bin** and an **obj** folder.

The .csproj File

This is one of the more important files that Visual Studio creates. This file essentially defines your project. It identifies what files are included in your project, as well as listing other assemblies, libraries, or other projects that this project uses. This file does for projects what the **.sln** file does for solutions.

Again, if you're using version control, this one belongs in the repository.

The .csproj.user File

This file, like the **.suo** file, contains user-based settings, but for your project instead of the solution. Like the **.suo** file, every user gets their own copy, so it should not go into version control.

The bin and obj Folders

You may or may not see folders called **bin** and **obj**. If you don't see them, open the project and compile it and these folders will appear. These are both used for building and storing executable versions of your projects and solutions (EXE and DLL files).

The difference between the **obj** folder and the **bin** folder is that when the program is compiling, at first it will place stuff in the **obj** directory. The stuff in here is not necessarily complete, and can be thought of as a staging area for the final executable code, which will get put into the **bin** folder. If you're going to hand off the EXE to anyone, or attempt to run the EXE from here, you should go with the one in the **bin** folder, not the one in the **obj** folder.

If you open up either of these directories, you'll likely see a folder in there called **Debug** or **Release**. Dig down further and you'll end up seeing the EXE file (or possibly a DLL file if the project is a class library) along with a random assortment of other files. You'll get a **Debug** or **Release** folder depending on how you've built your program. If you've told it to run in debug mode, you'll see a debug directory. If you've told it to run in release mode, you'll get a release directory.

Both the **obj** and the **bin** directories should be kept out of the version control repository, if you're using it. And if you're submitting an assignment for a class, or handing off your code to another programmer, these directories do not need to be included, because they can be rebuilt. (In fact, doing so is a good idea if you're giving it to someone else who has the capacity to compile it themselves. It reduces the total size of what you're sending and eliminates the executable files, which many email systems block.)

The Properties Folder

The **Properties** folder contains properties and resources that you've added to your project. For a large project, this can be quite large.

At a minimum, you'll probably see an **AssemblyInfo.cs** file in the Properties folder of the project. This file contains information about your project, including versioning information. This file is also quite important, and it is a shared file, so you should put it in version control.

You can open the file and see what it contains, but it is worth pointing out that everything in this file can be edited through Visual Studio by right-clicking on your project in the Solution Explorer and choosing **Properties**, and then on the **Application** tab click on the **Assembly Information...** button, where you can specify the information you want.

Source Files and Everything Else

At this point, we have covered everything except your actual code and other folders and files you may have added directly to your project in Visual Studio. If you look around, you're bound to find a bunch of **.cs** files, which are your C# code. Any directories that we have not discussed are directories that you have made yourself in Visual Studio, and each folder that you see likely represents a different namespace. This is the default behavior, but you can change your namespaces to be something else entirely. I wouldn't recommend doing that though, as it is very handy to have your namespaces match up with the folder structure that the files are in.

Part 6

Wrapping Up

In this final part, we're going to wrap up everything that we've been doing throughout this book, in more ways than one. Part 6 will cover the following:

- Give you several larger *Try It Out!* Problems for you to tackle, to help you be sure you've learned the things you needed to (Chapter 46).
- Discuss where you can go next, now that you've learned C#, including C#-based websites and web applications, desktop applications, and even video games (Chapter 47).

46

Try It Out!

In a Nutshell

- The best way to learn how to program is to do it. No amount of reading will make you learn.
- This chapter contains a variety of interesting problems that you can try out to help you learn. If you don't find any of these problems interesting, feel free to pick your own.
- This includes the following challenge problems to try out:
    - **Message from Julius Caesar:** Encryption using the Caesar cipher.
    - **Reverse It!:** A simple game that gives you practice with arrays and indexing.
    - **Pig Dice:** A simple multiplayer game involving randomness and dice.
    - **Connect Four:** Remake the classic game of Connect Four.
    - **Conway's Game of Life:** An exploration of a zero-player game.

The best way to learn to program is by programming. That's why I've included the *Try It Out!* sections throughout this book. It's also the reason why there's homework.

In this chapter, I'm going to present a collection of tasks that will hopefully be interesting to you, and give you something to work on that will help you be sure that you've learned the things you needed to.

Of course, if you've got your own project that you want to work on, you should go for that instead. As interesting as these projects are, if you've got one of your own in mind, that's a better choice for you to work on. You'll get the best results and learn more from something that you've personally chosen and are excited about.

I should point out that these challenges are not necessarily easy. They're like the final boss of a video game, where you'll use all of the tools and tricks that you've learned throughout the game. It may take hours of coding, maybe even spread out over days or weeks, to get the right answer. (Or not.) Many of these are based on the projects that I did while learning to program that I thought were the most interesting, fun, or memorable.

Each of these problems can be kept to a very simple minimum, or you can add extra features to make them more detailed and more interesting. Feel free to keep going with an idea until you get tired of it.

By the way, like all of the other *Try It Out!* problems throughout this book, I'm posting answers to each of these on the book's website and you can go there to see a complete working solution. See **http://www.starboundsoftware.com/books/c-sharp/try-it-out/**

Message from Julius Caesar

Encryption is the process of taking data and turning it into a form that is no longer readable. This keeps the information protected, so that people who aren't supposed to read it can't. Of course, the person who is *supposed* to read it needs to be able to decrypt it and recover the message.

When encryption is done, the algorithm used to encrypt stuff usually uses a special value called a key to perform the encryption. If a person has the key, they can usually decrypt the message as well.

The Caesar cipher is one of the simplest encryption schemes, and it is possibly something you used to send coded messages to friends when you were younger. It is an encryption method that supposedly Julius Caesar occasionally used when he wrote personal letters.

The basic idea is that for every letter in your message, you shift it down the alphabet by a certain number of letters. The amount you shift is the key for the algorithm. If you are using a key of 3, A would become D, B would become E, etc. Once you get to the end of the alphabet, it wraps back around, so Z would be C.

For example, with a key of 4, the message below is encrypted to look like this:

```
Plain text: EXPERIENCE IS THE TEACHER OF ALL THINGS
Encrypted:  IBTIVMIRGI MW XLI XIEGLIV SJ EPP XLMRKW
```

Write a program that will read in a message from a file (or from the user) and encrypt it, writing out the encrypted message to another file (or back to the console window). Don't overwrite the original file. Ideally, your program will ask for the name of a file to read, and a key (a number) to use in the encryption.

Anything besides a letter (punctuation, numbers, etc.) can either be skipped or passed along to the output without encrypting it.

Also create code that will do the reverse, decrypting a message given a decryption key.

If you want an extra challenge, try this. The Caesar cipher is really easy to crack. In fact, it is so basic that it provides little real protection in modern usage. Much more sophisticated algorithms are used now to encrypt data. To prove the point that the Caesar cipher can be cracked, we're going to try a "brute force" approach to crack the following code:

```
UX LNKX MH WKBGD RHNK HOTEMBGX
```

With the Caesar cipher, there are only 26 possible keys. Try each one, one at a time, until the decrypted message makes sense. The simple approach for this is to have a human (you) visually inspect the decrypted message to see if it makes sense. While it is more work, it is possible to have the computer figure out if the message has been decrypted by using a dictionary file (containing all or the words in the English language) and checking to see if all or most of the decrypted words are in it. A high percentage typically indicates a successful decryption.

Reverse It!

In this task, we're going to make a simple array-based game. We'll start by making an array that can hold 9 integer values, and we'll randomly put the numbers 1 through 9 in the array. Make sure you don't have any duplicates. Each number should appear exactly once.

The array will be printed out to the user like this:

```
5 3 8 6 4 1 2 9 7
```

Allow the user to type in a number 1 through 9. Whatever the user types in, we're going to reverse that many numbers, starting at the front. So from the randomized starting position shown above, if the user typed in the number 4, we'd end up reversing the first four numbers, and the resulting array would be:

```
6 8 3 5 4 1 2 9 7
```

The player can keep typing in numbers, and you will keep reversing the indicated part of the array until the array is in ascending order:

```
1 2 3 4 5 6 7 8 9
```

When this happens, the player has won the game, and the game ends. Before closing, show the number of moves it took the player to win.

Can you figure out an optimal strategy for this game? It is possible to win this game in no more than 16 moves, regardless of the starting arrangement.

Pig Dice

In the game of Pig Dice, players take turns rolling two six-sided dice. On a player's turn, they can roll the dice repeatedly until they either choose to hold (stop rolling) or roll a 1. As long as the player keeps rolling without getting a 1 on either die, the numbers on the dice are added to the player's turn score. At any point, a player can hold, which takes their turn score and permanently adds it to their overall score. If they roll a 1 on either die, they lose all of the points they had collected in their turn score, nothing extra is added to their overall score, and it becomes the next player's turn. When a player's overall score reaches 100, the game ends, and they have won.

To illustrate, let's say Alice (player #1) rolls a 3 and a 4. That gives her a turn score of 7. Alice chooses to roll again, getting a 2 and a 6. That gives her an extra 8 points, for a total turn score of 15 points. Alice then chooses to hold, and her current turn score is added to her overall score. Alice now has 15 points overall, and it is now Wally's turn (player #2). Wally rolls a 2 and a 3, giving him a turn score of 5. Wally chooses to roll again, but this time, gets a 1 and a 5. Since he rolled a 1, he loses his turn score (which was 5) leaving him still with 0 overall points, and it becomes Alice's turn again.

Our goal is to create a computer version of Pig Dice. On each player's turn, allow them choose to either hold or roll. When a player rolls, generate two random numbers between 1 and 6, and program the game to follow the logic described above. Between rolls, display the current status of the game, showing both player's overall score, and the current player's turn score. When a player gets up to 100, end the game and show who won. For an extra challenge, make the game playable by three or more players.

Connect Four

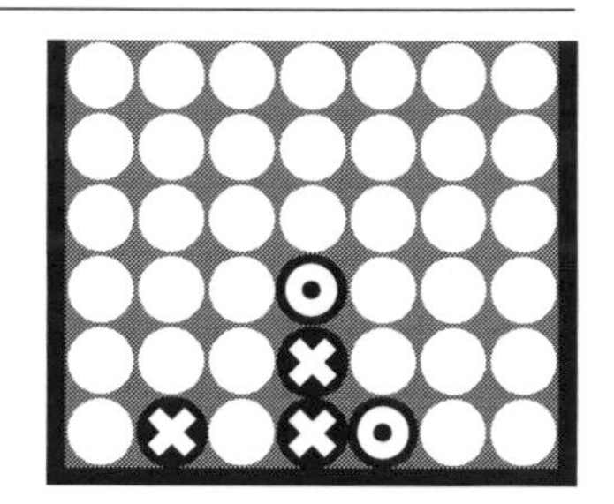

The game of Connect Four is a simple and interesting game played on a grid with six rows and seven columns. Two players take turns taking tokens and dropping them into one of the seven columns, where it falls down to either the bottom if the column is currently empty or to the top of the pile if it is not.

The goal of the game is to get four of your tokens in a row, up and down, side to side, or diagonally, before the other player does. Your task is to make a game where two alternating players can select a row (numbers 1 through 7) to place their token on and play the game of Connect Four.

There are a few pieces to this game that might be challenging. One will be writing the code to take a row and add a token to the top of it. On a related note, it may also be tricky to ensure that if a row is full, the game doesn't crash, but instead, notifies the player that the move they wanted to make was an illegal move, and give them another chance. Also, after each move, you'll want to check to see if that move caused the player to win.

By the way, while many games like this tend to look great with a GUI the game can still easily be made to look good by printing out the right things to the console window, as shown below:

```
. . . . . . .
. . . . . . .
. . . . . . .
. . . O . . .
. . . X . . .
. X . X O . .
```

Conway's Game of Life

In 1970, British mathematician John Conway invented a "zero player" game called the Game of Life. This is not the Milton Bradley board game called LIFE, but rather a mathematical game that simulates life.

The game is "played" on a grid, where the game advances in generations, with cells becoming alive or dying, based on certain conditions.

If a particular cell in the grid is empty (dead) then it can come to life if the following conditions are met:

- Exactly three neighbors that are alive.

If a particular cell in the grid is alive, then it dies in the following generation if any of the following are true:

- It has more than three neighbors. (Overcrowding.)
- It has 0 or 1 neighbors (Loneliness, I guess...)

For example, look at the little grid below:

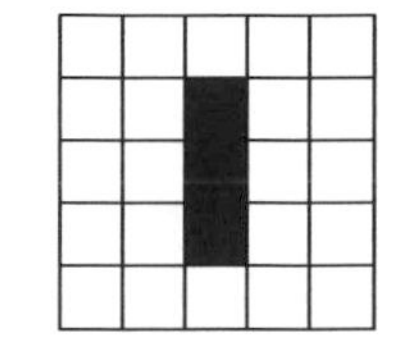

In the next generation (round) the top cell that is alive will die, because it only has one neighbor. The one in the middle will stay alive, because it has two neighbors. The one on the bottom will also die, because it only has one neighbor. Additionally, just off to the left and the right of the three "alive" cells, there is a cell with three neighbors, both of which will become alive. After one generation, this will turn into the following:

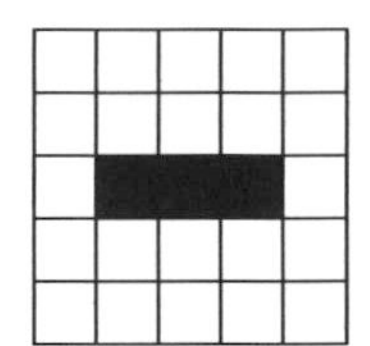

Interestingly, if you look at this and follow it through to the next generation, you'll end up back where you were initially. This happens to create what is called an oscillator in the Game of Life. It will keep repeating forever.

These simple rules for determining when cells in the grid come to life or die create very interesting results that are much more interesting to see than to read in a book.

Because of this, we're going to take upon ourselves the challenge of creating the Game of Life as a computer program.

There are several key parts to this. For one, we will need to store the current grid. There are lots of ways to do this, but one might be a two dimensional array of **bool**s.

We will also need a method for drawing our current grid. Again, it would look nice with a fancy GUI, but we can get away without needing to go beyond the console. If you're sticking with the console, your best bet is to use 40 columns in your grid, and have each cell take up two characters (an "X" or "." depending on if it is alive or dead, plus an empty space) because on the console, characters are twice as tall as they are wide, and there are 80 columns by default. Using 40 columns with two characters per column gives you nice, square shaped cells. About 12 rows fit on the console window by default, but you can go up to 20 rows pretty easily and then drag the console window to resize it and make it taller.

Additionally, you will need to create a mechanism to update the grid using the rules described earlier for cell death and generation. It is important to work with a copy for the next generation or updating one cell will mess up the calculation in another cell.

Finally, you'll need to put it all together in a loop that repeats forever, updating the grid, drawing the grid, and waiting for a while in between, to animate the game and let you see what's happening. For this, you might find the **Thread.Sleep** method useful. We talked about this method in Chapter 37. You will need to add a **using** directive to the top of your code for **System.Threading** to get access to this, as discussed in Chapter 26. You might also find it useful to be able to clear the console window before you draw the new grid, using **Console.Clear();**

The most interesting part about the Game of Life is being able to start with different starting configurations and watching it evolve. To make this happen, make it so you can load in initial configurations in a file. The specifics of the file format are up to you, but one possibility is to make them look like this:

```
........................................
........................................
........................................
........................................
........................................
........................................
...................X....................
...................X....................
...................X....................
........................................
........................................
........................................
........................................
.......................X................
........................X...............
......................XXX...............
........................................
........................................
........................................
........................................
```

You can create these files in any text editor, like Notepad.

When your game is up and running, try loading the following patterns and watch what happens:

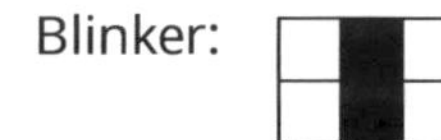

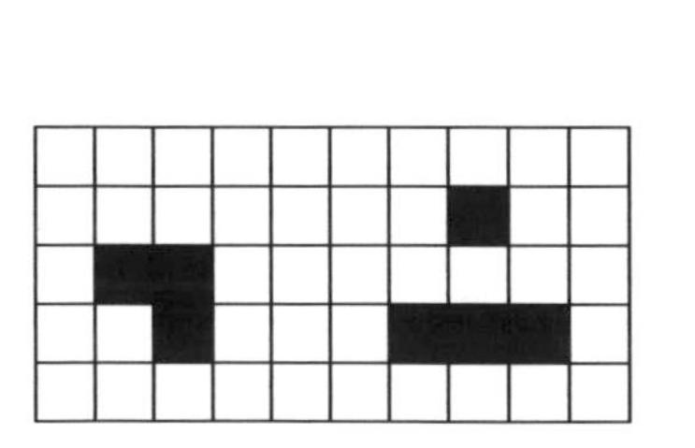

There are tons of interesting patterns that show up in the Game of Life, so when you get your game working, experiment with various configurations, and look online and see what other interesting patterns exist.

47

What's Next?

In a Nutshell

- Your next step might include learning Windows Forms or WPF if you want to make desktop applications, ASP.NET if you want to make websites or web-based programs, the libraries for Universal apps or the Windows phone, or MonoGame, XNA, Unity, or SharpDX if you want to try your hand at making video games.
- It is also worth your time to learn software engineering, as well as data structures and algorithms if you haven't already done so while learning another language.
- The best way to learn C# is to program in C#.
- You can get more information from MSDN, Stack Overflow, and this book's website.

As you finish learning everything this book has to offer, you may be wondering where you're supposed to go next. And it really depends. I want to take a little time and give you some pointers on where your next steps might take you.

Other Frameworks and Libraries

If you've gone through this book sequentially, you already know all of the basic things you need to know about the C# language. Much of the rest of your learning in the C# world will be more about learning additional libraries and frameworks, and understanding how to use the types and methods they contain, and less about learning new features of C# itself.

The specific libraries and frameworks that you'll want to learn depend on what you want to accomplish with C#. Below, I'll provide a brief discussion about some of the bigger, more useful ones.

WPF and Windows Forms

If you want to make desktop applications with windows, dialog boxes, check boxes, and the like, you'll want to take some time to learn Windows Forms (sometimes called WinForms) or Windows Presentation Foundation (WPF). Both of these do the same types of thing. Windows Forms is older, and generally considered to be a bit easier to understand. And while you can still make applications with it, Microsoft is no longer adding new features to it. WPF is designed to ultimately replace it, but there are lots of applications that still use WinForms, and even a lot of new ones that are still based on it. WPF is a little bit harder to learn, but it has far more power and is probably where the future of desktop-based Windows programming is. (But also consider Universal apps, below.) In the long run, WPF is probably a better investment of your time than Windows Forms.

ASP.NET

If you're more interested in making web-based software, ASP.NET is the next step. ASP.NET allows you to run C# code on the server side, in place of alternatives like PHP, Java (Java Server Pages), Python, or Perl. By the way, don't get this confused with Active Server Pages (ASP), which it replaced. ASP.NET allows you to run C# code on the server end of a website to generate dynamic content for the user. This doesn't take the place of client side code like JavaScript, nor does it eliminate the need for HTML. An ASP.NET program will generate a custom-made HTML page for the user of a website to look at.

Universal Windows Apps

Universal Windows Apps (usually shortened to Universal Apps, often called Windows Store Apps, and previously called Metro-Style Apps, among a dozen other names) is a framework that lets you make applications that run on all of the Microsoft platforms with only minor deviations for the different devices. That includes Windows on a desktop or laptop (often sold through the Windows Store), as well as Surface tablets, Windows Phones, and even the Xbox One.

Unfortunately, "universal" is probably a bit of a misnomer here. When creating universal apps, you'll be able to have a ton of shared code, but you'll inevitably end up with some custom code for each of the different target platforms. This is far better than trying to develop completely independent codebases for the different platforms, but there's still room for improvement. Fortunately, Microsoft is heading in the right direction with this, and the platforms are converging, rather than diverging.

The universal app platform relies heavily on XAML, one of the core technologies of WPF, so there's a lot of overlap when using these different frameworks.

Making universal apps requires Windows 8 or newer.

Windows Phone

Microsoft also provides tools to make it easy to make apps for the Windows phone. While Windows phones make up only a small percentage of smart phones worldwide, this is an area that is worth watching in the future. Microsoft's big push with universal apps, which will work on the next generation of Windows phones may bring a much larger collection of apps to the platform.

You can easily do app development for Windows 8 phones in Visual Studio, as long as you're developing on a Windows 8 computer and have the Windows Phone SDK installed.

Game Development

I've saved my personal favorite for last. C# is a very popular choice for game development, and you have a wide variety of options for doing so in C#.

To enumerate just a few, Microsoft built XNA, which has been a popular game development framework that runs on top of DirectX 9 and runs on a wide variety of Microsoft-based platforms. The XNA framework has since been ported under the name MonoGame, which targets the Mono framework (an open source port of the .NET Framework) and allows you to use either OpenGL or

DirectX 11 under the covers. In addition to Windows, it targets Linux, Mac, Android, iOS, web, PlayStation Mobile, and the Ouya console. XNA and MonoGame share essentially identical syntax, and you'll find yourself able to quickly make virtually any game you can dream up. XNA was effectively retired a few years back. While you can still make games with XNA, the momentum has shifted to MonoGame in recent years.

MonoGame and XNA tend to aim for the middle of the road when choosing between power and development speed. If you want more control or have a lot of very specific needs in your game, you can switch to raw OpenGL through OpenTK or Tao, or work with DirectX through SharpDX or SlimDX. These options require more low-level boilerplate code, but in exchange, you have unbound flexibility and control to build whatever you need to build. You can also often pull in libraries for networking, AI, physics, etc. to keep your development speed up.

On the other end, giving up some amount of flexibility and control for more rapid development, there are quite a few game engines that allow you to use C# for scripting. Probably the most well-known is Unity 3D, but other options include Paradox and Delta Engine. There are actually many game engines that use C#.

As you can see, C# has no shortage of game development options.

Other Topics

In addition to learning other frameworks, it is also important to learn things like the best way to structure your code, or various algorithms and data structures to use. If you are studying C#, and you're getting started with formal education at a university, you'll learn things like this in a software engineering class, or a data structures and algorithms class. Most universities require you to take these classes, but if it is optional, don't skip it. It will be worth it.

If you aren't at a university, you'll still want to learn these things. Take the time to learn software engineering practices. It will be worth the time. You'll know much better how to organize code into classes and how to make classes interact effectively. Doing it the right way makes it easier to make changes as your plans for your program change and grow over time. You'll learn things like how to best test your code to make sure it is bug free and working like it should. And you'll get a basic idea of how to manage your projects to finish things on time.

Also, be sure to learn about data structures and algorithms for things like sorting and searching. Making software is a whole lot more than knowing how to write lines of code. And to make truly awesome software, you'll need to understand how things will work at a higher level, and how to get the fastest results, or when to maximize for speed vs. memory usage.

Of course, if you've been programming for a while, especially if you had formal education in things like this, then you probably already know most of this stuff. Generally speaking, the best practices that you learned in other languages will carry over to the C# world.

Make Some Programs

By far the best thing you can do now is make some programs. Nothing will launch you into the C# world faster than making a real program. Or several. You'll quickly see what topics you didn't fully grasp (and you can then go back and learn them in depth) or you'll figure out what extra little pieces you'll need to learn still, and you'll be able to quickly hunt down those things on the Internet.

Of course, that's why I've included the *Try It Out!* chapter (Chapter 46), to give you some interesting problems to work through while testing your knowledge of various aspects of programming.

The programs you choose to work on do not need to be limited to the basics of C# that we've discussed in this book. If you want to try tackling something that also uses WPF, ASP.NET, universal apps, or game development, go for it. There's no better way to learn to program than by programming.

Where Do I Go to Get Help?

The best part about making software is that you're creating something that has never been done before. It is as much a creative process as it is mathematical or analytical. You're not just manufacturing the exact same thing over and over again. As such, you're going to run into problems that have never been encountered before. You're bound to run into trouble sooner or later, and so before we part ways, I want to point out to you a few meaningful things you can do to solve your problems when they come up.

First, obviously, a Google search will go a very long way. That's always a great place to start.

Second, make sure you fully understand what the code you are working with is doing. If you're having trouble with a particular class that someone else created (like the **List** class, for instance) and a method isn't doing what you thought it should, take the time to make sure you understand every piece that you're working with. I've learned over the years, that if you're running into a problem with code you don't quite understand, you're not likely to figure the whole problem out until you've figured out how to use all of the individual pieces that are in play. Learn what each parameter in a method does. Learn what exceptions it might throw. Figure out what all of its properties do. Once you truly understand the things you're working on, usually the problem solves itself.

Along those lines, the Microsoft Developer Network (MSDN) has tons of details about every type that the .NET Framework has, with plenty of examples of how to use them. It's a great resource if you're trying to learn how a specific type works. (http://msdn.microsoft.com/library/)

I probably don't need to say this, but if you're stuck, and you've got team members that might know the answer, it is always a good idea to talk to them.

If you don't have team members who can answer the question, Stack Overflow is one of the best sites out there for programming Q&A. (http://stackoverflow.com/) Stack Overflow covers everything from the basics up to some very specific, very detailed questions, and you're likely to get good, intelligent responses from people. Just be sure to do a Google search to make sure the answer isn't obvious, and search on the site as well, to make sure the question hasn't already been asked. (Because it probably has.)

Parting Words

You've learned the basics of C#, and ready to take the world by storm. There's so much that you can do with your new knowledge of C#. It truly is an exciting time!

From one programmer to another, I want to wish you the best of luck with the amazing software that you will create!

Glossary

.NET Framework
The framework that C# is built for and utilizes, consisting of a virtual machine called the Common Language Runtime and a massive library of reusable code called the Framework Class Library. (Chapters 1 and 40.)

Abstract Class
A class that you cannot create instances of. Instead, you can only create instances of derived classes. The abstract class is allowed to define any number of members, both concrete (implemented) and abstract (unimplemented). Derived classes must provide an implementation for any abstract members defined by the abstract base class before you can create instances of the type. (Chapter 22.)

Abstract Method
A method declaration that does not provide an implementation or body. Abstract methods can only be defined in abstract classes. Derived classes that are not abstract must provide an implementation of the method. (Chapter 22.)

Accessibility Level
Types and members are given different levels that they can be accessed from, ranging from being available to anyone who has access to the code, down to only being accessible from within the type they are defined in. More restrictive accessibility levels make something less vulnerable to tampering, while less restrictive levels allow more people to utilize the code to get things done. It is important to point out that this is a mechanism provided by the C# language and the .NET Framework to make programmer's lives easier, but it is not a way to prevent hacking, as there are still ways to get access to the code. Types and type members can be given an access modifier, which specifies what accessibility level it has. The **private** accessibility level is the most restrictive, and means the code can only be used within the type defining it, **protected** can be used within the type defining it and any derived types, **internal** indicates it can be used anywhere within the assembly that defines it, and **public** indicates it can be used by anyone who has access to the code. Additionally, the combination of **protected internal** can be used to indicate that it can be used within the defining type, a derived type, or within the same assembly. (Chapters 18 and 21.)

Accessibility Modifier
See *Accessibility Level*.

Anonymous Method
A special type of method where no name is ever supplied for it. Instead, a delegate is used, and the method body is supplied inline. Because of their nature, anonymous methods cannot be reused in multiple locations. Lambda expressions largely supersede anonymous methods and should usually be used instead. (Chapter 35.)

Argument
See *parameter*.

Array
A collection of multiple values of the same type, placed together in a list-like structure. (Chapter 13.)

ASP.NET
A framework for building web-based applications using the .NET Framework. This book does not cover ASP.NET in depth. (Chapter 47.)

Assembly
Represents a single block of redistributable code, used for deployment, security, and versioning. An assembly comes in two forms: a process assembly, in the form of an EXE file, and a library assembly, in the form of a DLL file. An EXE file contains a starting point for an application, while a DLL contains reusable code without a specified starting point. See also *project* and *solution*. (Chapter 40.)

Assembly Language
A very low level programming language where each instruction corresponds directly to an equivalent instruction in machine or binary code. Assembly languages can be thought of as a human readable form of binary. (Chapter 40.)

Assignment
The process of placing a value in a specific variable. (Chapter 5.)

Asynchronous Programming
The process of taking a potentially long running task and pulling it out of the main flow of execution, having it run on a separate thread at its own pace. This relies heavily on threading. (Chapters 37 and 38.)

Attribute
A feature of C# that allows you to give additional meta information about a type or member. This information can be used by the compiler, other tools that analyze or process the code, or at run-time. You can create custom attributes by creating a new type derived from the **Attribute** class. Attributes are applied to a type or member by using the name and optional parameters for the attribute in square brackets immediately above the type or member's declaration. (Chapter 39.)

Base Class
In inheritance, a base class is the one that is being derived from. The members of the base class are included in the derived type. A base class is also frequently called a superclass or a parent class. A class can be a base class, and a derived class simultaneously. See also *inheritance*, *derived class*, and *sealed class*. (Chapter 21.)

Base Class Library
The central library of code that nearly all C# programs will utilize, including the built-in types, arrays, exceptions, threading, and file I/O. (Chapter 40.)

BCL
See *Base Class Library*.

Binary Code
The executable instructions that computers work with to do things. All programs are built out of binary code. (Chapters 1 and 40.)

Binary Instructions
See *Binary Code*.

Binary Operator
An operator that works on two operands or values. Many of the most common operators are binary, such as addition and subtraction. (Chapter 7.)

Bit Field
The practice of storing a collection of logical (Boolean) values together in another type (such as **int** or **byte**) where each bit represents a single logical value. Enumerations can also be used as a bit field by applying the **Flags** attribute. When working with a bit field, the bitwise operators can be used to modify the individual logical values contained in the bit field. (Chapter 39.)

Bitwise Operator
One of several operators that operate on the individual bits of a value, as opposed to treating the bits as a single value with semantic meaning. Bitwise operators include bitwise logical operators, which perform operations like **and**, **or**, **not**, and **xor** (exclusive or) on two bits in the same location of two different values. It also includes bit shift operators, which slide the bits of a value to the left or right. In C#, the extra spots are filled with the value 0. (Chapter 39.)

Boolean
Pertaining to truth values. In programming, a Boolean value can only take on the value of true or false. Boolean types are a fundamental part of decision making in programming. (Chapter 6.)

Built-In Type
One of a handful of types that the C# compiler knows a lot about, and provides special shortcuts for to make working with them easier. These types have their own keywords, such as **int**, **string**, or **bool**. (Chapter 6.)

Breakpoint
While debugging, a line of code may be marked with a breakpoint, and when the program reaches that point, the program will pause and switch back to the Visual Studio debugger, allowing you to take a look at the current state of the program. (Chapter 44.)

C++
A powerful all-purpose programming language that C# is largely based on. C++ is generally compiled to executable code, and is not run on a virtual machine, with the exception of Visual C++, which is a variation designed for the .NET Framework. (Chapter 40.)

Casting
See *Typecasting*.

CIL
See *Common Intermediate Language*.

Class
One of several categories of types that exist in the C# language that can be custom designed. A class defines a set of related data that belongs to a single type or category of objects in the real world, or representation of a concept in the domain model, along with the methods that are used to interact with or modify that data. A class is as a blueprint for objects or instances of the class, defining what they store and what they can do. All classes are reference types. See also *struct*, *type*, and *object*. (Chapter 17.)

CLR
See *Common Language Runtime*.

Code Window
The main window, usually in the center of the screen, which allows you to edit code. The window can show multiple code files tabbed together. Any hidden window can be opened up through the **View** menu, or the **View > Other Windows** menu. (Chapter 41.)

Command Line Arguments
When a program is started, arguments may be provided on the command line of a program, which are passed into the program's main method where it can use them. These parameters are command line arguments. (Chapter 39.)

Comment
Additional text placed within source code that is designed to be read by any humans that are working with the code. The C# compiler ignores comments entirely. (Chapter 4.)

Common Intermediate Language
A special high-level, object-oriented form of assembly code that the CLR is able to execute. It includes instructions for things like type conversion, exceptions, and method calls. (Chapter 40.)

Common Language Runtime
A virtual machine that any .NET language including C# is built to run on. The Common Language Runtime, often called the CLR, converts CIL code that the C# compiler created into binary instructions for the computer to run on the fly. (Chapter 40.)

Compile-Time Constant
See *Constant*.

Compiler
A special program that turns source code into executable machine code. (Chapters 1 and 40.)

Compound Assignment Operator
An operator that combines a normal math operation with assignment, as in **x += 3;** (Chapter 7.)

Conditional Operators
The operators && (*and* operator) and || (*or* operator), which are used to perform checks for multiple conditions. (Chapter 10.)

Constant
A special type of variable that, once set, cannot be modified. Constants come in two variations. Compile-time constants are created when the program is compiled, using the **const** keyword. These are treated as global constants and cannot be changed without recompiling your program. Run-time constants, created with the **readonly** keyword, are variables that cannot be modified once assigned to. However, they can be assigned while the program is running. For instance, a type may have read-only instance variables, allowing instances of the type to be created with different values, but forcing the variables to be immutable. Run-time constants are preferred to compile-time constants, unless it is guaranteed that the value will never need to change. (Chapter 39.)

Constructor
A special type of method that initializes an instance of a type. The role of a constructor is to ensure that the new instance will be initialized to a valid state. Like a method, a constructor's definition may include any number of parameters. A constructor must have the same name as the type, and does not have a return type. A type may define multiple constructors. When creating a new instance of a type, a constructor is called, along with the **new** keyword. If a type does not explicitly include a constructor, the C# compiler will automatically generate a default parameterless constructor. (Chapters 17 and 18.)

Critical Section
A block of code that should not be accessed by more than one thread at once. Critical sections are usually blocked off with a mutex to prevent multiple simultaneous thread access. (Chapter 37.)

Curly Braces
The symbols **{** and **}**, used to mark blocks of code. (Chapter 3.)

Debug
The process of working through your code to find and fix problems with the code. (Chapter 44.)

Declaration
The process of creating something, specifying the important information about it. This is typically used to refer to variable declaration, but it is also applied to methods and type definitions. (Chapter 5.)

Decrementing
Subtracting 1 from a variable. See also *Incrementing*. (Chapter 9.)

Delegate
A way to treat methods like objects. A delegate definition identifies a return type and a list of parameter types. A delegate is created in a way that is similar to declaring a variable, and can be assigned "values" that are the names of methods that match the return type and parameter list of the delegate. The delegate can then be called, which will execute whatever method is currently assigned to the delegate and return the result. (Chapter 30.)

Derived Class
In inheritance, the derived class extends or adds on to another class (the base class). All members of the base class, including instance variables, methods, and events also exist in the derived class. Additional new members may also be added in a derived class. In the case of multiple levels of inheritance (not to be confused with multiple inheritance) a class may act as both a base class and a derived class. A derived class is also sometimes called a subclass. See also *base class* and *inheritance*. (Chapter 21.)

Divide and Conquer
The process of taking a large, complicated task and breaking it down into more manageable, smaller pieces. (Chapter 15.)

Division by Zero
An attempt to use a value of 0 on the bottom of a division operation. From a math standpoint, it is meaningless and isn't allowed. In programming, it often results in your program crashing. (Chapter 9.)

DLL
A specific type of assembly that contains no specific entry point but rather a collection of code that can be reused by other applications. See also *assembly* and *EXE*. (Chapter 42.)

Enum
See *Enumeration*.

Error List
A window in Visual Studio, usually located near the bottom, which displays a list of compiler errors and warnings. Any hidden window can be opened up through the **View** menu, or the **View > Other Windows** menu. (Chapter 41.)

EXE
A specific type of assembly that contains an entry point which can be started and run by the .NET Framework. See also *assembly* and *DLL*. (Chapter 40.)

Enumeration
A specific listing of the possible values something can take. In C#, enumerations are used to represent a type of data where there is a known, specific, finite set of options to choose from. (Chapter 14.)

Event
A mechanism, built on delegates, which allow one part of the code to notify others that something specific has occurred. Event handlers are methods which match a particular delegate and they can be attached or removed from the event, allowing it to start or stop receiving notifications from the event. (Chapter 31.)

Exception
An object that encapsulates an error that occurred while executing code. The object is "thrown" or passed up the call stack to the calling method until it is either handled ("caught") or is thrown from the Main method, causing the program to crash. (Chapter 29.)

Explicit
A term frequently used to mean something must be formally stated or written out. The opposite of "implicit." (Chapter 9.)

Extension Method
A special type of static method that appears to be a part of a class but is actually defined outside of it. This allows you to create additional methods for types that are sealed or that you do not have access to the source code for. Extension methods are essentially syntactic sugar, because when the code using the extension method is compiled, it will be rewritten to call the static method. (Chapter 34.)

FCL
See *Framework Class Library*.

Field
See *instance variable*.

Floating-Point Type
One of several built-in types that are used for storing real-valued numbers, such as fractional or decimal numbers. (Chapter 6.)

Framework Class Library
A massive collection of reusable code that is a part of the .NET Framework. This includes all of the Base Class Library, plus additional code for specific functional categories like database access or GUI programs. (Chapter 40.)

Fully Qualified Name
The full name of a type, including the namespace that it belongs in. (Chapter 26.)

Function
See *method*.

Garbage Collection
The process of removing objects on the heap that are no longer accessible. Garbage collection in C# is automated. Garbage collection makes it so that you do not need to worry as much about memory leaks, and you do not need to manually deallocate memory for your program, in most cases. See also *managed memory*. (Chapter 16.)

Generics
A mechanism that allows you to create classes that are type safe, without having to commit to specific types at compile time. Rather than creating something that can work with only a specific type (and then creating a very similar version for other types) and instead of using a very generalized type (like **object**) and casting to and from that to the wanted type, generics can be used to specify that at run-time, a certain specific type will be used, but it is currently unknown, and may be different in different instances. (Chapters 24 and 25.)

Heap
One of two main parts of a program's memory. The heap is unstructured, and any part of the program can access the information on the heap as needed. Because the heap is unstructured, memory allocation and deallocation must be handled carefully. The CLR manages the heap (see managed memory) using garbage collection. See also *stack* and *reference type*. (Chapter 16.)

IDE
See *Integrated Development Environment*.

IL
See *Common Intermediate Language*.

Implicit
A term frequently used to mean something happens without needing to be specifically stated. The opposite of "explicit." (Chapter 9.)

Incrementing
Adding 1 to a variable. See also *Decrementing*. (Chapter 9.)

Indexer
An indexer is a part of a type that defines what the indexing operator should do for the type. Indexers can use any number of parameters, and can use any type—they're not just limited to integers. (Chapter 33.)

Inheritance
The ability of one class to include all of the members that another class has, and add on additional members. Inheritance is designed to mimic an *is-a relationship*, or an *is-a-special-type-of* relationship—in other words, things where one type is a special type of another. For example, an octagon is a special type of polygon, and so an **Octagon** class may inherit from, or be derived from, a **Polygon** class. C# does not allow for inheritance from more than one base class (multiple inheritance) but a similar effect can be accomplished by implementing multiple interfaces. Inheritance can be many levels deep, where classes are derived from classes that are derived from yet other classes. Every type in C# is ultimately derived from the **object** type. Note that structs do not support inheritance. (Chapter 21.)

Instance
See *object*.

Interface
A listing of specific members that a type that implements the interface must have. It defines a specific contract that a type needs to present to the outside world. A type may implement multiple interfaces. All members defined in any interface a type implements must be defined in the type. (Chapter 23.)

Integer Division
A special kind of division, used when working with only integral types, in which any remainder or fractional part of the result is dropped. Using integer division, 7 / 2 is 3. (Chapter 9.)

Integral Type
One of several built-in types that are used for storing integers. (Chapter 6.)

Integrated Development Environment

A program that is built for the purpose of making it easy to create programs. It typically includes a source code editor and a compiler, along with many other features to assist with things like project management, testing, and debugging. (Chapters 1 and 41.)

IntelliSense

A feature of Visual Studio that performs auto-completion of various tasks such as name completion, but also provides a convenient way to view the XML documentation comments of types and their members. IntelliSense is brought up automatically when the dot operator (member access operator) is used, but it can also be brought up any time by using **Ctrl + Space**. (Chapter 41.)

Internal

See *accessibility level*.

Immutability

A feature of a type (or sometimes, just a member of a type) that prevents it from being modified once it has been created. In order to make changes to an instance, a new instance must be created with the desired changes. Immutability provides certain benefits, including simplicity, automatic thread safety, usable as keys in a dictionary or hash table, and you do not need to make defensive copies when returning them from a method or property. Value types should be made immutable in most cases. Classes should be immutable when possible. (Chapter 20.)

Implicitly Typed Local Variable

Using type inference, a local variable's type does not need to be specified in some cases. Instead, the **var** keyword is used to indicate that type inference should be used. Only local variables can be implicitly typed. It is important to remember that implicitly typed local variables actually do have a specific type (they're not loosely typed) but that type is determined by the compiler instead of the programmer. (Chapter 39.)

Instance Variable

A non-static variable that is declared as a member of a type. As such, it is available throughout the type, and depending on its accessibility level, it may be accessible from outside the type as well. Contrasted with a static class variable, where all instances of the type share the same variable, each instance of the type will have its own variable that functions independently of the variable with the same name in other instances. It is good practice to make sure instance variables are not accessible from outside of the type to adhere to the principle of encapsulation. (Chapters 17 and 18.)

Iterator

A mechanism for visiting or looking at each item in a collection of data. A **foreach** loop can "iterate" over (or loop over) items in any type that provides an iterator. (Chapter 39.)

Jagged Array

An array of arrays, where each array within the main array can be a different length. (Chapter 13.)

Java

A high-level, all-purpose programming language similar to C#. Like C#, it also runs on a virtual machine. (Chapter 1.)

JIT Compiling

See *Just-in-Time Compiling*.

Just-in-Time Compiling

The process of converting IL code to executable code right before it is first needed. The CLR uses Just-in-Time compiling (JIT compiling) as it runs C# code. (Chapter 40.)

Keyword

A special reserved word in a programming language. (Chapter 3.)

Lambda Expression

An anonymous method that is written with a simplified syntax to make it easy to create. Lambda expressions are powerful when used where a delegate is required, and is often very useful when using LINQ. If a lambda expression becomes complicated, it is usually better to pull it out into a normal method. (Chapter 35.)

Language Integrated Query

A part of the C# language that allows you to perform queries on collections within your program. These queries involve taking a set of data, filtering the result based on certain conditions, ordering, and returning the parts of the data that are relevant. Often called LINQ (pronounced "link"). LINQ provides a syntax that is similar to SQL and other query languages. LINQ is typically done using query expressions. (Chapter 36.)

LINQ

See *Language Integrated Query.*

Local Variable

A variable that is created inside of a method, and is only accessible within that method. Some local variables may have scope smaller than the entire method. A local variable that is declared inside of a loop, **if** statement, or other block will only have block scope, and cannot be used outside of that block. (Chapter 15.)

Loop
To repeat something multiple times. C# has a variety of loops, including the **for**, **while**, **do-while**, and **foreach** loops. (Chapter 12.)

Managed Memory
Rather than requiring you to allocate and deallocate or free memory on the heap, the CLR and the .NET Framework keep track of this for you, freeing you to concentrate on other things. Memory that is no longer used is cleaned up by the garbage collector. The CLR will also move memory around to best utilize the space. (Chapter 16.)

Member
Anything that can belong to a type, including instance variables, methods, delegates, and events. (Chapter 17.)

Method
A piece of code that does a single task. A method is given a specific name, while data can be handed to the method via parameters, and a single resulting value may be returned from the method upon completion. (Chapter 15.)

Method Body
See *method implementation*.

Method Call
Going from one method into another. When a method is called, data may be sent to the method by passing them in as parameters to the method. When a method call is finishing, information may be sent back or "returned" from the method. A method may call itself (recursion). Method calls are kept track of on the stack. As a new method is called, a new frame is placed on the stack. As the flow of execution returns from a method call, the top frame (which represents the current method) will be removed, returning execution back to the calling method. (Chapter 15.)

Method Implementation
The actual code that defines what the method should do. This is enclosed in curly braces right below the method declaration in most cases. A method without an implementation is an abstract method, and can only be made in an abstract class. (Chapter 15.)

Method Signature
The name of the method and types of parameters that a method has. This does not include its return type or the names of the parameters. This is largely what distinguishes one method from another. Two methods cannot share the same method signature. They must have either different names or different parameter types (or be declared in different types). (Chapter 15.)

Mutex
See *Mutual Exclusion.*

Mutual Exclusion
Setting up code so that only one thread can access it at a time. The mechanism that forces this is often called a mutex. (Chapter 37.)

Named Parameter
When passing values into a method, the name of the parameter they go with can be explicitly named, allowing the parameters to be given out of order. (Chapter 27.)

Namespace
A collection of related types, grouped together under a common label. Namespaces often represent broad features of a program, and contain the types that are needed to implement the feature. The types in a namespace are typically placed together in a single folder in the directory structure. (Chapters 3 and 26.)

Name Collision
A situation where two different types use the same name. At compile time, the compiler will be unable to determine which of the two types is supposed to be used. To resolve a name collision, you must either use fully qualified names or an alias to rename one or both of the types involved in the collision. (Chapter 26.)

Name Hiding
When a variable in method or block scope has the same name as an instance variable or static variable, making the instance variable or static variable inaccessible. The instance variable or static variable may still be accessed using the **this** keyword. (Chapter 18.)

NaN
A special value used to represent no value. It stands for "Not a Number." (Chapter 9.)

Nesting
Placing one statement or block of statements inside of another block. (Chapter 10.)

Null Reference
A special reference that refers to no object at all. (Chapter 16.)

Nullable Type
A mechanism in C# for allowing value types to additionally be assigned **null**, like a reference type. A nullable type is still a value type. A nullable type is simply one that uses the **Nullable<T>** struct, but the C# language allows for a

simplified syntax, by using the type, followed by a '?'. For example: **int? a = null;** (Chapter 39.)

Object
A collection of data and the methods that operate on that data. Objects are instances of a type (often a class) and represent a specific object or representation of something from the real world or the domain model. This is contrasted with the class (or other type definition) itself, which defines what kinds of data objects of that type will be able to keep track of and the methods that operate on that data. (Chapter 17.)

Object-Oriented Programming
An approach to programming or programming languages where code is packaged into chunks of reusable code that have data and methods that act on that data, bundled together. C# is an object-oriented programming language. (Chapter 17.)

Operator
A calculation or other work that is done by working with one, two, or three operands. Many of the C# operators are based on math operators, such as addition (**+**) or multiplication (*****). (Chapter 7.)

Operator Overloading
Within a type, providing a definition for what a particular operator (**+**, **-**, *****, **/**, etc.) should do. Not all operators are overloadable. Operators should only be overloaded if there is an intuitive way to use the operator. (Chapter 32.)

Optional Parameter
When a parameter is given a default value, allowing calling methods to not provide a value for the parameter if desired. (Chapter 27.)

Order of Operations
When an expression contains more than one operator, the order of operations dictates what operation is done first. (Chapter 7.)

Overflow
When the result of an operation exceeds what the data type is capable of representing. In a purely mathematical world, this doesn't happen, but on a computer, since all types have a certain size with set ranges and limits, overflow is when those bounds are surpassed. (Chapter 9.)

Overloading
Providing multiple methods with the same name but different parameter list. Not to be confused with overriding, an overload creates an entirely different method, though usually, the purpose is to do similar things with different inputs or parameters. For overloading operators, see *operator overloading*. (Chapter 15.)

Overriding
Taking a method, property, or indexer in a base class and providing an entirely different implementation of it in a derived class. In C#, in the base class, the method must be either **virtual** or **abstract**, and in the derived class, the method must be marked with **override** in order to override the member. (Chapter 22.)

Parameter
A type of variable that has method scope, meaning it is available from anywhere inside of a method. Unlike other local variables, the value contained in a parameter is determined by the calling method, and may be different for different method calls. (Chapter 15.)

Parent Class
See *base class*.

Parentheses
The symbols **(** and **)**, used for order of operations, the conversion operator, and method calls. (Chapters 7 and 15.)

Parse
The process of taking something and breaking it down into smaller pieces that have individual meaning. Parsing is a common task when reading data in from a file, or from input given by the user. (Chapter 28.)

Partial Class
A class that is defined in multiple files. This can be done to separate a very large class into more manageable pieces, or to separate a class into parts that are maintained by separate things, such as the programmer and a designer (a special window in Visual Studio, created to make it easy to build GUIs etc.). (Chapter 21.)

Pointer
See *reference*.

Pointer Type
A special type that can be used only in unsafe contexts, which uses pointers in a manner similar (but more limited than) C++. (Chapter 39.)

Polymorphism
In C#, a class may declare a method or property, including optionally providing an implementation for it, which can then be overridden in derived classes in different ways. This means that related types may have the same methods implemented with different behavior. This ability to have the same method that does different things when done by

different types is called polymorphism. The term comes from Greek, meaning "many forms," which reflects the fact that these different types can behave differently, or do different things, by simply providing different implementations of the same method. (Chapter 22.)

Public

See *accessibility level*.

Preprocessor Directive

Special commands embedded in source code that are actually instructions for the compiler. (Chapter 39.)

Primitive Type

See *Built-In Type*.

Private

See *accessibility level*.

Procedure

See *method*.

Project

In Visual Studio, a project represents the source code, resource files, and settings needed to build a single assembly (in the form of an .EXE file or a .DLL file). See also *solution* and *assembly*. (Chapter 41.)

Property

A member that provides a way for the outside world to get or set the value of a private instance variable, providing encapsulation for it. This largely replaces any **GetX** or **SetX** methods, providing simpler syntax, without needing to publicly expose the data. The instance variable that a property sets or returns is called a backing field, though not all properties need a backing field. Properties do not need to provide both a get and a set component, and when they do, they do not need the same accessibility level. Auto-generated properties can be used for very simple properties that require no special logic. (Chapter 19.)

Protected

See *accessibility level*.

Query Expression

A special expression in C# that allows you to make SQL like queries on data within a C# program. Query expressions are a fundamental part of LINQ. Query expressions have one or more **from** clause, which indicates the source of data, one or more **where** clauses which filter the dataset to find only the data of interest, and a **select** clause, which indicates what part of the data should be returned. Optionally, an **orderby** clause indicates the final ordering of the results. (Chapter 36.)

Rectangular Array

A special form of multi-dimensional arrays where each row has the same number of values. (Chapter 13.)

Recursion

Indicates a method that calls itself. Care must be taken to ensure that eventually, a base case will be reached where the method will not be called again, or you will run out of space on the stack. See also *recursion*. (Chapter 15.)

Refactor

The process of tweaking or modifying source code in a way that doesn't affect the functionality of the program, but improves other metrics of the code such as readability and maintainability. (Chapter 41.)

Reference

A unique identifier for an object on the heap, used to find an object that is located there. This is similar to a pointer, used in other languages, which is a memory address pointing to the object in question. However, a reference is managed, and the actual object may be moved around in memory without affecting the reference. (Chapter 16.)

Reference Semantics

When something has reference semantics, the identity of the object is considered to be the object, rather than the data it contains. When assigned from one variable to another, passed to a method, or returned from a method, while the reference is copied, it results in a reference to the same object or data. This means both variables are ultimately referencing the same data, and making changes to one will affect the other. In C# all reference types have reference semantics. See also *reference type* and *value semantics*. (Chapter 16.)

Reference Type

One of the two main categories of types in C#. Reference types are stored on the heap. A variable that stores a reference type will actually contain a reference to the object's data. Reference types follow reference semantics, and as they are passed into a method or returned from a method, while a copy of the reference is made, the reference still accesses or points to the same object. Because of this, the new reference will point to the same object on the heap, and modifying the object inside of the method will affect the object that was passed in. Classes are all reference types, as are the **string** and **object** types, as well as arrays. See also *value type*, *pointer type*, and *reference*. (Chapter 16.)

Reflection

The ability of a program to programmatically inspect types and their members. This includes the ability to discover the types that exist in an assembly and the ability to locate methods and call them. Reflection allows you to sidestep many of the rules that the C# language provides (such as no external access to private variables or methods) though performance is slower. (Chapter 39.)

Relational Operators

Operators that determine a relationship between two values, such as equality (==), inequality (!=), or less than or greater than relationships. (Chapter 10.)

Return

The process of going from one method back to the one that called it. It is also used to describe the process of giving back a value to the method that called it, upon reaching the end of the method. In this sense, it is said to "return a value" from the method. (Chapter 15.)

Run-Time Constant

See Constant.

Scope

The part of the code in which a member (especially variables) or type is accessible. The largest scope for variables is class or file scope, and it is accessible anywhere within the type. Instance variables or class variables have class scope. Method scope is smaller. Anything with method scope is accessible from within the method that it is used in, but not outside of the method. Parameters and most local variables have method scope. The smallest scope is block scope, which is for variables that are declared within a code block such as a **for** loop. Variables in block scope or method scope may have the same name as something in class scope, which causes name hiding. (Chapter 18.)

Sealed Class

A class that cannot be used as a base class for another class. This is done by adding the **sealed** keyword to the class definition. (Chapter 21.)

Signed Type

A numeric type that includes a + or - sign. (Chapter 6.)

Silverlight

A framework for making rich Internet applications using the .NET Framework, similar to Adobe Flash. This book does not cover Silverlight in depth. (Chapter 47.)

Solution

In Visual Studio, a solution represents a collection of related projects, and keeps track of how they reference each other and other external assemblies. See also *project*, *assembly*, and *Solution Explorer*. (Chapter 41.)

Solution Explorer

A window in Visual Studio, which outlines the overall structure of the code that you are working on. At the top level is the solution you are currently working on. Inside of that, any projects included in the solution are listed, and inside of that are any files that the project uses, including source code files, resource files, and configuration files. Any hidden window can be opened up through the View menu, or the **View > Other Windows** menu. (Chapter 41.)

Source Code

Human-readable instructions that will ultimately be turned into something the computer can execute. (Chapters 1 and 40.)

Square Brackets

The symbols **[** and **]**, used for array indexing. (Chapter 13.)

Stack

One of two main parts of a program's memory. The stack is a collection of frames, which represent individual method calls and that method's local variables and parameters. The stack is structured and is managed by adding frames when a new method is called, and removing frames when returning from a method. See also *heap* and *struct*. (Chapter 16.)

Static

A keyword that is applied to members of a type to indicate that they belong to the class as a whole, rather than any single instance. A static member is shared between all instances of the class, and do not need an instance to be used. A type may define a single, parameterless static constructor, which is executed just before the first time the class is used. This static constructor provides initialization logic for the type. If a type is static, all of its members must be static as well. (Chapter 18.)

String

A sequence of characters. In programming, you can usually think of strings as words or text. (Chapter 6.)

Struct

A custom-made type, assembled from other types, including methods and other members. A struct is a value type, while a class is a reference type. Structs should be immutable in most cases. (Chapter 20.)

Subclass

See *derived class*.

Superclass

See *base class*.

Subroutine
See *method*.

TAP
See *Task-based Asynchronous Pattern*.

Task-based Asynchronous Pattern
A pattern for achieving asynchronous programming (doing other things while you wait for a long-running task to complete) using tasks, primarily in the form of the **Task** class and the generic variant **Task<TResult>**. (Chapter 38.)

Ternary Operator
An operator that works on three operators. C# only has one ternary operator, which is the conditional operator. (Chapter 10.)

Thread
A lightweight "process" (contained within a normal process) which can be given its own code to work on, and run on a separate core on the processor. Depending on what you're trying to do, running your code on multiple threads may make your program run much faster, at the cost of being somewhat more difficult to maintain and debug. (Chapter 37.)

Thread Safety
Ensuring that parts of code that should not be accessed by multiple threads (critical sections) are not accessible by multiple threads. (Chapter 37.)

Type
A specific kind of information, as a category. A type acts as a blueprint, providing a definition of what any object or instance of the type will keep track of and store. C# has a number of built-in types, but custom-made value or reference types can be defined with structs or classes respectively. (Chapters 5 and 6.)

Typecasting
Converting from one type to another using the conversion operator: **int x = (int)3.4;** (Chapter 9.)

Type Inference
The ability of the C# compiler to guess the type involved in certain situations, allowing you to leave off the type (or use the **var** type). (Chapter 39.)

Type Safety
The ability of the compiler and runtime to ensure that there is no way for one type to be mistaken for another. This plays a critical role in generics. (Chapter 24.)

Unary Operator
An operator that works with only a single value, such as the - operator, used to represent negative numbers. (As in the value "-3".) (Chapter 7.)

Underflow
With floating point types, when an operation results in loss of information, because the value was too small to be represented. (Chapter 9.)

Unsafe Code
Unsafe code is not fully managed by the CLR, giving you some added abilities, like better performance, at the expense of giving up some of the benefits of the CLR. Pointer types can only be used in an unsafe context. Unsafe code is only rarely needed. (Chapter 39.)

Unsigned Type
A type that does not include a + or - sign (generally assumed to be positive). (Chapter 6.)

User-Defined Conversion
A type conversion mechanism that is defined by the programmer instead of the language. C# allows you to create implicit and explicit conversions for your types, but it is recommended that you use them sparingly, as they have a variety of complications that often arise from using them. See also *typecasting*. (Chapter 39.)

Using Directive
A special statement at the beginning of a C# source file, which identifies specific namespaces that are used throughout the file, to make it so you do not need to use fully qualified names. (Chapters 3 and 26.)

Using Statement
Not to be confused with **using** directives, a **using** statement is a special way of wrapping up an object that implements the **IDisposable** interface, which disposes of the object when leaving the block of code that it wraps. (Chapter 39.)

Value Semantics
Something is said to have value semantics if its value is what counts, not the thing's identity. When something has value semantics and is assigned from one variable to another, passed to a method as a parameter, or returned from a method, the value is copied, and while the two variables or sides will have the same value, they will be copies of the same data. Modifying one will not affect the other, because a copy was made. In C#, all value types have value semantics. See also *value types* and *reference semantics*. (Chapter 16.)

Value Type
One of two main categories of types in C#. Value types are stored on the stack when possible (when not a part of a reference type). Value types follow value semantics, and when they are passed to a method or returned from a method, are copied. Structs are value types, as are all of the built-in types except **object** and **string**. See also *reference type* and *pointer type*. (Chapter 16.)

Variable
A place for to store data. Variables are given a name, which can be used to access the variable, a type, which determines what kind of data can be placed in it, and a value, which is the actual contents of the variable at any given point in time. Once a variable is created, its name and type cannot be modified, though its value can be. Variables come in a number of different varieties, including local variables, instance variables, parameters, and static class variables. (Chapter 5.)

Virtual Machine
A special piece of software that can run executable code. The CLR is the virtual machine that the .NET Framework uses. Virtual machines usually provide controlled hardware access to the software that they are running and perform a number of useful features for the code that it runs, including a security model and memory management. Virtual machines are also usually responsible for the task of JIT compiling code to machine code at run-time. (Chapter 40.)

Virtual Method
If a method is virtual, any classes derived from the class that contains it may provide an alternate implementation of the method by overriding it. To mark a method as virtual, the **virtual** keyword should be added to it. See also *abstract method* and *overriding*. (Chapter 22.)

Visual Basic.NET
A programming language that is very different from C# in syntax, but has almost a 1-to-1 correspondence in keywords and features, because both were designed for the .NET Framework. (Chapter 40.)

Visual C++
See *C++*.

Visual Studio
Microsoft's IDE, designed for making programs in C# and other programming languages. (Chapters 2 and 41.) The latest version of Visual Studio is Visual Studio 2015.

Visual Studio 2015 Community Edition
A member of the Visual Studio 2015 family that is free and has identical features to Professional. It allows commercial development as long as you have no more than 5 developers, you don't have more than 250 computers, and your company doesn't gross more than $1,000,000 in revenue. There are additional exceptions that allow you to use this free, including for educational use and open source use. This is the version that most new C# developers will start with.

Visual Studio 2015 Professional
A member of the Visual Studio family that has identical features to 2015 Community, but comes at cost. If you have more than 5 Visual Studio users in your company, you have more than 250 computers in the company, or the company has gross revenue in excess of $1,000,000, then the company will have to find the money to pay for this.

Visual Studio Express Editions
In earlier versions of Visual Studio, there was no Community edition. Instead, there were Express editions that focused on a single specific functional area (such as web, phone, or desktop) and did not allow you to use add-ons or extensions. Express editions have been superseded by the Community edition, which combines all of the functional areas into one, and allows add-ons and extensions.

Windows Forms
A large collection of types designed for making Windows applications with the .NET Framework. Windows Forms is largely replaced by Windows Presentation Foundation (WPF) but Windows Forms is simpler and is still frequently used. This book does not cover Windows Forms in depth. (Chapter 47.)

Windows Presentation Foundation
A large collection of types that are designed to create Windows-based GUI applications. Windows Presentation Foundation is often abbreviated WPF. WPF is a replacement for Windows Forms, and is the recommended GUI framework for starting with new applications. This book does not cover WPF in depth. (Chapter 47.)

XML Documentation Comment
A special type of comment, placed immediately above a type or member of a type to define what it does and how it is used. These comments are started by using three forward slashes (///). The format can include certain XML tags. XML documentation comments are used by automated tools to generate documentation that can be displayed on websites for anyone else who uses your code. It is also used by Visual Studio to create IntelliSense for your code. (Chapter 15.)

Index

Symbols

A

B

C

D

E

F

G

H

I

J

K

L

M

N

O

P

Q

R

S

T

U

V

W

X

53814113R00204

Made in the USA
Lexington, KY
20 July 2016